Soil physics

T. J. MARSHALL
Formerly Division of Soils, Commonwealth Scientific and Industrial Research Organization

J. W. HOLMES
School of Earth Sciences, The Flinders University of South Australia

SECOND EDITION

CAMBRIDGE
UNIVERSITY PRESS

Published by the Press Syndicate of the University of Cambridge
The Pitt Building, Trumpington Street, Cambridge CB2 1RP
40 West 20th Street, New York, NY 10011-4211, USA
10 Stamford Road, Oakleigh, Victoria 3166, Australia

© Cambridge University Press 1979, 1988

First published 1979
Second edition 1988
Reprinted 1992

Printed in Great Britain at the
Athenaeum Press Ltd, Newcastle upon Tyne

British Library cataloguing in publication data

Marshall, T. J.
    Soil physics.—2nd ed.
    I. Title   II. Holmes, J. W.
    631.4'3   S592.3

Library of Congress cataloguing in publication data

Marshall, Theo John, 1907–
    Soil physics/T. J. Marshall, J. W. Holmes.—2nd ed.   p.   cm.
    Bibliography: p.
    Includes index.
    ISBN 0-521-35270-3. ISBN 0-521-35817-5 (pbk.)
    1. Soil physics.   I. Holmes, J. W. (John Winspere), 1921–
II. Title.
S592.3.M37 1988
631.4'3—dc19   87-35393CIP

ISBN 0 521 35270 3 hardback
ISBN 0 521 35817 5 paperback

Contents

vi

Contents

Preface to the second edition

In revising *Soil Physics*, we have retained the original framework of the book, keeping soil-water relations as the central theme. Much of the updating that has been necessary is in chapters that bear on field aspects of the subject where there has been much activity in research.

The intervening years have seen the wide adoption of SI units and the acceptance in the plant sciences of a terminology for water retention and movement that is closely allied to that used in soil physics. These welcome trends have made our task easier than it was when we prepared the first edition of this book.

We acknowledge our indebtedness to the libraries of Flinders and La Trobe Universities and of the Division of Soils, CSIRO.

June 1987   T. J. MARSHALL
                                                               J. W. HOLMES

Preface to the first edition

This book is written for the scientist who works with soils, plants or water resources, as well as for the student of soil physics. We have attempted to allow for the differing backgrounds of the soil scientist, the agronomist, the chemist, the physicist, the hydrologist or the civil engineer, who may be concerned with soil physics, by defining a number of terms in soil science, physics and mathematics that may seem elementary to some but can be unfamiliar to others. Similarly, we have treated units in some detail because the International System (SI) of Units that we use is not yet universally adopted in the field of soil physics. Here the differences in background are likely to be national rather than professional. To help in this we have, in many of the diagrams and tables, presented data in non-SI units as well. Units and conversion factors are listed in Appendix A. Vector notation is introduced to show the concise way in which flow equations are often expressed in the literature of soil physics, but knowledge of vector analysis is not required.

The central connecting theme is the state of water in a porous material. This is introduced early in the book in order to set out the principles of water retention that are basic to most of the topics that follow. The physical properties of soil are broadly dealt with in the opening chapter to provide an outline of the nature of the porous material with which water interacts and through which it moves. Properties concerned with soil behaviour are taken up again in greater detail later, after the relations of soil with water have been established. Wherever possible, we have related principles to field applications.

Even with a substantial bibliography, it has not been possible to make more than limited reference to early papers. The main original sources are cited and the reader seeking additional details has been referred to papers or books where more complete citations are to be found.

We wish to acknowledge the benefits that have come to us, in writing this book, from past and present association with colleagues in the Division of Soils, Commonwealth Scientific and Industrial Research Organization, Adelaide. In particular, one of us (T.J.M.) was given the opportunity to continue his writing in that environment after retiring from CSIRO, and he warmly acknowledges the help he received, especially from its Publications Section and Library. He wishes also to acknowledge the help of his wife, Ann Marshall, who transformed his unpromising index cards into an orderly set of references, and we are both grateful for her help in preparing the final list. Particular chapters have been read by Dr E. L. Greacen, Dr K. Norrish, Dr G. B. Allison and Professor G. W. Leeper, and we wish to thank them for their helpful comments. Finally we recall with gratitude the early influences of Dr G. B. Bodman of the University of California and the late Dr E. C. Childs of Cambridge University, which have in different ways contributed to the writing of this book.

July 1978   T. J. MARSHALL
Adelaide, Australia   J. W. HOLMES

1

Composition of soil

1.1 Description and classification of soil

The relatively thin mantle of soil over the land surface of the earth is a porous material of widely varying properties. Its solid phase consists of the inorganic products of weathered rock or transported material together with the organic products of the flora and fauna that inhabit soil. Some of these products are recognizable remnants in the form of stones, sand grains and leaf litter but others like clay minerals and humus result from the profound chemical changes that occur in both inorganic and organic material during the process of soil formation. The resulting soil can range in texture from coarse sands to fine clayey materials and it can range in its organic content from a usual amount that is less than 5 per cent by weight to about 80 per cent in peaty soil.

These and other properties can differ greatly from place to place over the earth's surface and from top to bottom through the succession of horizons or layers that constitute the soil profile. Hence it is useful to be able to recognize some order in the occurrence of soil materials in nature. The profile, extending from the soil surface to a depth that includes the zone explored by plant roots, forms the basis for studying soil distribution systematically. For a given set of conditions affecting its development, the profile exhibits predictable properties. Water penetrating downwards carries materials in solution or colloidal suspension to lower levels and so creates horizons of accumulation or depletion that characterize the profile. The more humid the climate the greater will be the leaching so that minerals including chlorides, sulphates and carbonates that can accumulate in arid areas (as in Fig. 1.1) are removed from the profile in drainage waters under wetter conditions. Similarly the type of parent material can leave its stamp so that for example sandy soils are more likely to develop from sandstone and granite, which contain abundant quartz of sand size, than from shale and basalt. Further, the type of vegetation affects the organic matter profile. The extensive grasslands of eastern Europe and North America

Fig. 1.1. A soil profile in a semi-arid locality (Renmark, Australia). The soil surface is at the second rung down the ladder, a horizon of carbonate accumulation commences at the fourth, and one of sulphate (gypsum) commences between the fifth and sixth rungs.

have for example allowed a deep horizon of dark soil to be formed as a consequence of the ramification and decay of fine roots. In contrast to these so-called chernozems or black earths, those soils developed under forest have their organic matter added mainly at the surface as leaf litter so that it occurs principally on relatively shallow surface horizons in the brown soils and podzols of forested regions. Also since soils can form to greater depth on flat lands than on slopes subject to erosion, topography influences soil development. Further, profiles as they appear today may have been thousands of years in developing to their state of maturity or they may, as in a drained swamp or an eroding hillside, be undergoing relatively rapid changes under the influence of a recent change in environment. Hence time, parent material, climate, vegetation and topography all have a part in determining the properties of soil.

Russian scientists demonstrated in the nineteenth century that broad zonal groups of soils occurred on a continental scale in patterns that conformed geographically with climate, vegetation and other environmental factors. The recognition of these great soil groups gave cohesion and impetus to the study of soils and their classification. However, this genetic approach to soil classification has practical limitations because soil origins are often imperfectly understood. In the early soil surveys undertaken in the United States of America at the level of detail of the farm, soils were identified by characteristic properties of the profile rather than by uncertain genetic criteria and this is current practice throughout the world in detailed surveys. The gaps in classification between the continental and farm scales are bridged in various ways in different classification systems (Buol, Hole and McCracken, 1973). No universally accepted system has yet evolved except that agreement has been reached on a basis for mapping the soils of the world by the United Nations organizations FAO and UNESCO. Otherwise the systems are largely national in character. The United States Department of Agriculture system of soil taxonomy (Soil Survey Staff, 1975) depends primarily on properties of the profile rather than on genesis. Northcote (1979) has developed a Factual Key by which Australian soils are mapped right up to the continental scale strictly on the basis of profile properties only. In the USSR, on the other hand, greater use is retained for soil genesis in separating the intermediate and high categories in soil classification (Rozanov, 1984).

A map showing the distribution of soils is made from inspections using hand auger, mechanical core-cutter, pits and road cuttings aided by surface observations and aerial photographs. Standardized descriptions of texture, structure, colour and other properties are made in the field. Measurements are then made in the laboratory on representative samples to provide quantitative data on the properties of the material in the various horizons of each type of soil. Large areas can be covered cheaply by soil surveys in this way and the resulting soil maps and profile details that are included in a report on the soils find applications in agriculture, hydrology, civil engineering, town planning and sewage disposal. Increasingly, soil survey reports include data and comments on non-agricultural applications (Bartelli, 1978). However, before expensive projects for roads, airfields and buildings go ahead, additional investigation including more intensive sampling and testing needs to be done. Classification systems for soil materials rather than soil profiles are employed in evaluating these test results for engineering purposes. Some of these tests will be dealt with later.

The user of a soil map has to allow for the fact that soils may vary continuously in their properties from one inspection site to the next and the discontinuities suggested by the mapped boundaries between different types of soil are to be treated cautiously. Also subsequent development of the land may introduce new conditions, such as a general groundwater table in an area brought under irrigation, and these changes may overshadow in importance any original differences between soil types. Aitchison (1973), in an otherwise favourable account of extensive experience on the use of soil survey for foundation engineering, shows that a change in the soil water regime at a depth of several metres, caused by urban development can seriously change the behaviour of a soil from that expected from the soil survey.

Soil physics deals with both the soil profile and the soil material. One aspect of the profile that will concern us is for example the distribution of water with depth through and beyond the plant rooting zone. In dealing with soil as a material we shall in the remainder of this chapter discuss the solid phase and the framework it provides for the storage and passage of water and air.

1.2 Soil particles

The individual particles making up the solid framework of soil can be divided arbitrarily into two classes according to whether or not they are smaller than 2 μm in diameter. The clay fraction (<2 μm) includes chemically and physically reactive clay minerals that have formed as secondary products from the weathering of rocks or have come directly from a parent material of transported deposits. The non-clay fraction consists of inert mineral and rock fragments and sometimes secondary concretions, and it can be further divided into silt, sand and gravel with size limits that vary between the different scales in current use. Three of these scales are set out logarithmically in Fig. 1.2b. Results of particle size analyses can be presented as cumulative curves as in Fig. 1.2a. From these the amount in the size fractions of any of the scales can be read off between appropriate size limits. More usually results are presented directly in size distribution tables showing the percentage of soil in each fraction. In soil survey reports, percentages are based on oven-dry soil passing a 2 mm sieve. Gravel and stones larger than 2 mm are reported separately if present. The analysis includes a pretreatment of the soil to ensure that its particles are separated so as to behave as individuals rather than clusters. The sizes are reported as diameters but, since the particles are of many shapes, they are to be regarded as effective diameters based on sizes of sieve openings in the case of the large particles and on settling velocities in the case of the smaller.

The size distributions of the three soils in Fig. 1.2 are plotted as points on a triangular diagram in Fig. 1.3a where the apices represent 100 per cent sand, silt and clay fractions respectively. The limiting size between silt and sand fractions is here 50 μm (US Department of Agriculture system) but other triangular diagrams are also available for the systems with limits of 20 μm (Marshall, 1947) and 60 μm (Hodgson, 1974). Areas marked off on these diagrams show the range of composition of soils

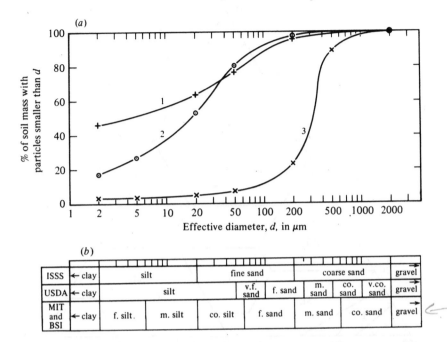

Fig. 1.2. Particle size analysis. (*a*) Cumulative curves for three soils with textures of (1) clay, (2) silty loam and (3) sand. (*b*) Limiting sizes of the fractions in systems of International Society of Soil Science (ISSS), United States Department of Agriculture (USDA), Massachusetts Institute of Technology (MIT) and British Standards Institution (BSI). (f. = fine, m. = medium, co. = coarse, v. = very.)

typical of each of the texture classes used for describing soils in the field. It will be noticed that the words 'clay', 'silt' and 'sand' are used both for the size fractions (where for example 'sand' means sand size) and for the texture names where 'sand' means soil of sandy texture. Texture is estimated in the field by working the moistened soil between the fingers to sense the coarseness or fineness of the non-clay material and the plasticity and strength imparted by clay. This provides useful qualitative information in the preliminary investigation of soil problems, in advisory services, and in soil surveys.

Particle size analysis can be represented also by plotting an average size of the non-clay material against the clay content. The resulting diagram has the advantage over triangular diagrams of being independent of all arbitrary size divisions except that at 2 μm. Texture classes have been marked out on such a diagram in Fig. 1.3*b* using symbols for five grades of clay content and three of the size of non-clay material. As in the triangular diagrams, these classes constitute the central part of the

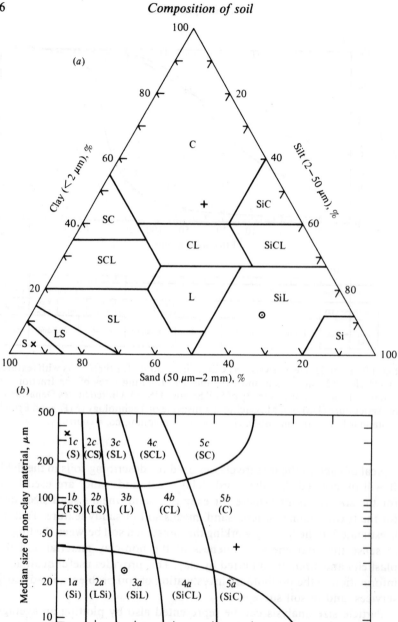

Fig. 1.3. Relation between texture and particle size distribution. (*a*) Texture triangle of the United States Department of Agriculture (Soil Survey Staff, 1975). (*b*) Texture diagram in which particles larger than 2 μm are represented by an average size (Marshall, 1947). (F = fine, S = sand(y), Si = silt(y), C = clay(ey), L = loam(y). Plotted points represent the three soils of Fig. 1.2*a*.)

areas where soils of each field texture description (shown in brackets) are found to cluster. The lines separating the grades of clay content are not vertical because soils of a given clay content feel less clayey as the average size of their non-clay material decreases (Marshall, 1947). This observation has influenced the design of triangular texture diagrams so as to allow a higher average clay content in a silty class than in the corresponding sandy class.

An upper limit for the clay fraction of 2 μm is a widely accepted but arbitrary one. It is possible to eliminate this from texture diagrams by plotting the geometric standard deviation (representing the spread of the distribution) against the geometric mean diameter. This has been done by Shirazi and Boersma (1984) in their adaptation of the texture triangle.

A texture diagram serves as a reference standard for field descriptions and further, in a number of national soil surveys, it is used for purposes of classification to define texture explicitly from the particle size analysis of soil samples. However, it needs to be recognized that when the particle size distribution of a sample is plotted on a texture diagram the texture class into which the point falls will not necessarily correspond to that given by field description. When the descriptions and analyses of large numbers of samples were compared by Marshall (1947) and by Foss, Wright and Coles (1975) a large scatter of points was found for each texture and only about half fell within the prescribed area of the diagram. Discrepancies arise not only because of the subjective nature of the field examination but because clay minerals, exchangeable cations, organic matter and cementing agents can strongly influence the field estimate of texture. Field descriptions, made by manipulating moist soil in the hand, reflect soil behaviour as well as particle size. They thus have a significance of their own and correspond with particle size analysis in an average way rather than on a one-to-one basis.

The minerals of which the particles are composed reflect the nature of the parent material and the degree of weathering that the soil has undergone. The minerals of the parent materials vary greatly in their stability and hence in their occurrence in soil. Quartz (SiO_2) often dominates the non-clay fraction because of its resistance to weathering, and its abundance in certain parent materials such as granite, sandstone and surface deposits. Among the primary silicate minerals, muscovite mica and potassium feldspar are more resistant than other felspathic minerals or the ferromagnesian minerals. The secondary minerals include some that are relatively soluble (e.g. gypsum, calcite, dolomite) and others that are not (e.g. oxides and hydroxides of iron, aluminium and silicon). Obviously the less resistant minerals are only common in soils not subjected to severe weathering and leaching or formed from parent material capable of providing them in relative abundance. For further

Composition of soil

information on these aspects of mineralogy of soils the reader is referred to Brewer (1964).

The above minerals occur chiefly in the non-clay fraction but they are not confined to it, since its lower limiting size of 2 μm is wholly an arbitrary one. The principal minerals of the clay fraction are secondary products of weathering referred to collectively as the clay minerals. These have characteristic crystal structures, as shown by X-ray diffraction patterns, and they occur mainly in the form of flakes, as shown by the electron microscope (Fig. 1.4). In addition there may be material present in the clay fraction that is amorphous to X-rays. The particles of the clay

Fig. 1.4. Electron microscope photograph of crystals of kaolinite. The length of the line represents 1 μm (K. Norrish, CSIRO, Adelaide).

fraction, ranging downward in size to about 5 nm, are the seat of phenomena such as swelling, plasticity and cation exchange that have profound influence on the physical behaviour of soil.

1.3 Pore space

The way soil behaves depends not only on the kinds and sizes of individual particles but also on how these are arranged and bonded together. Sometimes they occur as a collection of single individual grains (as in sands) but usually they are linked into clusters or aggregates of varying stability. The properties of the constituent particles are masked by this clustering which thus has a marked effect on soil behaviour. Between the particles, arranged singly or in aggregates, there is an intricate system of pore space on which plant roots, micro-organisms and soil animals depend for the storage and movement of water and air.

These topics will be dealt with more fully later but it will be useful to outline here some broad structural features of the three-phase system of solid particles, soil water and soil air. Taking V_s, V_l, V_g and V_t to represent respectively the volume of solid, liquid, gas and the total or bulk volume of a quantity of soil and taking m_s, m_l, m_g and m_t to represent the corresponding masses, we define

$$\text{density of solid particles,} \quad \rho_s = m_s / V_s \tag{1.1}$$

$$\text{bulk density of soil,} \quad \rho_b = m_s / V_t \tag{1.2}$$

$$\text{porosity,} \quad \varepsilon = (V_l + V_g) / V_t. \tag{1.3}$$

Writing Equation (1.3) as $\varepsilon = (V_t - V_s)/V_t$ and combining this with Equations (1.1) and (1.2), we obtain

$$\varepsilon = (\rho_s - \rho_b)/\rho_s. \tag{1.4}$$

In soil undergoing change of volume as a result of compaction, swelling, or shrinking, V_t in Equation (1.3) is not constant and there are then advantages in using the void ratio, e, instead of porosity. This is defined as

$$e = (V_l + V_g) / V_s. \tag{1.5}$$

The relation between ε and e, obtained from Equation (1.3) by dividing the numerator and denominator by V_s, is

$$\varepsilon = e/(1+e) \tag{1.6}$$

or

$$e = \varepsilon/(1-\varepsilon). \tag{1.7}$$

In the above equation, ε and e are dimensionless ratios and density has the dimensions ML^{-3}. The SI units for density are $kg\,m^{-3}$ but the fractional SI units of $g\,cm^{-3}$ are also used ($1\,g\,cm^{-3} = 1\,Mg\,m^{-3}$). In most soils the average density of the particles, ρ_s, is in the range 2.5 to 2.8 $Mg\,m^{-3}$ but the density of soil organic matter is much less than this, and that of the ferromagnesian minerals, iron oxides and other so-called heavy minerals is greater than 3 $Mg\,m^{-3}$. Bulk density, ρ_b, increases with the degree of compaction and tends to increase with depth in the profile because of increasing overburden and decreasing disturbance. In a clay soil it decreases with the degree of swelling. Fine textured soils tend to be less dense than sands especially if the size distribution of the sand is such that the gaps between the larger particles are occupied by smaller ones.

Selected values of ρ_b are given in Table 1.1 covering the range ordinarily found in soils and from these ε and e are calculated taking ρ_s as 2.65 $Mg\,m^{-3}$. The values are appropriate to the described soil conditions and to the porous rock (sandstone) but they are not specific to them and serve only as representative examples. On the other hand, those for spheres in open (cubical) and closest (rhombohedral) packing given for comparison are quite specific to the geometry for spheres of any uniform size (Slichter, 1899).

Table 1.1. *Range of values found in bulk density and associated properties illustrated by particular cases. Particle density is taken as 2.65 Mg m^{-3}*

Description	Bulk density, ρ_b Mg m^{-3}	Porosity, ε	Void ratio, e
Surface soil of wet clay	1.12	0.58	1.37
Surface soil of loam texture	1.28	0.52	1.07
Spheres of uniform size in open packing	1.39	0.48	0.91
Subsoil of sandy texture	1.61	0.39	0.65
Sandy loam soil compacted by heavy traffic	1.90	0.28	0.39
Spheres of uniform size in closest packing	1.96	0.26	0.35
Sandstone	2.12	0.20	0.25

Methods for measuring particle density and bulk density are described by McIntyre and Loveday (1974a) and by Blake and Hartge (1986a, b). The most usual method for bulk density is to cut out a cylindrical core of known volume and find the mass of the dried soil. It can also be measured on a clod whose volume is found by displacement in water after its surface has been sealed with a resin. This is a good method for many purposes, but the bulk density of a clod will be greater than that of the soil at large if there is much pore space between the clods. Bulk density can be measured in situ from the attenuation of gamma rays transmitted through the soil as discussed in Section 3.1.

1.4 Water content

The porosity consists of a portion V_g/V_t occupied by soil air and another V_l/V_t occupied by soil water. The amount of water in soil is expressed in this and other ways as

$$\text{water content, volume fraction,} \quad \theta = V_l/V_t \tag{1.8}$$

$$\text{water content, mass basis,} \quad \theta_m = m_w/m_s \tag{1.9}$$

$$\text{degree of saturation,} \quad S = V_l/(V_l + V_g). \tag{1.10}$$

Water content is most commonly measured by finding the loss of mass, m_w, on drying in an oven at the arbitrary temperature of 105 °C to a constant mass, m_s, although there are also other methods, as described in Chapter 3. On combining Equations (1.8) and (1.9), the following relation is obtained for converting from the mass basis to the volume fraction which is generally more useful in field studies:

$$\theta = \theta_m \rho_b/\rho_w. \tag{1.11}$$

In obtaining Equation (1.11), it is assumed that the density of water is unaffected by being adsorbed in soil so that m_w/V_l is equal to ρ_w, the density of pure free water. Values of ρ_w at different temperatures are given in Table 1.7 but, for most applications of Equation (1.11), it can be taken with sufficient accuracy to be 10^3 kg m^{-3}. It may also be noted that the volume fraction, θ, is equivalent to a depth fraction representing the ratio of the depth of water to the depth of the soil profile that contains it. This form is used when examining gains and losses of water in the field, since precipitation and evaporation are also expressed as depths of water.

In a swelling soil whose volume is not constant, water content is sometimes expressed as a ratio of the volume of water to the volume of solids. This so-called liquid ratio has the same form as the void ratio. It is obtained when ρ_s, the density of the solids, is substituted for ρ_b in Equation (1.11), and is equal to $\theta(1+e)$.

Water content as a volume fraction ranges from zero at oven dryness to a value ε at pore space saturation. For agronomic purposes, two intermediate stages are commonly recognized during the drying of wet soil. The wetter of these is the 'field capacity' which is the water content found when a thoroughly wetted soil has drained for about two days. It is determined in the field under conditions that prevent evaporation and allow good drainage (a condition that cannot always be met). The drier stage, the 'permanent wilting point', is the water content found when test plants growing on the soil wilt and do not recover if their leaves are kept in a humid atmosphere overnight. These terms lack critical physical definition (see Marshall, 1959a, and Sections 5.2 and 10.1) but they have an established utility.

Field capacity and permanent wilting point are used for marking the upper and lower levels of the water content of a soil at which water is ordinarily available for plants. As illustrated by data for selected non-organic soils in Table 1.2, they both tend to increase with increasing clayeyness of the soil. The data also show that at field capacity the degree of saturation of the sandy soil is much lower than that of the clayey soil. This is because a larger amount of pore space in the sandy soil is made up of relatively large pores that drain readily. As will be shown in subsequent chapters the size distribution of pores influences water retention, water movement and aeration and hence is often more important than size distribution of particles. A method for measuring it will be discussed in Chapter 2.

Table 1.2. *Water content as a volume fraction of three surface soils of coarse, medium, and fine texture at three soil water conditions*

Texture	Clay content %	Saturated	Field capacity	Permanent wilting point
Sand	3	0.4	0.06	0.02
Loam	22	0.5	0.29	0.05
Clay	47	0.6	0.41	0.20

1.5 Clay minerals

The common clay minerals in soils belong to three main groups – kaolinite, illite (or hydrous mica) and smectite. The smectites include montmorillonite, a mineral much used in experiments on the behaviour of clay. Details of crystal structure are given by Grim (1968),

C. E. Marshall (1964), van Olphen (1977) and only an outline will be given here. The clays are silicates in which the negatively charged atoms (ions) O^{2-} are coordinated around positively charged silicon Si^{4+}, aluminium Al^{3+}, magnesium Mg^{2+} or other positively charged atoms (ions). The principles applying to this arrangement established by L. Pauling in 1930 require that the oxygens are arranged in positions of closest packing around the positively charged atom so that charges are balanced over the shortest possible distance.

By considering atoms in crystals as rigid spheres in contact with one another, atomic radii have been worked out that apply generally in different compounds. Radii of some of the atoms of interest in clays are listed in Table 1.3 and a full discussion, covering coordination numbers other than that quoted in the table, is given by Megaw (1973, p. 23). Because of the relative sizes, as shown in Table 1.3, four oxygen atoms can be coordinated around one silicon atom, which is small enough to fit into the position between them. Lines joining the centres of the oxygens in this four-fold coordination represent a tetrahedron or pyramid. Similarly six oxygens (or hydroxyls of about the same size) can be coordinated around one aluminium or one magnesium atom which being larger than silicon requires the larger gap between the oxygens. In this six-fold coordination the oxygens represent an octahedron. These two coordinations form the basis for the structures of the principal clay minerals. Since each of the four oxygen atoms of a tetrahedron carries two negative charges while the central silicon atom carries four positive charges, the oxygens can be shared with other groups. Three at the base are shared with a neighbouring silicon in other tetrahedral groups to form a continuous sheet of tetrahedra with their bases all in the same plane. Similarly a continuous sheet of linked octahedra is formed by the sharing of oxygen atoms. These two kinds of sheets unite to form a unit layer of the crystal lattice of a clay mineral by sharing the fourth oxygen

Table 1.3. *Ionic radii, in nm (from Kaye and Laby, 1973, for coordination number 6)*

Cation	Radius	Anion	Radius
Na^+	0.102	Cl^-	0.181
K^+	0.138	O^{2-}	0.138
Mg^{2+}	0.072	—	—
Ca^{2+}	0.100	—	—
Al^{3+}	0.053	—	—
Fe^{3+}	0.065	—	—
Si^{4+}	0.040	—	—

atom of the tetrahedron with the octahedral sheet. This unit layer is made up of one of each type of sheet in kaolinite, a so-called 1 : 1 mineral, and of two tetrahedral to one octahedral sheet in the 2 : 1 minerals illite and montmorillonite. Commonly a clay crystal consists of a stack either of 1 : 1 or 2 : 1 unit layers bonded together by different means in different minerals.

The structure of some clay minerals of importance in soils is illustrated in Fig. 1.5. In the kaolinite crystal, the unit layers are held together by hydrogen bonding. Hydrogen of the hydroxyl ion of one layer is bonded with oxygen in the next layer. The repetition of the layers is shown by X-ray diffraction to occur at intervals of 0.72 nm. This distance between corresponding planes in successive layers is called the basal spacing of

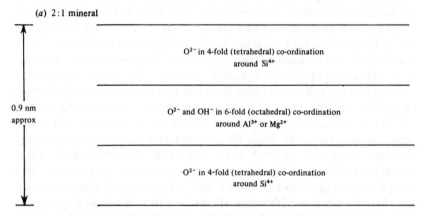

(*a*) 2 : 1 mineral

0.9 nm approx

O^{2-} in 4-fold (tetrahedral) co-ordination around Si^{4+}

O^{2-} and OH^- in 6-fold (octahedral) co-ordination around Al^{3+} or Mg^{2+}

O^{2-} in 4-fold (tetrahedral) co-ordination around Si^{4+}

Illite: Fixed distance to next layer and bonded to it through non-exchangeable K^+. Basal spacing 1 nm.
Montmorillonite: Variable distance to next layer in an expanding lattice; bonded to it by exchangeable cations. Basal spacing variable but generally about 1.4 nm.

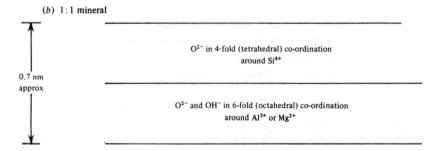

(*b*) 1 : 1 mineral

0.7 nm approx

O^{2-} in 4-fold (tetrahedral) co-ordination around Si^{4+}

O^{2-} and OH^- in 6-fold (octahedral) co-ordination around Al^{3+} or Mg^{2+}

Kaolinite: Bonded to next layer at a fixed distance by hydrogen bonding. Basal spacing 0.72 nm.

Fig. 1.5. Unit layer of common clay materials.

the crystal and is a distinguishing characteristic of each mineral. In illite, the two tetrahedral sheets are combined with one octahedral sheet between them to form a unit layer as in Fig. 1.5. One layer is bonded to the next through non-exchangeable potassium ions and the stack built up in this way has a basal spacing of about 1 nm. In montmorillonite, the bond between the layers is through cations which can be exchanged for those in any outer solution that bathes the clay. Water molecules can also enter between the layers and separate them further as first shown by Hofmann, Endell and Wilm (1933). Measurement by X-ray diffraction shows that the montmorillonite crystal expands in steps corresponding to the degree of hydration of the exchangeable cations. Water molecules attracted to the cations cause the basal spacing to increase progressively from 0.96 to 2 nm to accommodate them as shown in the first stage of swelling in Fig. 1.6 from Norrish (1954, 1972). During this stage, swelling is opposed by electrostatic attraction between the cations and the negatively charged layers. As shown by Norrish (1954), swelling beyond 2 nm is possible only if the hydration energy of the cation is sufficient to overcome the attraction. In the case of Ca-montmorillonite (or montmorillonite with other polyvalent cations) the layers separate no further unless work is done on the system to shear them apart. But lithium and sodium ions are not prevented from hydrating further and can proceed to the second stage of swelling. The clay changes from a crystalline solid

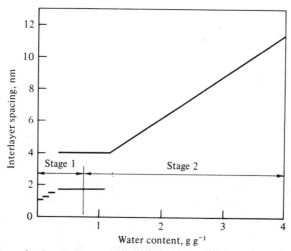

Fig. 1.6. Stages in the swelling of Na-montmorillonite. Stage 1: crystalline solid; swells due to hydration of cations (opposed by attraction between interlayer cations and negatively charged layers). Stage 2: plastic paste with irregularly spaced layers; swells due to osmotic effect of exchangeable cations (opposed by edge-face attraction of layers). (After Norrish, 1972.)

to a plastic paste with irregularly spaced layers on entering stage 2. Swelling is now caused by osmotic pressure arising from the concentration of exchangeable cations near the clay surface. Stability is maintained by the opposing attraction between negatively charged faces and positively charged edges of clay layers. The swelling behaviour of Na-montmorillonite provides a model for a theory of swelling of clays to be discussed in Chapter 2.

The effect of crystalline swelling on the bulk volume and plasticity of sodium-montmorillonite is illustrated in Table 1.4. Obviously the swelling shown there and in Fig. 1.6 progresses much more freely than it does in soil under field conditions where it is usually restricted by the presence of polyvalent exchangeable cations and by the restraint of neighbouring and overlying soil particles. But volume can change greatly enough in soils to affect their morphology, cause periodic cracking, and provide unstable conditions for the construction of buildings and pavements. The smectite group of clay minerals with an expanding crystal lattice and illite with a small crystal size are responsible for substantial volume changes in soils with sufficient content of these clay minerals.

Table 1.4. *Crystalline swelling of an oriented flake of Na-montmorillonite by adsorption of water vapour (Mering, 1946)*

Relative vapour pressure	Water adsorbed	Average basal spacing	Macroscopic swelling
	$g\,g^{-1}$	nm	
0.0	0.0	0.96	No swelling
0.90	0.4	1.62	Beginning of visible swelling and plasticity
0.96	0.6	1.85	About 30% swelling
0.99	1.0	1.95	About 100% swelling

The relative abundance of the different clay minerals in a soil can be estimated approximately from the intensity of their characteristic lines in the X-ray diffraction pattern. Interpretation is complicated by the presence of material that is amorphous to X-rays and by the interlayering of one type of mineral with another to form randomly interstratified material. The average mineral composition of the clay fraction of 1600 samples of Australian soils was found by Norrish and Pickering (1983) to be 6, 20, and 45 per cent respectively in smectite, illite, and kaolinite. A number of other minerals may also be present in the clay fraction including chlorite, vermiculite, palygorskite (in deserts) and allophane (in volcanic regions). Oxides and hydroxides of iron and aluminium and oxides of silicon are also common in the clay fraction.

1.6 Exchangeable cations

The ability of soils to exchange cations with those contained in solutions that bathe them has important effects on soil behaviour. The exchange complex of soils normally consists of a mixture of the cations Ca^{2+}, Mg^{2+}, K^+, and Na^+ in which Ca^{2+} comprises a large part and K^+ and Na^+ only a small part. H^+ also enters into exchange in acid soils but the reaction may be complicated by breakdown of the crystal structure of the clay mineral and the release of Al^{3+} to the exchange complex. Soil can be treated so as to make the effect of one particular cation dominant and in so doing the physical properties of the soil can be greatly changed. For example, if Na^+ replaces about 15 per cent or more of the Ca^{2+} in the exchange complex of a calcium-saturated soil of medium or clayey texture, the soil structure may lose its stability, as will be discussed in Chapter 8, and the permeability to water will then decrease drastically.

Exchangeable cations are located at accessible sites on the surfaces of particles and also between layers of the expanding crystal lattice of montmorillonite and vermiculite. They are attracted by negative charges on the particles the existence of which can be demonstrated by the migration of clay particles towards the positive pole when a current is passed through a suspension of clay in water, a behaviour referred to as electrophoresis.

The charges arise in the following ways. One source is on the edges of clay crystals where oxygen atoms, lacking a neighbouring positively charged atom, have their negative charges satisfied by exchangeable cations. Since positive as well as negative charges arise from these 'broken bonds' at the edge of crystals, a basis also exists for anion exchange which can in fact occur. This edge source of cation exchange is relatively more important in kaolinite than in other clay minerals where the following mechanism is mainly responsible. As demonstrated by C. E. Marshall in 1935, negative charges arise from isomorphous substitutions of one atom for another of about the same size, but with a lower positive charge. From the sizes given in Table 1.3, it follows that Mg^{2+} can take the place of Al^{3+} at the centre of octahedra and Al^{3+} can substitute for Si^{4+} in tetrahedra and so leave unsatisfied negative charges that are balanced by exchangeable cations.

The resulting capacity for exchangeable cations varies with the type of clay and, if expressed as milliequivalents per kg of clay, it can be about 100 in kaolinite, 300 in illite and 1000 in montmorillonite. It may, however, be about 50 per cent greater or less than these values according to the range given by Grim (1968, p. 189) for the cation exchange capacity of these (and other) clay minerals. Variation occurs because of effects of crystal size, isomorphous substitution and pH. Oxides and organic matter also have a cation exchange capacity. In both these cases it depends on pH and in humic material at pH 7 it is about 4000 mequiv. kg^{-1} (Kononova, 1966, p. 62). The non-clay fraction of a soil seldom contributes significantly to the soil's cation exchange capacity and hence, in soils that are not highly organic, this depends upon clay content and type of clay and commonly varies with the specific surface area of the particles.

The exchange of one cation for another is governed in a general way by Coulomb's law which states that the force of attraction, F, between two point charges, q_1 and q_2, varies directly with the product of the charges and inversely with the square of the distance, l, between them or $F \propto q_1 q_2 l^{-2}$. Cations of a higher valence (higher charge) are attracted more strongly to the negatively charged clay than those of a lower valence. Hence Ca^{2+} replaces Na^+ readily in a reaction that is of practical importance in reclaiming sodic soil as will be discussed in Chapter 8. There is also a tendency for larger cations to enter more readily in exchange than small ones of the same valency. This is attributed to greater hydration of the smaller ions which, with a larger hydrated radius, are thus more distant from the negative charges on the clay. These generalizations, originated by G. Wiegner around 1930, have been confirmed by later workers although exceptions occur especially for cations that have specific reactions with particular minerals. Among monovalent cations the exchanging power is in the order $Li < Na < K < Rb < Cs$ and in the divalents it is $Mg < Ca < Sr < Ba$. These series may be compared with radii given in Table 1.3. The two groups overlap; but for cations of a similar size the one with the higher valence will in general be the more powerful in exchange. In addition to these consequences of Coulomb's law there is also an effect of concentration on exchange equilibrium so that an increase in the concentration of a cation in the outer bathing solution will cause an increase in the amount of it entering into the exchange complex of the clay in competition with other cations.

1.7 Distribution of ions at clay surfaces

When adsorption of water has progressed far enough for swelling and plasticity to become apparent (as at the end of stage 1 in Fig 1.6), clay has enough water for a diffuse layer of exchangeable cations to extend out from the negatively charged surface of the mineral sheets under suitable circumstances. Their positions are governed by the opposing effects of electrostatic attraction towards the surface and diffusion away from it in the direction of decreasing concentration much as gas molecules of the atmosphere are distributed under the opposing influences of gravitational attraction and diffusion. The clay surface with its localized negative charges and the swarm of cations of opposite charge are then said to constitute a diffuse electrical double layer. If an outer solution bathes the clay, this will affect the distribution of the cations. Also because of the charge on the clay, the concentration of anions near its surface will be lower than in the outer solution ('negative adsorption' of anions). The resulting distribution of ions is shown diagrammatically in Fig. 1.7.

These principles are given quantitative expression in the diffuse double layer theory proposed by G. Gouy in 1910 and further treated by D. L. Chapman in 1913. From this the distribution of ions out from a negatively charged surface can be calculated. Calculations in Table 1.5 show that the potential extent of the cation layer from the surface varies inversely with the valence and concentration of the cation in the outer solution. This theory is discussed further by Yong and Warkentin (1975, p. 55) and by van Olphen (1977).

A number of modifications to the double layer concept have been proposed because of serious anomalies between the theory and experimental results. Stern does not consider the first layer of cations to be

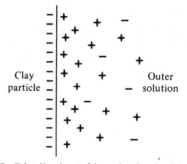

Clay particle — Outer solution

Fig. 1.7. Distribution of ions in the double layer.

Table 1.5. *Potential extent of double layer as affected by concentration of electrolyte in the outer solution and valence of exchangeable cations (Bolt and Koenigs, 1972). The extent is given in units of nm*

Valence of exchangeable cations	Concentration of electrolyte		
	10^{-3} mol l^{-1}	10^{-2} mol l^{-1}	10^{-1} mol l^{-1}
Monovalent	20	6	2
Divalent	10	3	1
Trivalent	6	2	0.6

part of the diffuse layer. Shainberg and Kemper (1966) suggest that cations in this so-called Stern layer lack part of their hydration shell and are directly on the clay surface. The fraction of the exchangeable cations in the Stern layer was believed by them to be about 20 per cent for Li^+ and about 40 per cent for Na^+ in accordance with the difference in their hydration energy which as we have noted varies inversely with radius for cations of a given valence. Divalent cations such as Ca^{2+} with their higher charge are more strongly attracted to the surface and were considered to be mainly in the Stern layer. Hence the diffuse double layer theory cannot be applied quantitatively to clays in soils.

1.8 Heat of wetting

When water is added to dry soil heat is evolved because water molecules lose kinetic energy in changing from bulk water to the hydration shell of a cation and to other possible adsorption sites on the clay. Up to three water molecules may become associated with a monovalent cation and about six with a divalent, but there is some tendency for the smaller cations of a given valence to attract more water than the larger ones. The heat of wetting one gram of a clay material to saturation from different initial water contents is shown in Fig. 1.8 from Janert (1934). The reaction depends largely on the polar character of the water molecule as will be discussed in Chapter 2. This is shown by the much lower heat of wetting when a non-polar liquid was used. From the decreasing slope of the curve in Fig. 1.8 with increasing initial water content, it can be

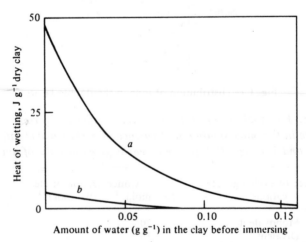

Fig. 1.8. Heat of wetting for a clay material in (*a*) water and (*b*) carbon tetrachloride (Janert, 1934).

seen that the most energetic reaction between water and soil occurs when the soil is at its driest.

The reaction of the solid with water varies with the cation exchange capacity and the specific area of surface, as is illustrated in Table 1.6 where data of Greene-Kelly (1962) have been summarized for the heat of wetting when dry clay minerals were immersed in water. These minerals all had the same exchangeable cation (Na^+) in order to avoid any difference in effect from type of cation. Kaolinite with its relatively large crystals has the lowest specific area of surface, exchange capacity and heat of immersion. Montmorillonite with small crystals and an expanding lattice that exposes a large internal surface to these reactions has high values for all properties. Hydrous micas (which include illite) lie between these two. The ratio of cation exchange capacity to surface area which represents the density of charge on the surface was found by Greene-Kelly to be of the same order in the three types of mineral, i.e. 2.0, 1.7, 1.4 for kaolinite, hydrous micas, and montmorillonite, respectively, in units of 10^{-3} mequiv. m^{-2} as can be checked from the data of Table 1.6.

Table 1.6. *Heat of wetting* (H), *surface area* (S), *and cation exchange capacity* (CEC) *per gram of clay for three types of sodium clays* (*from data of Greene-Kelly, 1962*)

Clay mineral	H $J g^{-1}$	S $m^2 g^{-1}$	CEC mequiv. g^{-1}
Kaolinite	7	16	0.03
Hydrous mica (includes illite)	39	167	0.28
Montmorillonite	78	677	0.96

In Chapter 2 we shall deal with the interaction of water with soil as it affects water retention. The evolution of heat when water is added serves to illustrate the energetic nature of this interaction which is the basis for treating water retention. The methods to be described for measuring this do not however include heat of wetting, which has only occasional use as an energy characteristic of the soil that bears on its texture, the specific area of its particles and other properties as outlined by Dimo and Utkayera (1984).

1.9 Particle size analysis

Size of particles can be expressed directly as the distribution of the mass of soil particles with respect to their effective diameters or indirectly as the area of the surface exposed by particles in unit mass of soil. Both types of measurement will be discussed. The first, called particle size or mechanical analysis, provides data about the composition of a soil over the whole range of its particle sizes for use as a starting point for investigation of soil problems and as a quantitative basis for soil classification. The second type of measurement, specific area of surface, is useful for examining the size of particles within the clay fraction.

Particle size analysis usually employs sieves for the coarser particles and the rate of settling in water for the finer. Methods used with soils are described in detail by Gee and Bauder (1986) and by McIntyre and Loveday (1974b) and only general principles of these will be dealt with here. It is first necessary to ensure that the particles are separated from one another so that single particles rather than clusters are measured in sieving and settling. Treatments that bring this about are described for general and particular cases by McIntyre and Loveday. These include hydrogen peroxide treatment when necessary to remove organic matter, washing when necessary to remove soluble material and vigorous mechanical action upon the soil in water to which a dispersion agent has been added. In many soils this last step is sufficient without the others. The mechanical action shears the particles apart and the dispersion agent tends to prevent reclustering of the clay particles by providing a monovalent cation (Na^+) to replace divalent cations from the exchange complex of the clay. This results in a more extensive diffuse layer of cations around each of the clay particles since the extent of the layer varies inversely with the valence (charge) of the cation according to double layer theory. The clay particles are prevented by this extensive layer of like charge around each of them from coming close enough together to cohere and hence they settle as separate individuals.

When particles cohere, they settle rapidly in clusters from a clay suspension and the clay is said to be flocculated. This occurs when particles, approaching one another in the course of Brownian movement, come under the influence of van der Waals forces. These arise from the attraction the atoms of one particle have for those of another at short range and they are a cause, for example, of deviations in the behaviour of real gases from the ideal gas law. The relative protection from flocculation given by double layer repulsion is indicated in Table 1.5 where the effects of type of cation and electrolyte concentration on the extent of the layer are shown. It can be seen there that the extent of the layer and hence the stability of a suspension will be greatest with monovalent cations in the exchange complex and with the lowest concentration of electrolytes in the suspending water.

In making up soil suspensions for particle size analysis, sodium tripolyphosphate or sodium hexametaphosphate together with sodium hydroxide is commonly used as dispersing agent. The mechanical treatment usually consists of stirring the soil rapidly in water with an electric mixer or shaking or tumbling it vigorously in water. Sonic or ultrasonic vibration of soil in water was used by Edwards and Bremner (1967) as an alternative to these. They found this effective even without chemical treatments but Emerson (1971) demonstrated that a dispersion agent was still needed for certain soils. An ultrasonic probe does less damage than the other mechanical treatments to minerals that fragment readily, accordingly to Walker, Woodyer and Hutka (1974).

Since an analysis may be worthless unless the soil is fully dispersed, short cuts that omit some of the treatments in the dispersion procedure can only be taken with caution. But on the other hand it is desirable also to avoid unnecessarily harsh treatments in particular cases. For example severe grinding to break down lumps when preparing the sample may break particles that fragment readily; and acid treatment, that may be necessary for some soils (McIntyre and Loveday, 1974*b*), destroys carbonate particles that would otherwise be included in the analysis. Although some particles may have an ill-defined ultimate size that depends on the treatment, analyses of most soils are highly reproducible by routine procedures of soil laboratories.

When the necessary treatment is completed, the soil is ready for sieving or sedimentation. Details about sieves, sieving procedures and sieving errors are given by Krumbein and Pettijohn (1938), Dallavalle (1948) and Herdan (1960). Wire mesh sieves with uniform square apertures are used in soil analysis to separate the dried particles according to their size. They are available with aperture sizes from about 50 μm upwards. Sedimentation measurements can be done in a variety of ways as described in the books referred to above and in that by Keen (1931). We shall confine ourselves to those that require a sample to be taken or a density measurement to be made at a particular depth and time in a suspension of soil in water. These are the types mainly used for soils and the principles to be discussed for them can readily be applied to other sedimentation methods.

The dispersed soil is washed into a glass cylinder and made up with water to a known volume. It is stirred into a uniform suspension and the particles are then allowed to settle. From the amount of material remaining in the suspension at a depth, l, after a settling time, t, the fraction of the soil with a settling velocity less than l/t is determined. The concentration, C, of a pipette sample (mass of dry soil per unit volume of suspension) taken at this depth and time is expressed as a percentage of the concentration, C_0, corresponding to the total mass of dry soil taken for analysis divided by the total volume of suspension. The percentage of the soil having a settling velocity, v, less than l/t is thus given by $100C/C_0$. In the pipette method C_0 is usually 20 g dry soil per litre of suspension and, in assessing C, allowance is made for the amount of dispersion agent in the pipette sample.

Settling velocity is related to the radius r, of a spherical particle in the following way. The force acting downward on each particle due to its weight in water is $\frac{4}{3}\pi r^3(\rho_s - \rho_w)g$ where ρ_s is the density of the particle, ρ_w is the density of water and g is the acceleration due to gravity. Because of the viscous resistance of the water, this is opposed by a force which Stokes found in 1845 to be given by $6\pi\eta rv$ where η is the viscosity of water and v is the velocity of fall. This resisting force is zero when $v = 0$ at $t = 0$ and it increases with increasing v until it is equal to the downward force. The particle has then reached a constant terminal velocity. On equating these two forces we obtain Stokes' law relating radius and terminal velocity,

$$v = 2(\rho_s - \rho_w)gr^2/9\eta. \tag{1.12}$$

In applying this to particle size analysis, we assume with negligible error that the particles move at their terminal velocity as soon as settling commences. We can safely assume also that there is no slip between soil particles and the water so that the resistance offered by the water is properly represented in the equation. Also the settling velocity must be so slow that no turbulence occurs and resistance is wholly due to the viscosity of the water. This requirement is met by soil particles with a diameter smaller than about 80 μm. The deviations from Stokes' law that occur for larger particles are illustrated by data of Gibbs, Matthews and Link (1971) in Fig. 1.9. It will be noted that this upper limit for applying Stokes' law is larger than the lower limit for sieves (about 50 μm) so that there is some overlap possible between sedimenting and sieving. One further assumption in applying Stokes' law is that the particles are smooth and spherical. Since this is not the case in soils, r is to be regarded as an equivalent rather than an actual radius. These and other assumptions in Stokes' law and its application are discussed fully by Krumbein and Pettijohn (1938, p. 96).

As an example of the application of Equation (1.12), we may need to know what settling velocity corresponds to the upper size limit of the clay fraction (2 μm diameter). At 20 °C and assuming a particle density of 2.65×10^3 kg m^{-3}, we have

$$v = 2(2.65 \times 10^3 - 1.00 \times 10^3) \times 9.80 \times (1.00 \times 10^{-6})^2/(9 \times 1.00 \times 10^{-3})$$

$$= 3.6 \times 10^{-6} \text{ m s}^{-1}.$$

In this example, ρ_s and ρ_w are in units of kg m^{-3}, g is in units of m s^{-2}, r is in metres, and η is in units of pascal second (Pa s). (Alternative units are listed in Appendix A tables.) For this calculated velocity the

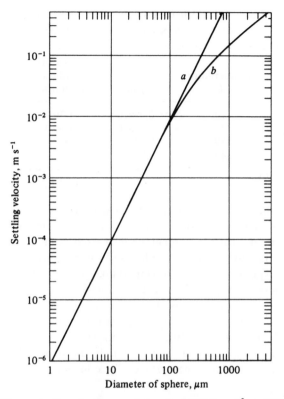

Fig. 1.9. Settling velocity of spheres of density 2.65 Mg m^{-3} at 20 °C according to (*a*) Stokes' law and (*b*) data of Gibbs *et al.* (1971) for diameters between 50 μm and 5 mm.

pipette sample can be taken for example at a depth, *l*, of 10 cm after a settling time, *t*, of 7 h 43 min. This sample will contain no particles larger than an equivalent diameter of 2 μm but all the particles of the clay fraction will be present in their original concentration. Sedimenting columns are best kept at constant temperature to avoid the effects on density and viscosity of water that are shown in Table 1.7. Viscosity is strongly dependent on temperature. Density is less so but if it is not controlled there is a risk that convection currents will arise from thermal gradients within the suspension.

The results of an analysis can be plotted as a cumulative curve as in Fig. 1.2 in which the abscissa represents the particle size on a logarithmic scale and the ordinate the percentage of materials smaller than any particular size. Alternatively results can be given in a size distribution table showing the percentage in each size class or fraction. Results are

Table 1.7. *Density and dynamic viscosity of*
water (from Kaye and Laby, 1973)

Temperature °C	Density Mg m^{-3}	Viscosity mPa s or centipoise
10	0.99970	1.3037
20	0.99820	1.0019
30	0.99565	0.7982
40	0.99222	0.6540

obtained for the table either directly by measurements made at the precise
size limits of the classes or indirectly from the cumulative curve by
reading off the increase in percentage over each interval of size.

Hydrometers and plummets can be used for measuring the amount of
material in suspension without having to take pipette samples, evaporate
them to dryness and weigh. These are illustrated in Fig. 1.10. Hydrometers
were introduced by Bouyoucos (1927) and, after some refinements were
made by A. Casagrande for estimating settling depth from the variable
depths at which they float, they came into wide use. They measure the
density, ρ, of a suspension at approximately the depth of the centre of
volume of the bulb below the surface. The concentration, C, of the soil

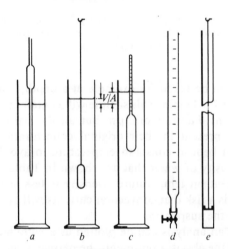

Fig. 1.10. Some settling methods of particle size analysis: (a) pipette, (b) plummet
hanging from balance, (c) hydrometer, (d) bottom withdrawal tube, (e) settling
tube for coarse particles with pan hanging from balance. (V = volume displaced;
A = cross-sectional area of cylinder.)

in the suspension can be calculated from

$$\rho = \rho_w + C(1 - \rho_w/\rho_s),\tag{1.13}$$

where ρ_w is the density of water and ρ_s is the density of the soil particles. The stem of the hydrometer is calibrated at a particular temperature to read in density units 0.995 to 1.030 Mg m^{-3} or alternatively in percentages (100 C/C_0) when the initial concentration $C_0 = 0.05$ Mg m^{-3} and $\rho_s = 2.65$ Mg m^{-3}. Details of hydrometers and their use are given by Gee and Bauder (1986) and by the British Standards Institution (1975).

A plummet hanging from a balance arm can be used instead of a hydrometer. If this weighs as a mass W when immersed in the suspension at a chosen depth and time, the concentration, C, can be calculated from

$$W_0 - W = VC(1 - \rho_w/\rho_s),\tag{1.14}$$

where W_0 is the mass representing the weight in water at the same temperature as the suspension and V is the volume of the plummet. A plummet can be used with a simple balance developed by Marshall (1956) whose pointer reads percentage of material directly when the plummet is immersed at the desired depth in a suspension. Details of its use in particle size analysis are given by Hutton (1955) and McIntyre and Loveday (1974b). Of the conventional balances that have been used with a plummet, a torsion balance of small capacity as used by Slager and Koenigs (1964) is probably the most convenient.

Adjustments are applied to plummet and hydrometer readings for the effect of temperature other than that of calibration on ρ_w and for the amount of dispersion agent present. The depth, l, at which the readings are assumed to apply is the distance from the surface to the centre of the plummet or of the bulb of the hydrometer of volume, V, less $V/2A$ where A is the cross-sectional area of the cylinder containing the suspension. This reduction arises from the displacement of the suspension and the raising of the surface level by the immersed body, as explained by Dallavalle (1948, p. 81).

In all three methods, pipette, plummet and hydrometer, it is assumed that the measured reading applies to the horizontal plane through the centre of the volume that is to be sampled or displaced, and hence that the lower density in the top half of this volume of suspension is compensated for by the higher density in the bottom half. The pipette usually takes a spherical sample of 20 cm^3 in a manner discussed by Keen (1931, p. 32) from the work of Köhn, and the plummet and hydrometer displace roughly cylindrical volumes of about 35 and 60 cm^3, respectively. Since the top and bottom halves will rarely compensate exactly, it follows that the depth to which the reading is assumed to apply will usually be in error. The error will be greatest in the hydrometer because of its long bulb and a method for correcting for this was proposed by Day (1950). However, any error in the assumed settling depth, l, and hence in settling velocity, v, will be small relative for example to the 100-fold range of v that separates the limiting sizes of silt and clay fractions (20 and 2 μm) on the ISSS scale (Fig. 1.2).

For precision, the three methods rank in decreasing order pipette → plummet → hydrometer. The pipette method is regarded as the standard one because of the following advantages over one or both of the more rapid density methods. The first is that it measures C independently of ρ_s which as we have seen can vary between soils and between different minerals in the one soil. Correction for ρ_s is seldom attempted for the density methods and it can be shown from Equations (1.13) and (1.14) that, for the fairly representative range 2.5 to 2.8 Mg m^{-3}, the maximum error is 3 per cent of the value of C when the density is assumed to be 2.65 Mg m^{-3}. It may be noted that the estimate of radius from settling velocity through Stokes' law is also affected by ρ_s but, within the above range, the maximum error is 5 per cent of r which is small relative to the large range of size found among soil particles. The second advantage of the pipette method is that it is free from error caused by sediment collecting on the shoulders of hydrometers and plummets. This only affects early readings when the larger particles can accumulate rapidly on the shoulders during the time of observation. It can be minimized by having the instrument in the suspension for as short a time as possible.

Other advantages of the pipette are also shared by the plummet. Both can operate with the low initial concentration of 20 g soil per litre instead of the 50 g required for the hydrometer because of its low sensitivity. This means there is less interference between settling particles and less risk of flocculation using the pipette or plummet. A further advantage for them both is that the depth of measurement is selected in advance in contrast to the hydrometer which floats at a depth that depends on the suspension density. Hence readings taken with the pipette or plummet at a predetermined time can be made to correspond to fraction sizes but hydrometer results have to be obtained for particular sizes by interpolation. There is also an advantage for the pipette and plummet in having a greater depth for their measurements than the hydrometer has when it floats high in dense suspensions. This reduces the relative error that arises in the assumed settling depth because, as already discussed, the measurement is not made precisely at a horizontal plane. It also results in a longer settling time for a given velocity, so that for early readings there will be less effect from any eddies that continue after the suspension has been stirred. However, these several disadvantages of the hydrometer are usually marginal in their effects so that in most soils reasonable agreement can be expected with the pipette. The grosser inaccuracies in particle size analysis arise more particularly from inadequate dispersion which can affect all settling methods.

Other settling procedures are also in use. In the ISSS method of particle size analysis (McIntyre and Loveday, 1974*b*), the sand fraction is separated by decanting all the fine material with a settling velocity less than that corresponding to 20 μm diameter. This is done repeatedly (and automatically in the method of Hutton, 1955) until the sand in this decantation vessel is virtually free of finer material. Another settling method, used chiefly in water laboratories (Guy, 1969) for examining sediment loads in streams where the concentration of the suspension is relatively small, makes use of the so-called bottom withdrawal tube (Fig. 1.10). In this method the bottom part of a suspension column is allowed to run out of the tube and the total mass of the particles that have accumulated there in a given settling time is determined. The particle size distribution can be determined from the masses successively withdrawn from the suspension over a range of settling times.

All the settling methods dealt with so far depend upon Stokes' law. The settling tube method of water laboratories (Guy, 1969) makes use of the empirical relations demonstrated in Fig. 1.9 to exist between d and v for large particle sizes where Stokes' law does not apply. The material to be analysed is released at the upper end of a column of water and the amount settled out is recorded as a function of time either from its volume (visually) or from the mass accumulating on a pan suspended from a balance at a given depth. Since the density of the suspension is necessarily greater above the pan than below it, convection currents make this weighing method unsuitable for fine material (Keen, 1931, p. 66).

1.10 Surface area of particles

Within the clay fraction there are particles small enough to undergo erratic Brownian movement arising from bombardment by water molecules in thermal motion. For this order of size, settling procedures are uncertain and use is made of methods that estimate size in terms of the surface area per unit quantity of the material. This is called the specific surface area and in soils it is usually given in units of $m^2 g^{-1}$. For a spherical particle of radius, r, and density, ρ_s, this is $4\pi r^2/\frac{4}{3}\pi r^3 \rho_s$ or $3/r\rho_s$. Thus spherical particles 2 mm in diameter have a specific area of 1.13 $m^2 kg^{-1}$ when $\rho_s = 2650$ kg m^{-3} while for those of 2 μm diameter and the same density it is 1.13×10^3 m^2 kg^{-1}. However, the shape of most clay particles is better approximated by a disc of radius, r, and thickness, a, than by a sphere. In the case of a platelet consisting of a single layer of fully dispersed montmorillonite in which $r \gg a$, the specific area of surface from this model is $(2\pi r^2 + 2\pi ra)/a\pi r^2 \rho_s$ or, when the edge area of the thin disc is neglected, $2/a\rho_s$. Taking $a = 1$ nm and $\rho_s = 2650$ kg m^{-3} we have for the specific surface area 750×10^3 m^2 kg^{-1}. This is comparable in magnitude with values obtained experimentally for montmorillonite as in Tables 1.6 and 1.8. An alternative way of obtaining this calculated value is given by van Olphen (1977, p. 254).

The electron microscope enables shapes and sizes of clay particles to be examined and with the aid of a shadow-casting procedure the study is made three-dimensional, as in Fig. 1.4 for kaolinite. However, other soil minerals do not provide as clear a picture as do the relatively large crystals of kaolinite, and hence their surface area is not readily determined in that way. An adsorption method developed by Brunauer, Emmett and Teller (1938) is widely used to determine the specific surface area of clays. Nitrogen is adsorbed at low temperature in the neighbourhood of the boiling point of liquid nitrogen on to the surface of the porous material. The heat of adsorption of the nitrogen decreases with each successive layer of molecules adsorbed but, by making the simplifying assumption that for layers subsequent to the first it is equal to the heat of condensation of the gas, the following equation was developed

$$\frac{p}{V(p_0 - p)} = \frac{1}{V_m c} + \left(\frac{c-1}{V_m c}\right)\left(\frac{p}{p_0}\right), \qquad (1.15)$$

where V is the total volume of gas adsorbed, V_m is the volume of gas adsorbed when the entire adsorbent surface is covered with a complete unimolecular layer, p is the pressure of the gas, p_0 the saturation pressure of the gas and c is a constant involving the heat of adsorption in the first layer and the heat of condensation. Data obeying the Brunauer, Emmett and Teller (BET) theory will fall on a straight line when $p/V(p_0 - p)$ is plotted against p/p_0 and the value of the constants V_m and c can be obtained from its slope and intercept. The surface area of the material can then be calculated from V_m on the basis that the molecules adsorbed in the first molecular layer pack uniformly over the surface with a known area for each. Since nitrogen does not enter between the layers of an expandable crystal lattice, it gives only the external surface area of clays including montmorillonite and vermiculite. Other gases and vapours can be used with the BET method and water vapour has accordingly been used in work reviewed by van Olphen (1975) to measure the total (internal as well as external) surface of clays. But since water has a polar molecule that is strongly attracted by exchangeable cations, it was not expected that it would pack uniformly over the surface into a true monolayer, as nitrogen is presumed to do. Hence other methods have been proposed for measuring total surface more critically.

Dyal and Hendricks (1950) introduced a different type of method for surface area measurement, in which ethylene glycol is added and allowed

Table 1.8. *External and total surface areas of clay minerals determined by adsorption of nitrogen (BET method) and cetyl pyridinium bromide, respectively*

| | Specific surface area, $m^2 g^{-1}$ | |
	External	Total
Kaolinite	20	19
Illite	112	117
Montmorillonite	66	733

to evaporate until the rate of loss decreases, at which point a monolayer is presumed to exist. Later ethylene glycol monoethyl ether was used as the absorbate, making the method more rapid and precise because it evaporates more rapidly. Carter, Mortland and Kemper (1986) have described this method which is widely used for measuring specific surface area of clays including the internal as well as external surfaces. Greenland and Quirk (1962) have used cetyl pyridinium bromide absorbed from solution in a further method for measuring the total surface area. Results obtained by them in this way are compared in Table 1.8 with others they obtained for external surface area only by adsorption of nitrogen. The large internal area of montmorillonite is represented by the difference of more than $600 \, m^2 g^{-1}$. The data are averages of the original results after omitting from the six kaolinite samples one that was noted as exceptional by the authors.

2

Interaction of soil and water

2.1 Retention of water by the matrix

Water is held within the soil matrix by adsorption at surfaces of particles and by capillarity in the pores. We have seen in Chapter 1 how surface adsorption is influenced by the specific surface area and the exchangeable cations. Clay particles in particular are able to adsorb water actively and swelling can occur as a consequence. Surfaces of quartz grains are not so reactive and only limited surface adsorption can occur in a bed of sand. However, water can be drawn into the pores between the grains by capillarity. It is not always possible to distinguish which of these two mechanisms – surface or capillary adsorption – controls water retention and it is not necessary for our present purposes to do so. Their combined effect on water retention can be measured using a tensiometer as in Fig. 2.1 in which water at A in the soil is hydraulically continuous through a porous ceramic cup with a column of water in the glass manometer tube. At equilibrium, the height of the open end of a hanging water column relative to A is h and the gauge pressure of water in the tensiometer is given by

$$p = \rho g h \qquad (2.1)$$

where ρ is the density of the water and g is the acceleration due to gravity. Gauge pressures are zero at atmospheric pressure and have negative values when h is negative as is ordinarily the case in tensiometers. For the convenience of dropping the sign from negative values of p and h in unsaturated soil, their numerical values $|p|$ and $|h|$ are often referred to as matric suction, s.

When the open end of the water column is higher than the point of measurement in the soil as at B in Fig. 2.1, p and h have positive values. The point B is below a water table and in such cases the measurement can be made by a piezometer which is a simple device fitted with a screen over its opening at B to hold the soil in place. The tensiometer cup at A differs from the screen at B in having pores that are sufficiently small to prevent air from entering the saturated cup when p is less than

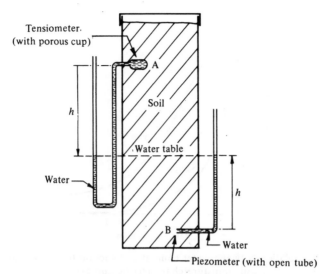

Fig. 2.1. A tensiometer located above a water table (where $h < 0$) and a piezometer below a water table (where $h > 0$) in a column of soil at static equilibrium.

atmospheric pressure. The terms pore-water pressure and pressure head are often used for p and h respectively.

The relation between water content, θ, and suction, s, is a basic property of a soil that is important enough to warrant the special name 'moisture characteristic' or 'water retention curve' sometimes given to it. Examples of curves $\theta(s)$ for two soils drying from saturation are given in Fig. 2.2. From these it can be seen that sandy soils release more of their water at low suctions than do more clayey soils. Further a sand of fairly uniform particle size releases most of its water over a small range of suction. Commonly suctions of 1 and 150 m are selected as useful reference points on the curve corresponding in an approximate way to the water content at field capacity and the permanent wilting point, respectively, in many soils. But they differ physically from these in being well-defined equilibrium values.

The range of matric suction in Fig. 2.2 far exceeds that covered by tensiometers which break down because of air entry when s reaches about 8.5 m. Higher values can be reached with a pressure membrane apparatus using a gas pressure rather than a suction to force water from the soil sample. Also adsorptive forces for water within the matrix lower the vapour pressure of soil water which therefore offers another way of determining matric suction, provided allowance is made for the effect that solutes in the soil solution also have upon the vapour pressure. These methods, which together enable the full range from wet to dry to be covered, will be described in Chapter 3. The data on which the curves

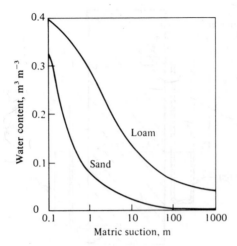

Fig. 2.2. The relation between water content and suction for a sand and a loam soil (moisture characteristic curves).

of Fig. 2.2 are based are from progressively drying soils. The curves for wetting soils would differ from these for reasons that are given later in this chapter.

The relation between suction, s, and water content, θ, obtained experimentally by these methods can be represented approximately by an empirical equation of the form used by Visser (1969) and others,

$$s = a\theta^b,$$

where a and b are parameters that can be determined for a particular soil from the intercept and slope of a straight line fitted to the plot of log s against log θ. It may be noted that b has negative values since θ varies inversely with s. Equations such as this for $s(\theta)$ or $\theta(s)$ with parameters that are evaluated by curve fitting are used to represent the moisture characteristic in the analytical solution of problems of water movement. Further examples are given in Chapter 8 as Equations 8.11 and 8.13.

The amount of water held by a soil at a given suction is influenced by a number of properties of the soil including its texture, structure, organic content and the nature of its clay minerals. Regression equations have been established showing the effect of these properties on water content at various suctions. Gupta and Larson (1979) have tabulated regression coefficients for predicting water content from the sand, silt, clay and organic matter contents, and bulk density, at each of several suctions. Hence the moisture characteristic of a soil can be drawn up when these properties are known. Haverkamp and Parlange (1986) used the cumulative curve for the particle size distribution of a soil (together with other soil properties) as a basis for predicting it. In a different method for employing soil properties, Williams *et al.* (1983) arranged soils into eight groups according to similarities in their moisture characteristics and determined how each of several soil properties influenced the grouping. This enabled them to predict the moisture characteristic group to which a soil belonged from its known properties. Among the properties used to identify the grouping, field texture and structure descriptions as well as quantitative data from particle size and clay mineral analysis were found to be important.

These and other approximate methods have been proposed for estimating the moisture characteristic with the aim of avoiding direct measurements of it. Details and tests of some of the methods are presented by Ahuja, Naney and Williams (1985). One of the obstacles to methods that make use of soil properties is that pore space geometry is not well represented in the criteria used. Clay and organic content and type of clay mineral relate well to surface adsorption. But size distribution of pores, which greatly affects water retention by capillary adsorption, is inadequately represented. We shall show later how this property is in fact best obtained from the moisture characteristic itself.

2.2 Potential energy of soil water

Water in soil is seldom in the static state represented in Fig. 2.1. Under non-equilibrium conditions there is a tendency for it to move at a given point in a direction that results from the combined effect of gravity, hydrostatic pressure and other possible forces. The movement is towards a position of lower energy since energy is used up in overcoming the viscous resistance to flow. The force resulting from the various water-moving forces can be obtained in both magnitude and direction from the gradient of the potential energy at the point. This will be seen in Chapter 4 to be basic to the treatment of water transport in soil.

The influence of these forces is therefore treated in terms of the amount of work that must be done against any one or any particular combination of them when water from a reference pool is added to the soil water. When work is performed on the water under certain specified conditions, its potential energy is changed by an amount that is equal to the work done. The change in potential energy from that of water in the reference pool is called its potential. The following potentials of soil water are defined using definitions of a committee of the International Society of Soil Science (Aslyng 1963), modified to accommodate decisions by a later committee of the Society (Bolt, 1976).

The total potential, Φ, of soil water is the amount of useful work that must be done per unit quantity of pure water to transfer reversibly and isothermally an infinitesimal quantity of water from a pool of pure water at a specified elevation at standard atmospheric pressure to the soil water at the point under consideration. The required condition that the process is reversible means that no energy is lost in friction and the second condition requiring it to be isothermal means that none goes into changing the temperature. Also the transferred quantity has to be infinitesimally small so that the soil water remains unaffected by the transfer. Measureable components of total potential will be defined in a similar way.

Pressure potential, P, is the amount of useful work that must be done per unit quantity of pure water to transfer reversibly and isothermally to the soil water an infinitesimal quantity of water from a pool at standard atmospheric pressure that contains a solution identical in composition to the soil water and is at the elevation of the point under consideration. Before use can be made of this, it is necessary to be able to express P in terms of a measureable gauge pressure, p, as given by a tensiometer or piezometer. If a volume, V, of water is transferred from a body of water where the gauge pressure is zero to one where it is p, the work done against p under the prescribed conditions is pV, as explained by Edlefsen and Anderson (1943, p. 103). This can be illustrated (after Rose, 1966) by assuming this water to be displaced from a tube of length, l, and cross-sectional area, A, into water at pressure, p. The work done in this hypothetical transfer against p will then be pAl or pV. The work per unit volume transferred is therefore p which thus represents the pressure potential, P_v, per unit volume of water. The pressure may be represented as in Equation (2.1) by the height, h, of the open end of a water column in a manometer relative to the point of measurement. Hence

$$P_v = p = \rho g h, \qquad (2.2a)$$

where ρ is the density of water, g is the acceleration due to gravity, and h is called the pressure head. Since the almost incompressible volume on which the work was done contains a mass ρV and a weight $\rho g V$ of water, the potential can also be expressed per unit mass, P_m, and per unit weight, P_w, as

$$P_m = p/\rho = g h \qquad (2.2b)$$

and

$$P_w = p/\rho g = h. \qquad (2.2c)$$

P, p and h are greater or less than zero depending on whether or not the point is submerged.

The above definition of pressure potential can be taken to cover the state of water (1) under a positive pressure resulting from submergence below a water table (as measured at B in Fig. 2.1) or (2) under the influence of attractive forces which are represented as an equivalent negative pressure or suction in the tensiometer at A in Fig. 2.1. The second of these alternatives is the matric potential which depends on water content and is discussed separately below because of its importance for dealing with retention and movement of water in soil that is not submerged. However, since positive and negative pressures can both be measured by one instrument, the tensiometer, it is convenient to define matric potential as part of P (Bolt, 1976). The definition includes also within pressure potential the effect (if any) of having an external gas pressure other than atmospheric pressure. If there is a difference in gas pressure between the reference pool and the soil, the liquid pressure as measured at A or B will be affected by an amount that is equal to this difference. This component of pressure potential, called pneumatic potential (Rose, 1966), will be used only in dealing with the pressure membrane apparatus.

Matric potential, ψ, is a pressure potential that arises from the interaction of water with the matrix of solid particles in which it is embedded. Water added to soil is subject to forces of capillary and surface adsorption that vary with water content and, because these forces are attractive, matric potential has negative values under the terms of our definition. Matric potential is defined in the same way as pressure potential except that any effect from a non-standard gas pressure is excluded. It is the amount of useful work that must be done per unit quantity of pure water to transfer reversibly and isothermally to the soil water an infinitesimal quantity of water from a pool that contains a solution identical in composition to the soil water and is at the elevation and the external gas pressure of the point under consideration.

The forces against which work has to be done in adding unit quantity of water can be represented by the gauge pressure, p, of a tensiometer (as for pressure potential) and we can therefore re-write Equation (2.2) for the matric potential:

$$\psi_v = p = \rho g h \qquad (2.3a)$$

$$\psi_m = p/\rho = gh \qquad (2.3b)$$

$$\psi_w = p/\rho g = h. \qquad (2.3c)$$

The subscripts v, m, and w refer, respectively, to the volume, mass and weight basis and h is the pressure head as before. Here, the quantities h, p and ψ vary with water content as illustrated in Fig. 2.2 and have negative values. The corresponding suction, s, is given by $|p|$ or $|h|$. The contribution (if any) to ψ from overburden will be discussed in Section 2.6 on swelling. Bolt (1976) used the name 'envelope pressure potential' for this as a subcomponent of matric potential but attempts to evaluate it separately are seldom made.

Matric potential was introduced in classical theoretical and experimental work of Buckingham (1907) under the name capillary potential. Since capillarity is not always the sole responsible mechanism, Marshall (1959a, p. 14) proposed instead a term that would associate water retention with the solid phase and its internal geometry. From names considered for this, he chose the term 'matrix' suggested to him by C. G. Gurr. The roles of both capillarity and surface adsorption in water retention will be discussed later in this chapter.

Osmotic potential, Π, arises from the presence of solutes in the soil water. The work of adding water is done against a pressure, p_π, as given by the gauge pressure to which water must be subjected in order to be in equilibrium through a semipermeable membrane with a sample of the soil solution free of the soil. This gauge pressure can be represented by the height, h_π, of a hanging water column with respect to the point of measurement in the solution. As in the case of the tensiometer in unsaturated soil, it is always negative. It is equal but of opposite sign to the osmotic pressure, π, which is the pressure to which the solution (rather than the water) is subjected in order to be at equilibrium through a semipermeable membrane with pure water. Osmotic pressure can be regarded as a solute suction (Richards, 1960) which when added to matric suction gives the total suction, $|h_\pi| + |h|$, as in Fig. 2.3.

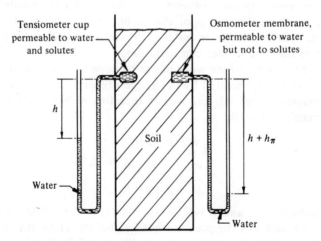

Fig. 2.3. A tensiometer and an osmometer in a column of soil showing the effects of unsaturation ($h < 0$) and solutes ($h_\pi < 0$) on retention of water by soil.

Osmotic potential is defined as the amount of useful work that must be done per unit quantity of pure water to transfer reversibly and isothermally an infinitesimal quantity of water from a pool of pure water at a specified elevation at atmospheric pressure to a pool containing a solution identical in composition with the soil water at the point under consideration but in all other respects identical to the reference pool. In this case the work is done against a solute suction and the values will be negative as in the case of matric potential. If we consider the transfer of unit quantity of water from a body of water at zero gauge pressure to one at a gauge pressure of $-\pi$, we find by the same process as was used for pressure and matric potentials that osmotic potential can be represented as

$$\Pi_v = -\pi = \rho g h_\pi \qquad (2.4a)$$

$$\Pi_m = -\pi/\rho = g h_\pi \qquad (2.4b)$$

$$\Pi_w = -\pi/\rho g = h_\pi. \qquad (2.4c)$$

Gravitational potential, Z, is defined as the amount of useful work that must be done per unit quantity of pure water in order to transfer reversibly and isothermally an infinitesimal quantity of water from a pool containing a solution identical in composition to the soil water at a specified elevation at atmospheric pressure to a similar pool at the elevation of the point under consideration. This represents work done against the gravitational field and hence gravitational potential will be positive if the point is above the specified reference plane and negative if below. In practice, the soil surface or the water table is commonly used as a reference plane. If a volume V of water is moved from zero height at the reference plane to the height, z, of the point, the amount of work done is $V\rho g z$. Hence the gravitational potential per unit volume of water is given by

$$Z_v = \rho g z. \qquad (2.5a)$$

Since the volume contains a mass ρV and a weight $\rho g V$, the gravitational potentials per unit mass and unit weight are

$$Z_m = gz \qquad (2.5b)$$

$$Z_w = z. \qquad (2.5c)$$

The height z is also called the elevation head.

Pressure, osmotic and gravitational potentials are used in the following combinations:

$$\text{total potential, } \Phi = P + \Pi + Z$$

$$\text{hydraulic potential, } \phi = P + Z.$$

For soil that is not submerged, ψ can be substituted for P in these combinations. Total and hydraulic potentials will be used in discussing water movement. Water is moved by a force that is given in magnitude and direction by the gradient of an appropriate combination of potentials. For water moving as a liquid through soil, the force is equal to the gradient of the hydraulic potential. One further combination in much use called the water potential sums the effects of the matrix and solutes on water retention, that is

$$\text{water potential} = \psi + \Pi.$$

This provides a basis for examining the release of water from soil to plants through semipermeable membranes of the root system and from soil to the atmosphere by evaporation. This combination is measured from the vapour pressure of soil water which, because it is adsorbed within the matrix and contains solutes, is lower than that of free, pure water.

It may be noted here that (1) methods quoted above for measuring potentials are not necessarily the best available as will be seen in Chapter 3; (2) suffixes used in this section will be omitted when the meaning of symbols is clear without them; and (3) heads represented as heights of columns of pure water here will not commonly be so in actual measurements because of solutes (Bolt, 1976). However, the effect this has on density is of little practical consequence except in saline groundwaters for which it is discussed by Lusczynski (1961).

2.3 Units of potential

The unit quantity of water on which potential is based can be a volume, mass or weight. All three bases are in use in soil physics. Potential per unit volume is expressed in joules per cubic metre ($J\,m^{-3}$) or, since this has the dimensions of pressure, in the equivalent pressure units, newtons per square metre ($N\,m^{-2}$) or pascals (Pa). Potential per unit mass of water is expressed as joules per kilogram ($J\,kg^{-1}$) and per unit weight it is a length (head) in metres. A potential per unit weight of h metres is equivalent to $gh\,J\,kg^{-1}$ on a mass basis or ρgh Pa on a volume basis. In these conversions, g can be taken as $9.80\,m\,s^{-2}$ and ρ as $10^3\,kg\,m^{-3}$ with sufficient accuracy for most purposes.

For comparison of these units, reference can be made to a guide given in Table 2.1 to the order of magnitude of matric potential under different soil water conditions. Values are given there for potential per unit volume, mass, and weight in SI units of kPa, $J\,kg^{-1}$, and m, respectively. Potential per unit volume is also given in bars. The bar ($= 10^2$ kPa) with magnitude of about one atmosphere (1 bar $= 0.99$ atmosphere $= 10.22$ m column of

Table 2.1. *Matric potential in different units showing its order of magnitude over a broad range of soil water conditions*

Condition at the quoted potential	Matric potential			
	kPa or J kg^{-1}	m	bar	pF
Saturated or nearly so	-10^{-1}	-10^{-2}	-10^{-3}	0.0
Near field capacity	-10	-1	-10^{-1}	2.0
Near permanent wilting point	-1.5×10^3	-1.5×10^2	-1.5×10	4.2
Air dry at relative vapour pressure of 0.48	-10^5	-10^4	-10^3	6.0
Conversion from a matric potential of h metres	$9.8h$	h	$0.098h$	$\log 100|h|$

water at 20 °C) is a pressure unit that was much used in soil physics and meteorology but it is not part of the SI system and its use is declining. The term pF meaning the logarithm to the base 10 of the suction expressed in cm remains also in some use and values for this are listed in Table 2.1. Conversion factors in the table show that the apparent 10-fold difference between the potential values given in metres and those in other units is a rounded factor. This approximation will be used occasionally. For details of units see Appendix A.

2.4 Water in relatively dry soil

The definition of matric potential does not require a distinction to be made between the different mechanisms responsible for attracting water to the matrix. There is not much point in doing so when methods like the tensiometer that are used for measuring matric potential are not able to separate them. However, those mechanisms influence the way soil behaves. In particular water attracted by reactive clay minerals will cause swelling but when water is attracted by capillarity into the pores of a sandy soil there is no swelling at all. It is therefore necessary to consider the interaction between water and the solid phase and also some properties of water itself that affect this interaction.

Water has unusual properties and behaviour because of the structure of the molecule H_2O. The single electron of each hydrogen atom is involved in bonding it to the oxygen atom thus leaving a positive charge on both hydrogen atoms. Since the two hydrogen atoms are arranged towards one side of the oxygen atom, as shown in Fig. 2.4a, the water molecule acts as an electrical dipole with a positive pole due to hydrogen

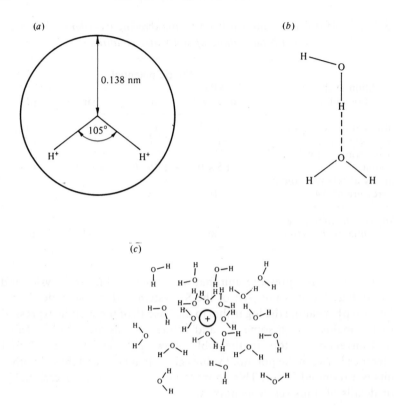

Fig. 2.4. Model of the water molecule. (a) Water molecule showing the position of the two hydrogen protons. (b) Hydrogen bonding between two water molecules. (c) Effect of a charged ion on the arrangement of the polar water molecules.

atoms and a negative pole due to the oxygen atom. Furthermore one molecule of water can link up with another through hydrogen bonding thus allowing some degree of association between molecules in liquid water, as established by J. D. Bernal and R. H. Fowler in 1933 (Fig. 2.4b). Many structures have been proposed for water. According to the flickering-cluster structure of Frank and Wen (1957), at any instant some of the water molecules are hydrogen-bonded together in clusters while the others are free. These clusters form only momentarily and immediately vanish to reform in different combinations. A structure such as this helps to explain anomalies in the behaviour of liquid water. Horne (1968) in a review of work on the structure of water, cites as one example the effect of hydrostatic pressure upon viscosity of water. Due to increased crowding of molecules, the viscosity of a liquid should increase with pressure. But in water it first decreases and only after the pressure reaches about 100 MPa does it increase. This is explained by considering that the clusters of hydrogen-bonded molecules (which are more openly spaced than 'free' molecules as is also the case in ice structure) are progressively eliminated by increasing the pressure until the water behaves like a normal liquid. Among the many unusual properties of water are two of especial importance in soil physics – its high surface tension and heat capacity.

Because their centres of positive and negative charges are separated, water molecules are attracted and oriented by the electrostatic field of a charged ion and this results in the hydration of solute ions (Fig. 2.4c). As evidence of the rearrangement of water molecules that accompanies hydration, there is commonly a reduction in the overall volume (electro-striction) when a salt is added to water. Hydration of ions can obviously occur in soil too, when the polar water molecules interact with the exchangeable cations and this was discussed in Chapter 1 as a principal mechanism in water adsorption at the first stage of wetting of soil. Other possible mechanisms of adsorption at this stage are (1) intermolecular attraction between the solid surface and water over a short range due to van der Waals forces and (2) hydrogen bonding of water molecules to oxygen atoms on the solid surface. Low and White (1970) consider that hydrogen bonding gives rise to a partially bonded network of water molecules originating at the oxygen atoms of the mineral surface and extending outward with decreasing effectiveness for a distance of 5 nm or more. However, the existence of any long range structure in water at interfaces is open to question (Conway, 1977).

The vapour pressure of water adsorbed in soil provides a means for examining the interaction between water and soil under relatively dry conditions. The water content is measured in equilibrium with various relative vapour pressures, e/e_0, where e is the vapour pressure of the soil water which because of the attraction of the solid matrix is lower than the vapour pressure of pure free water, e_0, at the same temperature. Each measurement is made by exposing a soil or clay mineral sample to water vapour in equilibrium with an aqueous solution of known vapour pressure, e. Thomas, in papers published in the period 1921 to 1928 showed that the amount of water adsorbed by a soil at a given value of e/e_0 increased with clay content of the soil and was affected also by the type of exchangeable cations (Thomas, 1928). It was later shown that type of clay mineral was also important. Montmorillonite with its expanding crystal lattice adsorbs much more water at a given value of e/e_0 than kaolinite with its relatively large crystals and smaller specific surface area available for adsorption. Illite has an intermediate position with small crystals but without the expanding lattice of montmorillonite. These effects of clay properties on adsorption are illustrated in Fig. 2.5.

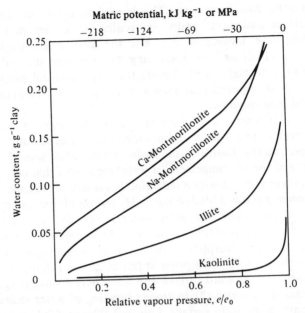

Fig. 2.5. Adsorption of water vapour by different clays. (Data for illite and kaolinite from Orchiston, 1954.)

Soluble salts also lower the vapour pressure of soil water but, if they can be neglected as in the materials discussed above, e/e_0 can be used to calculate the matric potential, ψ, through the relation

$$\psi_m = RTM^{-1} \ln (e/e_0) \tag{2.6}$$

where R is the gas constant $(8.3143 \text{ J K}^{-1} \text{mol}^{-1})$, T is the temperature $(^\circ\text{C}+273.16)$ in kelvin (K), M is the mass in kg of a mole of water (0.018015), and ln represents the natural logarithm. If for example the relative vapour pressure of water in soil at a temperature of $20\,^\circ\text{C}$ is $e/e_0 = 0.85$, we have,

$$\psi_m = 8.3143 \times 293.16 \times (1/0.018015) \times (-0.162516)$$
$$= -21988 \text{ J kg}^{-1}.$$

Values for ψ per unit mass of water obtained in this way are shown in Fig. 2.5 which thus becomes a moisture characteristic diagram over a range of low water contents for these materials. The relative dryness of the range can be illustrated by the fact that when e/e_0 is as great as 0.989, ψ is approximately -1500 J kg^{-1} (or 15 bars suction) which as indicated in Table 2.1 is in the region of permanent wilting point of plants in soils. Hence the range of interest in agriculture lies almost wholly between 0.989 and 1. It will be shown in Chapter 3, where Equation (2.6) is explained and applied more fully, that ψ (or $\psi + \Pi$ if solutes are present) can be measured within that narrow range of e/e_0 using hygrometers.

In Fig. 2.5 the curves for Na- and Ca-montmorillonite tend to cross over at $e/e_0 = 0.90$ which is the stage of wetting shown in Table 1.4 at which macroscopic swelling and plasticity become apparent in Na-montmorillonite. Beyond that stage of wetting, at $e/e_0 = 0.975$, Anderson, Brown and Buol (1967) found the water film in montmorillonite to be sufficiently developed for diffusion of dyes to be measurable. With further wetting to $e/e_0 = 0.989$, where the matric potential is -1.5 MPa and soil becomes just moist enough to support plant life, water can move as a liquid in soil (Marshall and Gurr, 1954; Scotter, 1976) and it is presumed in this book to have the usual properties of water in bulk.

2.5 Capillarity

The molecules within a body of water (as in any liquid or solid body) are subjected to attractive forces from their neighbours in all directions; but those at an air-water interface are not and they are consequently attracted inward. Because of this, energy has to be expended on the system if the area of the interface is to be enlarged. The amount required per unit of new area is the surface tension which for water is 73 mJ m^{-2} at 20 °C. This may also be expressed as a force per unit length of 73 mN m^{-1}.

When water is placed on a clean glass surface it will spread readily under the influence of the attractive forces that have been discussed in the previous section and the angle of contact, α, between the water and the glass is then said to be zero. But if the glass is not clean the water molecules may not be attracted so strongly. They tend to cohere to one another rather than adhere to the glass; and the water will then contact the glass at some angle α that is greater than zero. Organic matter can cause angles of contact between water and dry soil particles to be greater than zero and when this happens the entry of water especially in dry sandy soils becomes affected as will be discussed later.

The water held in a glass tube of a small radius, r_t, at an equilibrium height, s, above the free water surface as in Fig. 2.6, provides an example of capillary retention. In this three-phase system of glass, water and air, the column of water is held up by a force $2\pi r_t \gamma \cos \alpha$ where γ is the surface tension of water and α is the contact angle of water with the tube. This expression arises from the force $\gamma \cos \alpha$ acting vertically upward per unit length of the circumference of the circle of length $2\pi r_t$ along which the air-water interface contacts the wall of the tube. The

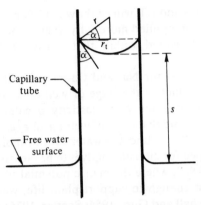

Fig. 2.6. Water in a capillary tube showing the relation between the radius of the tube, r_t, and the radius of the air–water interface, r, when the angle of contact is α. In this central vertical section through a tube, the width of the tube is greatly exaggerated relative to the height, s, of the water.

cylinder of water of length s and mass $\pi r_t^2 s\rho$ that is held up in this way exerts a force downward of $\pi r_t^2 s\rho g$, where ρ is the density of water and g is the acceleration due to gravity. Neglecting the mass of the air displaced by the column of water and the mass of the water lens of the meniscus, we have, on equating the magnitudes of the upward and downward forces,

$$2\pi r_t \gamma \cos \alpha = \pi r_t^2 s\rho g \quad \text{or} \quad (2\gamma \cos \alpha)/r_t = \rho g s = -p, \qquad (2.7)$$

where p is the pressure of water at the top of the column with respect to the free water surface as datum and s can be taken as the suction at the top of the column. From the construction shown in Fig. 2.6, it can be seen that Equation (2.7) can be written as

$$p = 2\gamma/r, \qquad (2.8)$$

where r, the radius of curvature of the water–air meniscus, is given negative values when the centre of curvature is on the air side of the interface.

The water–air interface in a soil pore is not in general part of a spherical surface and it cannot ordinarily be described completely by a single radius of curvature. Two radii of curvature, r_1 and r_2, lying in planes normal to one another are required to describe the shape at any point on the interface. The pressure difference across the interface is then given by

$$p = \gamma(1/r_1 + 1/r_2), \qquad (2.9)$$

where r_1 and r_2 are taken as positive or negative according to whether the centre of curvature lies on the water or on the air side of the interface. Equation (2.9) is derived by equating the amount of work done in moving a small rectangular portion of an interface a small distance under pressure p with the amount done in changing the area of this portion in the course of this movement. Details are given in books on the properties of matter and in that by Kirkham and Powers (1972). Innumerable combinations of r_1 and r_2 correspond to any one pressure p in soil water. For certain shapes, such as an annular ring of water around the contact zone of two adjoining spheres, the centres of curvature are on opposite sides of the interface and the value of r_1 will therefore be opposite in sign to that of r_2. All pairs will give a negative value for $(1/r_1 + 1/r_2)$ in conformity with a negative value for p.

From Equation (2.8) or (2.9) it can be seen that, for a given configuration of the water–air interface (and hence for a given water content corresponding to that configuration), the pressure p will vary with surface tension. Since surface tension decreases with temperature as shown in Table 2.2, the moisture characteristic, $\theta(p)$, must be temperature dependent whenever capillarity influences adsorption of water by soil. This is shown to be the case by a number of workers. Sometimes the increase of p (i.e. decrease in suction) with increase in temperature is greater than that expected from the decrease in surface tension with temperature. This is explained by the increase in pressure of air trapped in pores when temperature is increased. Haridasan and Jensen (1972) give data and list papers dealing with the effect of temperature on the moisture characteristic. Surface tension of water is also affected by solutes. Tschapek, Scoppa and Wasowski (1978) found that it was lowered by about $8 \, mJ \, m^{-2}$ in soil solutions from surface soils and they found that organic rather than inorganic solutes were mainly responsible.

Table 2.2. *Effect of temperature in °C on surface tension of water in units of $mJ \, m^{-2}$ or $mN \, m^{-1}$ (Kaye and Laby, 1973)*

Temperature	0	10	20	30	40
Surface tension	75.7	74.2	72.75	71.2	69.6

The geometry of the air–water interface throughout an unsaturated porous material with particles of irregular shapes and various sizes is too complex for precise quantitative treatment. To determine pore size, resort is made to Equation (2.7) for calculating an effective radius, r_t, of the largest pore to remain full of water when a matric suction, s, is used to drain the soil. Assuming the contact angle to be zero

$$s = 2\gamma/\rho g r_t. \tag{2.10}$$

The volume of water withdrawn on increasing the suction from s_1 to s_2 in soil with a rigid matrix represents the volume of pores having an effective radius between r_{t_1} and r_{t_2}. A moisture characteristic curve (Fig. 8.5) can in this way be used for soil that does not shrink on drying to show the cumulative amount of pore space that is smaller than a given effective radius, r_t. The determination of an effective radius for pores from Equation (2.10) is a simplification that may be compared with that made for particles of irregular shape in determining particle size from sedimentation rates using Stokes' law. Despite this rather gross simplification and the complication of hysteresis in drying and wetting which is discussed later, the distribution of pores according to size obtained this way was found by Swanson and Peterson (1942) to compare satisfactorily with that obtained when pores were measured and counted microscopically. Applications and calculations are given in Chapter 8.

Water retained by capillarity is under tensile stress as is illustrated in Fig. 2.6 and Equation (2.7) and the tensile strength of water is therefore concerned. It is found in practice that tensiometers break down when a suction of about 8.5 m has been reached because air bubbles develop in the column of water. For this reason the use of tensiometers is limited to rather wet soils and other devices to be described later have to be used to measure greater suctions. However, despite these practical limits, much higher tensile strength can be demonstrated in water if strict precautions are taken to avoid nucleation of air to form bubbles. Using a centrifugal method, Briggs (1949) measured a tensile strength of 22.6 MPa equivalent to a water column of 2230 m and Temperley (1947) on theoretical grounds considered the order of magnitude of the tensile strength of water to be 50 to 100 MPa. In applying capillary theory to the water held in thin films and small pores at high suctions it is often assumed that it is capable of withstanding tensile stresses about as great as the limit measured by Briggs under controlled conditions.

2.6 Swelling

In a soil that contains much of the clay fraction, a change in water content may be accompanied by a change in volume. The shrinking of a soil on drying and its swelling on wetting have very marked effects on soil structure, on water movement and on the stability of buildings and pavements, as will be shown in later chapters.

Haines (1923) observed that during the drying of a remoulded block of a clay soil, the decrease in volume was at first equal to the volume of water lost so that the block remained saturated as drying proceeded. This first stage, called 'normal' shrinkage by Keen (1931, p. 138), was followed by what Haines called 'residual' shrinkage, when the volume decreased less rapidly than the water content as air entered the soil. These stages are represented in Fig. 2.7 in the drying curves for blocks moulded by Holmes (1955) from soil containing 64 per cent of clay. The degree of saturation remains fairly constant during the first stage of drying because the rate of decrease in bulk volume of the plastic soil is approximately the same as the rate of loss of water (normal stage). It then decreases rapidly because of increasing resistance of the soil to deformation as the clay loses its plasticity (residual stage). Holmes found that the matric potential at which the transition from normal to residual shrinkage occurred in his remoulded clay soils was of the order of -10^4 kPa. Macroscopic shrinkage ceased at about -10^5 kPa. Stirk (1954) showed that the lower the clay content, the higher the potential for these transitions. Inactive grains of sand and silt provide a rigid framework that arrests the shrinkage process in a drying soil. With enough of them present, macroscopic shrinkage is of course prevented altogether.

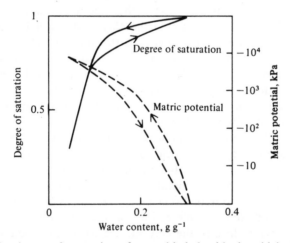

Fig. 2.7. The degree of saturation of remoulded clay blocks which change their volume during wetting and drying. The change of matric potential with water content is also shown (Holmes, 1955).

Lauritzen and Stewart (1941), Lauritzen (1948) and Stirk (1954) working with naturally structured soil rather than remoulded blocks noted that, at the start of drying, there was an initial stage, called 'structural' shrinkage by Stirk, in which the decrease in soil volume was less than the volume of water removed and air entered large pores and cracks.

Clay soil with stable porous structure associated for example with grasslands exhibits structural shrinkage markedly. If normal shrinkage follows
structural shrinkage it will necessarily be at a lower degree of saturation
than in Fig. 2.7 because air has already entered the soil. A large structural
stage in soil of stable structure or an early residual stage in soil of low
clay content will restrict the range of the normal stage which is therefore
dominant mainly in soil of massive structure and high clay content, such
as a clay subsoil.

The mechanism of shrinking and swelling of clays is not fully understood but the behaviour of the clay mineral montmorillonite, as discussed
in Chapter 1, provides a partial guide to the swelling of the other clays.
Water molecules adsorbed at active sites on the layers of the crystal
lattice of Na-montmorillonite cause it to expand. Hydration of the
exchangeable cations is largely responsible for this adsorption and for
the accompanying increase in the interlayer spacing. As a consequence
of this process the clay swells macroscopically. With continuing adsorption it becomes plastic and its further expansion is explained as follows
by the development of a diffuse layer of ions (Fig. 1.7). If, because of
the exchangeable cations in the diffuse layer, the solution midway
between two parallel surfaces has a higher electrolyte concentration than
that of an outer solution bathing the clay, water will be attracted in
osmotically, and the distance between the surfaces will consequently
increase. The exchangeable cations, unable to move freely out to the
bathing solution, act as if retained within a semipermeable membrane.

Clays that do not have an expanding crystal lattice are without the
regular arrangement of parallel sheets capable of separating on swelling
as in Na-montmorillonite. However, the plate-like crystals of most clay
minerals in soils are arranged in packets or domains within each of which
there is a degree of parallel orientation. Swelling is then ascribed to
increasing separation of crystals forming these packets as water is adsorbed. Adsorption probably depends primarily on hydration of exchangeable cations as in the system discussed above. But, as discussed in Section
1.7, the divalent exchangeable cations of ordinary soils may be mainly
in the Stern layer close to the clay surface where they neither hydrate
fully nor form part of a diffuse layer of ions. The swelling of clays in
soil is therefore not of the magnitude of that of Na-montmorillonite.

We may conclude that the swelling behaviour of soils depends on the
clay content of the soil, the cation exchange capacity of the clay minerals
and the types of exchangeable cations associated with them. Swelling is
greater with smectite (which includes montmorillonite) and with illite
than it is with kaolinite because of their greater cation exchange capacity
and it is greater with monovalent than with divalent or trivalent exchangeable cations which are mainly in the Stern layer.

An alternative theory of swelling proposed by Terzaghi (1931) has received some support as a possible supplementary cause of swelling (Murray and Quirk, 1980). According to this theory, the platelets are distorted by capillary forces during drying and recover their shape elastically on re-wetting. Terzaghi's concept of effective stress which underlies this theory will be discussed in Section 9.3.

2.7 Effect of load on water retention

A soil that swells while water is being adsorbed will exert a pressure on any body that tends to confine it. Conversely a load applied to a swollen soil, previously brought into equilibrium with a supply of water under a suction, will force water out of it in much the same way as if an equivalent increase in suction had been applied to it. This behaviour of swelling material contrasts markedly with that of rigid material where the external pressure is supported by a framework of constant bulk volume and where air enters when suction is applied. In practice any pressure arising from overburden of soil lying above the point under consideration or from any building or other body supported by the soil will tend to reduce the equilibrium water content of swelling soils. The degree to which loads are transmitted to the soil water depends on the behaviour of the soils. In a non-swelling soil they can be carried by the solid phase through intergranular contacts and not at all by the water; but in a swelling soil of high clay content and of sufficient water content they may be wholly carried by the water. A compressibility factor, β, ranging in value from 0 to 1 for these two extreme cases was used by Croney and Coleman (1953) for the fraction of the applied pressure supported by the water. It was determined experimentally by measuring the pressure p in the water of the loaded soil with a tensiometer and then measuring the pressure p_u in the water of an unloaded sample immediately after removal from the soil and without change of water content. The value of β for a particular soil at a given water content is then obtained from the relation between these quantities:

$$p = p_u + \beta L, \tag{2.11}$$

where L is the pressure from the overburden and from any other load carried by the soil at the point of measurement and βL is called the envelope pressure (Bolt, 1976). Croney and Coleman also measured β on a sample in a consolidometer from the change in p caused by a change in the applied load L when p_u was kept constant. In a third method it was measured indirectly from the shrinkage curve for the soil taking β to be the slope $dV/d\theta$ of the curve relating volume, V, with water content, θ, in a sample of drying soil. Thus from Fig. 2.7 it can be seen that $\beta = 1$

in moist clay and decreases with decreasing water content to 0. The effect of texture upon β has not been explored greatly but Croney, Coleman and Black (1958) quote 1 for a clay, 0.3 for a silty clay and 0.02 for a sand, under the moist conditions that correspond to an unloaded suction of less than 40 cm. Talsma (1977) measured β on a number of swelling clay soils in the field and on undisturbed samples in the laboratory. He found that, even at high water content, $\beta < 0.25$.

It should be remarked that the compressibility factor of Croney and Coleman applies to volume changes associated only with shrinkage and swelling and not with other cases in which the structure fails under a load greater than any to which it has been previously subjected. This aspect of soil strength will be dealt with in Chapter 9.

The three pressure terms in Equation (2.11) are equivalent to potentials per unit volume of water and the equation can be written as

$$\psi = \psi_u + P_e$$

where ψ is the matric potential of the loaded soil, ψ_u is the matric potential of the unloaded soil and P_e is the envelope pressure potential (equal to βL) arising from the load (Bolt, 1976). In the absence of any load imposed at the surface, the load, L, at a given depth is due to the overburden of soil lying above, as given by

$$L = g \int_z^0 \rho_{wb}\, dz \qquad (z \leqslant 0) \qquad (2.12)$$

where ρ_{wb} is the wet bulk density of the soil, g is the acceleration due to gravity and z is the height relative to the soil surface.

2.8 Hysteresis in the moisture characteristic

The moisture characteristic of a soil has already been discussed for the important case where the water content, θ, decreases as the matric suction, s, increases in a soil drying from saturation to air-dryness. The $\theta(s)$ relation is, as shown by W. B. Haines in 1930, not so simple because, on re-wetting, a different curve is traced. Also if the process is reversed at any intermediate stage of wetting or drying between saturation and air-dryness the curve will follow a different course within the limits set by these two boundary curves. Such hysteresis is illustrated in Fig. 2.8 where the arrows show the order followed in making successive measurements during drying and wetting. It will be seen that after any reversal of direction the drying soil has a higher water content than the wetting soil at the same suction. It might be thought that this is simply a consequence of equilibrium not being reached between water content

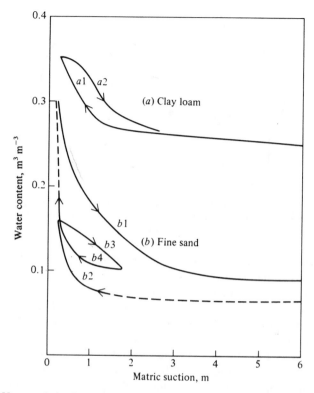

Fig. 2.8. Hysteresis in the relation between water content and suction during the wetting and drying of two soils in the field. (*a*) Soil at 8–9 cm depth wetted (in curve *a*1) by irrigation (Watson, Reginato and Jackson, 1975). (*b*) Soil at 40 cm depth wetted (in curves *b*2 and *b*4) by rain (adapted from Royer and Vachaud, 1975).

and suction when the measurements are made due, for example, to the time taken for air to enter isolated pores during drainage, as noted by Corey and Brooks (1975) and others. Certainly equilibrium is approached slowly but true hysteresis is independent of delay and it is explicable on quite different grounds.

Hysteresis can commonly be explained by the fact that many pores are larger than their openings. The pore illustrated in Fig. 2.9 will remain full of water according to Equation (2.10) until the suction exceeds kr_t^{-1} where r_t is the tube radius and k is a constant. On the other hand on re-wetting, water can refill the pore only when the suction falls to the lower value kR_t^{-1} corresponding to the larger radius, R_t, of the body of the pore. It can be seen on the basis of this behaviour that, at any given suction between these two, the water content of the drying pore exceeds

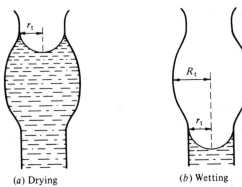

(a) Drying   (b) Wetting

Fig. 2.9. Hysteresis in the amount of water contained in a pore at a given suction due to the 'ink-bottle' effect. R_t is the maximum and r_t the minimum radius of the pore.

that on wetting. This so-called ink-bottle effect is a cause of hysteresis in most porous materials.

Variation in the contact angle of water with soil is also a source of hysteresis under some conditions. The pore shown in Fig. 2.10 will not fill and empty reversibly if the contact angle, α, differs during wetting (α_w) and drying (α_d). Commonly α is presumed to be zero during drying so that cos α in Equation (2.7) equals 1; but in soil that is being wetted this is less likely to be true and with $\alpha_w > 0$ we have cos $\alpha_w < 1$. A higher contact angle for wetting can be due to the presence of surface active substances on the solid and to surface roughness, neither of which is likely to affect the contact angle for drying. From Equation (2.7), the tube radius, r_t, that will support a column of height, s, is given by

$$r_t = (a/s) \cos \alpha$$

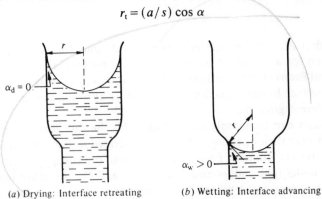

(a) Drying: Interface retreating   (b) Wetting: Interface advancing

Fig. 2.10. Hysteresis due to the effect of contact angle during drying (α_d) and wetting (α_w) on the amount of water contained in a pore at a given suction. The radius of curvature, r, is the same in each case.

where a is a constant. Hence if $\alpha_w > \alpha_d$ we have $\cos \alpha_w < \cos \alpha_d$ and at equilibrium with a given suction, s, r_t must then be smaller for wetting than for drying conditions. This means that the pore shown in Fig. 2.10 will contain less water on wetting than on drying at a given suction and hysteresis will therefore occur.

In clays, hysteresis can occur even when the soil remains saturated during a cycle of wetting and drying. Moisture characteristics for blocks made of compressed clay by Holmes (1955) illustrate this in Fig. 2.7. The volume change in these blocks was equal to the water content change at water contents above $0.17 \mathrm{~g~g}^{-1}$ as shown by the degree of saturation being constant (near saturation). The hysteresis that is shown to occur in that range cannot therefore be explained by the ink-bottle effect or by contact angle differences both of which apply only to three-phase systems where the amount of air-filled pore space is changed during wetting and drying. Hysteresis in this clay is apparently associated with particle rearrangement during swelling and shrinking instead of with the filling and emptying of pores. A plausible explanation of hysteresis in these circumstances is that the internal resistance to particle rearrangements keeps the volume higher on shrinking than on swelling when the clay is coming to equilibrium with a given suction.

Air trapped in soil during wetting reduces the amount of water able to enter the pore space. The curves for the first drying and wetting cycle of an initially saturated soil may not close as a consequence, although subsequent cycles will do so. If desired this effect can be avoided in the laboratory by wetting under reduced air pressure but for some purposes it is more realistic to accept the fact that ordinarily a small part of the pore space remains air-filled at 'saturation'. Removal of this air by the natural processes of solution and diffusion is slow and uncertain.

Hysteresis in the relation between water content and suction affects the use that can be made of moisture characteristics. If we are concerned with soils that are drying from an initial wet condition, the boundary drying curve can be used. In practice this is the curve most often used and unless we are told otherwise we can assume that a single moisture characteristic curve given for a soil is a boundary drying curve. Commonly, in the field, soils gain water more quickly than they lose it and so for much of the time the drying curve applies. This is particularly true of irrigated soil where rapid wetting is followed by a relatively long drying period. However, in field experiments exposed to changing weather conditions of wetting and drying it may be necessary to measure both suction and water content if both are wanted, since one cannot always be inferred from the other with assurance through a moisture characteristic curve. The problem is greatest in sandy soils at low suction, as illustrated in Fig. 2.8. The question also arises of whether a drying or a wetting curve should be employed for estimating pore size distribution using Equation (2.7). It might appear from the preceding discussion of the ink-bottle effect that the drying curve is unduly influenced by the size of openings rather than the size of pores. However, data for the drying curve are more easily obtained and are more reliable than wetting data, which may be affected by the contact angle being greater than zero. Hence they are used consistently in pore size distribution estimates.

Any curve lying between the boundary moisture characteristics can be established approximately for a soil by an application of the independent domain concept of D. H. Everett, A. J. Enderby and others, as described by Poulovassilis (1962) and Childs (1969, p. 129). The assumption is made that the pore water can be divided into volume elements each of which is consistently withdrawn between certain limits of suction during drying and is added between another particular pair of limiting suctions during wetting no matter what the previous history of wetting and drying of the soil may be. According to Poulovassilis and others, agreement with experimentally determined data is reasonably good unless, as observed by Topp and Miller (1966), the pores are of a narrow range of size. Other models of hysteresis have been proposed by Mualem (1984) and Shcherbakov (1985).

3

Measurement of water content and potential

3.1 Water content

The gravimetric method

The water content of soil can be measured by finding the mass, m_w, of water lost upon drying a sample in an oven at 105°C to a constant mass, m_s. The water content on a mass basis is

$$\theta_m = m_w/m_s. \qquad (3.1)$$

Since the amount of water lost increases with the drying temperature in any soils that contain clay or organic matter, the oven temperature must be controlled within a range of about 100–110 °C for routine work. The procedure is set out by Gardner (1986). Despite arbitrary features of the method, it is the standard with which others are compared. The water content as a volume fraction, θ, is obtained from θ_m by means of Equation (1.11),

$$\theta = \theta_m \rho_b/\rho_w \qquad (3.2)$$

where ρ_b is the dry bulk density of the soil and ρ_w is the density of water which for most purposes can be taken to be $1 \, \mathrm{Mg \, m^{-3}}$.

A disadvantage of the method is that sampling of experimental areas interferes with continuing experiments. When holes are bored, roots are cut and infiltration and drainage behaviour are affected. A large number of samples may be required because water can be distributed unevenly in the field due to the effects of variable texture and structure on its retention and movement and due to the variable demands of roots and the interception of rain by leaves of plants. A coefficient of variation of about 10 per cent or more has been found in field samples by a number of workers quoted by Holmes, Taylor and Richards (1967). The effect of such variation upon ability to determine changes of water content in the field may be assessed in the following way. Suppose a soil that is to be studied for seasonal change of water content has a total water content

of 450 mm between the surface and 1.5 m depth. If the water content
were determined on the basis of 15 samples, each of depth increment
100 mm, and the coefficient of variation of each individual sample was
10 per cent, then it may be shown that the standard deviation (SD) of
the total water content, 450 mm, would be 12 mm. Therefore, on the basis
of comparison of two sampled profiles, a difference of 24 mm (2 SDs)
would be significant at approximately the 95 per cent level. On the basis
of comparison of the means of 10 sampled profiles on each of two
occasions, a difference of 7.6 mm could be resolved at the 95 per cent
confidence level. Considering that an evapotranspiration rate of 1–4 mm
day^{-1} is a common range of experience, the variability of soil makes the
moisture sampling approach unattractive.

If texture varies with depth, variation can be reduced by sampling
within morphological horizons rather than within fixed depth intervals
(Keisling *et al.*, 1977).

The neutron moisture meter

The neutron moisture meter as an instrument to measure the water content
of soils in the field was developed in the late 1940s. It was one example
of the appearance at that time of non-destructive methods in industry,
based upon penetrating radiation from radioactive sources. Of all the
elements, hydrogen is by far the most effective in slowing down fast
neutrons, because the mass of a proton and the mass of a neutron are
closely similar. On average, fast neutrons emitted from a point source
lose energy principally by elastic collisions with the hydrogens of the
soil water. The other main constituents of the soil, oxygen, aluminium
and silicon are less effective in slowing the fast neutrons because their
atomic masses are larger. As a consequence, about nine times as many
collisions of a neutron with the oxygen nucleus are needed to reduce its
energy to ambient, thermal energies (and about 14 for Al and 16 for Si)
compared with collisions with the hydrogen nucleus. Once thermalized
the neutrons diffuse through the soil in a way that is similar to gaseous
diffusion, except that they are very soon absorbed by an elemental
nucleus. Although a free neutron decays to a proton and an electron
with a half-life of about 13 minutes, the actual life of a thermal neutron
in even a weakly-absorbing medium such as soil is only of the order of
milliseconds.

These processes and the special properties of hydrogen were exploited
by Belcher, Cuykendall and Sack (1950), Spinks, Lane and Torchinsky
(1951) and Gardner and Kirkham (1952), all apparently unaware of the
others' work. The neutron moisture meter resulted from their work, and
many others, and commercial equipment became available about 1956.
The equipment comprises a probe (see Fig. 3.1) containing a source of

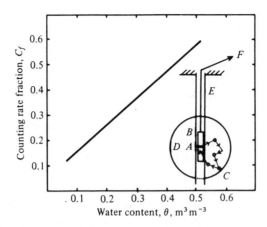

Fig. 3.1. Calibration curve of the neutron moisture meter, C_f versus θ, with C_f the counting rate in the soil as a fraction of the counting rate in water. The inset is a schematic diagram of the process of neutron scattering in the soil. A = annular source; B = counter tube; C = hydrogen nucleus; D = approximately spherical zone of measurement; E = access hole lining; F = to scaler and counting rate recorder.

fast neutrons (^{241}Am mixed with Be) and, close to the source, a detector of slow neutrons. The latter may be a gas-filled proportional counter tube (filled with BF_3 or ^{3}He) or a scintillation detector. The detector does not need to be shielded from the source because it is sensitive only to thermal neutrons. Proportional counting mode is required, however, because there is a background counting rate due to gamma ray emission from the source that must be discriminated against. The probe is suspended on a cable that transmits the neutron-induced pulses to a suitable amplifier, discriminator and counting device.

When the probe is lowered down the access hole in the soil, which is commonly about 50 mm in diameter, a concentration distribution of slow neutrons is established immediately in the soil surrounding the hole. The neutrons populate the space within the borehole as well, where the detector of slow neutrons samples their density. The latter is proportional to the volumetric water concentration in the soil. An example of the final calibration of an instrument is shown in Fig. 3.1, where the response to water content variations is expressed as the ratio of the counting rate with the probe in the soil to the counting rate in a drum of pure water, in each case the access to the medium being in a dry hole with identical casing. The water content profile of the soil can be obtained by lowering the probe to successively greater depths, at increments of about 0.3 m. Because of the large sample volume, any sharp change in water content which may occur at a horizon boundary tends to be smoothed out in the moisture profile, but this does not introduce significant error in the integrated, total water content. However, near the soil surface the probe should be sufficiently deep at the first location to ensure that the volume measured is effectively infinite for the process of slowing down and diffusion of neutrons to be fully developed in the soil itself, and that none escape through the soil surface to the air before they are absorbed in the soil.

The access hole for the probe should have a liner of aluminium or polythene tube, sealed at the bottom, and it is important to avoid any cavities between the soil and the lining material. A power-auger may be used to construct a hole into which the lining tube is subsequently pushed. Sometimes the site could be damaged by bringing up a drilling-rig. Then hand-augering must be resorted to, with reaming of the hole to size. If the installation must go deeper than about 2 m, polythene tubing proves to be convenient because it can be easily welded in the lengths of about 2 m required in the field.

The relation between counting rate and water content is best established for each soil material of interest. This can be done in the field by taking samples for gravimetric measurement of water content, together with the soil bulk density, at locations close to the access hole. Alternatively, the calibration may be determined by packing the soil into a drum in the laboratory, to the appropriate field bulk density and at known water contents. The drum needs to be about 1.2 m diameter by 1.4 m deep to be large enough at low water contents. The relationship between the counting rate and water content is approximately linear and commercial instruments are usually designed by their manufacturers to approach a linear calibration as closely as the basic processes of slowing down and scattering will permit. The relationship is not universal for all soils because it is influenced by properties other than the volumetric water fraction. For example, the counting rate is increased by the hydrogen atoms in organic matter, and those in the bound $-OH$ groups of clay minerals, neither of which would contribute to the water content as determined by the standard gravimetric method. The counting rate is also increased as bulk density increases for a given water content. On the other hand it is reduced by the increased presence of slow neutron absorbers, such as boron, chlorine, iron, gadolinium and other exotic elements which absorb neutrons more strongly than the commonly abundant elements.

A useful source book on good technique to be followed was prepared by Greacen (1981). The safety precautions and procedures that should be adopted in handling the radioactive source are discussed in one of its chapters. The questions of calibration and accuracy of the derived water content are treated thoroughly. Greacen *et al.*, (1981) remark that the linearised calibration equation of the form

$$\theta = a + bC_f, \tag{3.3}$$

should not imply that the proper procedure is to regress θ on C_f. On the contrary, the least squares fitting should be done by regressing C_f on θ, because of the larger variance of θ than of C_f (see also Sprent, 1969). Then the convenient estimation of θ should be obtained by

$$\theta = \frac{a'}{b'} + \frac{1}{b'} \cdot C_f, \tag{3.4}$$

where a' and b' are the intercept and slope of the regression C_f on θ, and a'/b' and $1/b'$ are equal to a and b in Equation (3.3).

Soils are so variable that there is no universal calibration curve and the slopes of different calibrations may vary by up to 25 per cent. Since calibration by the method of packing soil or by field sampling are both arduous and time-consuming there have been a number of attempts to derive a calibration curve by the application of neutron transport theory when the principal elements have been determined by chemical analysis. Wilson and Ritchie (1986) used a multi-group diffusion theory code to show that calculation is feasible and accurate. A calibration curve determined experimentally by Greacen and Schrale (1976) was closely reproduced by their calculation, but it is questionable whether the facility for calculation would be widely available.

The precision in water content determination can be as good as 2 mm standard deviation in a total water content of 450 mm in the soil profile to a depth of 1.5 m, so a 4 mm change could just be resolved significantly (Holmes and Colville, 1964). However, caution needs to be exercised in transferring such an estimate of error to the estimation of precision in large changes of water content in a time-series, because the calibration slope itself may have an undetected bias (Williams and Sinclair, 1981).

The neutron moisture meter is widely used in agriculture, forestry, hydrology and soil engineering to follow changes in water content. Since results are obtained on a volume basis without need to measure bulk density, it can be used directly to follow gains and losses expressed as an equivalent depth of water in a given depth of soil. In this form the water content change as expressed above is compatible with other components of the water balance such as rainfall and evaporation.

The gamma ray method

A beam of gamma rays, emitted from a source such as caesium-137 becomes attenuated during its transmission through soil and the degree of attenuation increases with the wet bulk density. This was exploited by Bernhard and Chasek (1953) and later by Vomocil (1954) in a method that became well established for measuring bulk density in the field. Since water contributes to the attenuation of a gamma ray beam passing through soil, the method was used by Danilin (1955) and Ashton (1956) to follow changes in water content in the root zone. Then Gurr (1962) and Ferguson and Gardner (1962) independently refined the method for critical measurement of water content in laboratory experiments on water flow.

Water content is measured in the laboratory with a source and detector on opposite sides of a soil column. In the field the usual procedure is to lower a source and a detector down parallel holes as a double probe (Fig. 3.2). Details are given by Gardner (1986) and Nerpin and Chudnovskii (1967). If N_0 is the count rate for gamma rays transmitted from the source through air to the detector and N_m is the rate through moist soil, then

$$N_m/N_0 = \exp\{-x(\mu_s\rho_b + \mu_w\theta)\}, \qquad (3.5)$$

where x is the path length through soil of dry bulk density, ρ_b, and water content, θ (as mass per unit volume of soil), and μ_s and μ_w are mass attenuation coefficients for soil and water, respectively. In dry soil ($\theta = 0$),

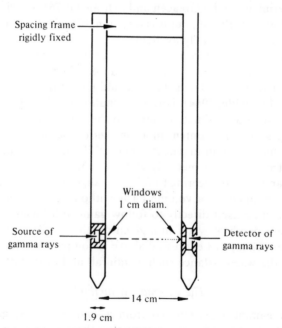

Fig. 3.2. Double-tube probe for measuring bulk density and water content in the field by gamma ray attenuation. The dimensions are those given by Gurr and Jakobsen (1978) for use with a 300 mCi [241]Am source.

Equation (3.5) becomes

$$N_d / N_0 = \exp{(-x\mu_s\rho_b)}. \qquad (3.6)$$

From Equations (3.5) and (3.6) it follows that

$$\theta = \frac{\ln N_d / N_m}{x\mu_w}. \qquad (3.7)$$

This gives a theoretical basis for determining water content from gamma ray attenuation in soil whose density does not change with water content (Gurr, 1962). If dry soil is not available to enable N_d to be determined directly, a one-point calibration of counts with water content is undertaken. An established value is available for μ_w but this is subject to some modification when low energy gamma rays are used as when [241]Am is the source (Groenevelt, de Swart and Cesler, 1969). Alternatively a calibration curve is established for the soil (Ferguson and Gardner, 1962).

If two sources of gamma rays with different energies such as ^{137}Cs and [241]Am are used instead of one, two sets of values are available for the parameters N_m, N_0, μ_s and μ_w in Equation (3.5) which can then be solved for the two unknowns ρ_b and θ. Thus, bulk density and water content can be determined in a swelling soil in which bulk density varies with water content. The two-source method was developed by Gardner and Calissendorff (1967), Soane (1967), Bridge and Collis-George (1973) and others and has proved successful especially for laboratory use on swelling soils. Gurr and Jakobsen (1978) have described equipment suitable for use in the field.

The gamma ray transmission method of determining water content has the advantage over the neutron scattering method of giving better depth resolution and of being more readily used near the surface of the soil. It has become a valuable tool in laboratory experiments enabling water content to be measured in transient systems with considerable precision (Gardner, 1986). However, it is not widely used in the field because of difficulties in collimation and discrimination under field conditions. Error in spacing the parallel holes for the dual probes can also affect readings. For their equipment, with the centre lines of their holes spaced 14 cm apart, Gurr and Jakobsen (1978) state that an error of 1 mm in the spacing causes a 2 per cent error in wet density.

Time-domain-reflectometry

The electrical capacitance of two conductors placed in the soil depends upon the water content of the soil because the dielectric constant of water (80) is much larger than that of the dry soil (5) or the air (1) which replaces the water as the soil dries. This fact encouraged numerous

attempts to develop a capacitance method, but without success until
about 1980. Childs (1943a) remarked that the attempted correlation was
fraught with dangers. The dielectric is leaky and the capacitance, as
measured, is dependent upon the resistance between the condenser plates,
that is, the soil electrical conductivity, as well as the apparent dielectric
constant. He stated that capacitance measurements should therefore be
made at frequencies larger than 20 MHz to avoid the uncontrolled
influence of varying electrolyte concentration of the soil water.

A new technique known as time-domain-reflectometry (TDR) has
circumvented a direct capacitance measurement but it depends upon
exactly the same soil properties. Electrical transmission-line theory leads
to the expression for the propagation velocity of an electromagnetic wave
in a transmission line as

$$V = c[\tfrac{1}{2}K'\{1 + (1 + \tan^2 \delta)^{1/2}\}]^{-1/2}, \qquad (3.8)$$

where c is the velocity of light, and $\tan \delta = \{K'' + (\sigma_{dc}/\omega\varepsilon_0)\}/K'$. The
remaining symbols are explained as: K' is the real part of the complex
dielectric constant, K'' is the imaginary part, σ_{dc} is the zero-frequency,
electrical conductivity of the medium, ω is the angular frequency and
ε_0 is the free-space permittivity. Measurement at very high frequency
makes $\tan \delta$ tend to zero, so that

$$V = c/(K')^{1/2}. \qquad (3.9)$$

Assigning the symbol L to the length of a segment of the transmission
line, Equation (3.9) can be recast as

$$t = L(K')^{1/2}/c, \qquad (3.10)$$

where t is the time taken for the propagation of an electromagnetic pulse,
or perturbation, along the length L. Putting realistic values of the constants into Equation (3.10), $L = 0.3$ m, $K' = 25$, $c = 3 \times 10^8$ m s^{-1}, the time
of propagation would be about 5 ns.

Topp *et al.* (1982) have used this theory and electronic apparatus
developed in a different, industrial context, to measure water content of
soil. They used parallel rod transmission lines, made of 12.7 mm diameter
brass rods spaced at 50 mm centre-to-centre and up to 0.5 m long, embedded vertically in the soil. Either open circuit or short circuit termination
of the line allows a satisfactory reflection from its end. The frequency
used was ~ 0.1 GHz, and the small changes in soil water content created
by their laboratory set-up could be measured accurately by the apparatus,
which employs an electronic analogue of the optical stroboscopic method.
Topp and Davis (1985) tried the method in the field and reported that

the precision in measuring water content appeared to be superior to gravimetric sampling, as is to be expected of a method that uses a larger sample volume.

Other methods

As we have seen water content can be determined from the relation it has with other measureable properties of the soil. A further example of this is provided in Fig. 2.2, in which the moisture characteristic curve for a soil constitutes a calibration curve for determining water content from matric potential, but because of hysteresis in this relation its use for this purpose is greatly restricted. Since water content affects electrical conductivity and thermal conductivity, these properties can be used once a calibration curve has been established for a particular soil. Here there is a difficulty common to all devices placed in the soil to measure these properties. It is that a good electrical or thermal contact can hardly be maintained with the soil when its bulk density changes with water content change. In addition, electrical conductivity is affected by solutes whose concentration in the water may vary with time. These difficulties are met by measuring the chosen property in absorbent blocks buried in the soil (as will be discussed in a following section) but calibration of the property is then best made with matric potential of the water rather than with water content. Cornish, Laryea and Bridge (1973) found the thermal method to be satisfactory when they used a buried thermistor both as a heat source and as a temperature sensor in relating the rate of dissipation of heat to water content. However, in the field, a thermal method is more likely to be useful in sandy than in clayey soils, as de Vries (1953) observed.

An estimate of the change in water content of soil within the root zone in the field can be obtained by drawing up a balance sheet of gains in water from precipitation and irrigation and losses by evaporation, transpiration, run-off and drainage below the root zone (see Chapter 12). The changes can alternatively be measured directly in a weighing lysimeter. A commonly used size is 2 m diameter by about 1.5 m depth. Such a volume of soil together with its soil container and attached weighing mechanism would weigh about 8 tonnes. The precision of weighing claimed for a good installation is about 0.3 kg corresponding to 0.1 mm change of water content of the lysimeter. The lysimeter body has to be carefully isolated from the surrounding soil so that it can be weighed without removing it from its field situation in a growing crop. It may be noted that, to avoid the expense and difficulty of installing weighing devices, the water content of the soil in a lysimeter can be measured using the neutron moisture meter (McGuiness, Dreibelbis and Harrold, 1961; Holmes and Colville, 1964) or the gamma ray method (Danilin,

1955). Tanner (1967) has reviewed work done on the construction and use of lysimeters.

3.2 Matric potential

The gravitational, pressure, matric and osmotic components of the total potential of water in soil can be measured separately or in certain combinations. Examples of these measurements have already been quoted in Chapter 2. Gravitational potential is determined immediately from the elevation of the point of measurement with respect to an arbitrarily chosen reference height and this needs no further discussion. The pressure potential corresponding to a positive hydrostatic pressure is readily measured by means of a piezometer as illustrated in Fig. 2.1. In the field, the height of water in open auger holes sunk below the water table provides this information for drainage studies. If water enters a hole from one horizon and not from others, it may be necessary to seal off part of the profile with a liner in order to identify the water-bearing horizon and measure the pressure potential of water in it.

In unsaturated soil the pressure (or matric) potential is not so easily measured. The early work of Buckingham (1907) on retention and movement of water found only limited application until tensiometers came into use (L. A. Richards, 1928) for measuring matric potential. The present discussion will deal mainly with matric potential measured alone or in combination with osmotic potential.

Field and laboratory methods for measuring matric potential are based upon the same principles, though the apparatus and procedure must be designed for the purpose in hand. When a sample is taken from the field to obtain, by laboratory measurement, the relation between its matric potential and water content, its natural 'undisturbed' structure should be preserved. Unless this is done, as by cutting cores or digging out blocks of soil, the moisture characteristic curve that is obtained will be in error at its wetter end. In this connection it is found that while a naturally structured soil, at a water content corresponding to field capacity, may be at a matric suction of about 10 kPa, the same soil if pulverized and wetted would require a suction of about 33 kPa (330 mbar) to bring it to the same water content (the so-called $\frac{1}{3}$ bar water content). However, the moisture characteristic curve is not much affected at the dry end because the smaller pores are little changed by pulverization.

For greater detail on the methods than can be given here, the reader is referred to *Methods of Soil Analysis* edited by Klute (1986).

The tensiometer

The tensiometer (Fig. 3.3) is widely used for measuring matric potential in the field and in the laboratory. It has a porous cup, usually ceramic,

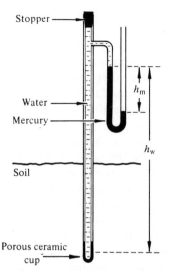

Fig. 3.3. Tensiometer with a mercury manometer. For symbols, see text.

which when buried in the soil allows water to move freely through its walls between the soil and the water-filled systems of the tensiometer, but excludes air when wet. The negative pressure or suction of the water is measured by a mercury manometer or a vacuum gauge. When the unbalanced length of water column in Fig. 3.3 has been allowed for, the matric suction is obtained in units of length as

$$s = |h| = (\rho_m/\rho)h_m - h_w \qquad (3.11)$$

and the matric potential in units of energy per unit volume or of pressure as

$$\psi = \rho g h_w - \rho_m g h_m \qquad (3.12)$$

where ρ and ρ_m are the density of water and mercury, respectively, and the lengths h_m and h_w are as shown in Fig. 3.3. Solutes can pass through the walls of the cup into the water of the tensiometer but the error involved in assuming this to have the density of pure water is ordinarily unimportant.

Tensiometers have two limitations. The first of these is that, as the suction increases towards 100 kPa, air comes out of solution and the water column in the tensiometer breaks. Their useful range is confined to suctions less than 85 kPa. At this stage of drying not much water is left in sandy soils but in soils of other textures much still remains available for plants after the tensiometer ceases to be of use. This can be seen to be the case by comparing the retentivity curves of Fig. 2.2. Hence as a guide to irrigation, tensiometers are chiefly useful for plants such as green vegetables for which the soil must be kept moist. The nuisance of having to recharge them with water after each breakdown has restricted their use by farmers under other conditions. It will be seen also from Equation (3.11) that the range of suction that can be measured decreases with increasing depth of installation because of the accompanying increase in h_w.

The second limitation is one that chiefly concerns the tensiometer as a research tool, both in the laboratory and in the field, when the matric potential of the soil to be measured is changing rapidly. The response of the tensiometer to such change is determined by two characteristics of its construction, the conductivity of the porous ceramic cup and the sensitivity of its pressure measuring device. Suppose that p_0 is the initial water pressure inside the tensiometer body at the time zero, and that for $t \geq 0$ the pressure in the soil water outside the tensiometer is maintained at p_1. The pressure inside the tensiometer body, at sufficiently large t will approach p_1. At any intermediate time the pressure inside the tensiometer will be intermediate between p_0 and p_1, let us say p.

Water enters through the porous cup in response to the pressure difference $(p_1 - p)$ and the rate of entry is given by

$$dQ/dt = K'(p_1 - p), \qquad (3.13)$$

where K' is the cup conductance, a property that is related to hydraulic conductivity (see Chapter 4). The increment of volume of water inside the tensiometer body, dQ, is also given by

$$S \, dQ = dp, \qquad (3.14)$$

where S is the pressure gauge sensitivity. Equation (3.13) therefore becomes

$$dp/dt = K'S(p_1 - p). \qquad (3.15)$$

The solution of Equation (3.15) for $p = p_0$ at $t = 0$ is

$$p = p_1 - (p_1 - p_0) \exp -(K'St). \qquad (3.16)$$

Equation (3.16) is the tensiometer response equation, first given by L. A. Richards (1949), and the parameter $1/(K'S)$ is the time constant, τ, of the tensiometer. For good response τ should be as small as possible.

Under some circumstances, the rate of transfer of water between tensiometer and soil may not be fast enough to allow an accurate measurement to be made of the changing matric potential in the soil. Furthermore, the volume of the exchange may itself upset the matric potential of the soil at the site of measurement. To overcome this, pressure transducers are sometimes employed to allow equilibrium to be maintained with negligible interchange of water between soil and tensiometer cup. A hydraulic selector switch can be used to enable a number of tensiometers to be connected in turn to a single pressure transducer that has a very large gauge sensitivity. Cassel and Klute (1986) give further details.

Suction plates

Porous plates containing water under suction are used for finding the relation between water content and matric potential for soil samples in the laboratory (Fig. 3.4). A series of bubbling towers connected to a vacuum line can provide any constant suction up to 85 kPa (Holmes, 1955). Ceramic plates large enough to take several samples at once are usually used. The structure of the original soil needs to be maintained in the samples by cutting cores or using lumps rather than pulverized material. Continuity of water between the sample and the plate is aided by smoothing the underside of the sample and interposing a thin layer of fine-grained material that will retain water at the greatest of the required suctions.

Gas pressure devices

The useful range of a ceramic plate can be extended to matric potentials below -100 kPa by applying a pressure, p_g, to gas in a chamber that encloses the soil on the plate (Fig. 3.5). Water from a wet soil then passes out through the plate into a body of water in contact with the underside of the plate. This water is at atmospheric pressure (zero gauge pressure) so that, when flow to it ceases, the pressure potential, P, of the soil water in the chamber will be zero. P has two components, the matric potential, ψ, associated with water retention by the soil matrix, and the pneumatic

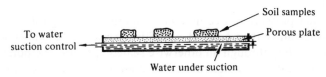

Fig. 3.4. Suction plate. A cover, to prevent evaporation, is not shown.

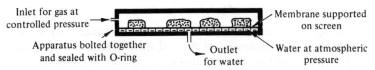

Fig. 3.5. Pressure membrane apparatus.

potential arising from the imposed gas pressure. From the definitions given in Chapter 2, the pneumatic potential is equivalent to p_g when potentials are expressed per unit volume of water, and hence at equilibrium

$$P = \psi + p_g = 0, \qquad (3.17)$$

and $\psi = -p_g$. It is thus possible by means of the pressure chamber to determine the relation between matric potential and water content of a soil over a greatly extended range. Ceramic plates are limited by gas escape to gas pressures of about 1 MPa but cellulose acetate membranes suitably supported can take a pressure up to 10 MPa without leaking air. Pressure plate and pressure membrane apparatus are in general use in laboratories and are available commercially. Their development is due to S. J. Richards (1938), L. A. Richards (1941) and L. A. Richards and Fireman (1943) and their use is described fully by Klute (1986).

The equivalence of an external gas pressure and a water suction of the same magnitude, in influencing the curvature of a water film and hence the water content of a capillary porous system, is evident from Equation (2.9). This equation shows the relation between film curvature and the pressure difference across the film, irrespective of how this difference arises. However, some disagreement has been observed in moist soil between the pressure applied and the suction measured in the soil by a tensiometer after the pressure is released. This is ascribed to air bubbles forming or enlarging and so causing the geometry of the liquid phase to change when the water is no longer under pressure. This has the effect of decreasing the matric suction below that expected from the pressure that had been used in bringing the soil to its equilibrium water content (Peck, 1960; Chahal and Yong, 1964). The deviation occurs in moist soil and is not apparently of consequence for drier conditions with suctions greater than 100 kPa, where the pressure methods are required.

One other pressure device may be mentioned although it is used for plant tissue rather than soils. If a leaf is placed in a chamber with its petiole projecting through an otherwise sealed aperture, the potential of water in the leaf can be determined by finding what gas pressure in the chamber is required to make sap appear at the cut end of the petiole (see Section 12.1).

Absorbent blocks

The water content of an absorbent block buried permanently in the soil varies with the matric potential of the soil water. Hence matric potential can be determined indirectly from the water content of the block or from some electrical or thermal property that varies with its water content.

Any calibration curves set up experimentally for the blocks are subject to error due to the occurrence of hysteresis in the relation between block water content and matric potential. For field use of these blocks, calibration curves are usually based on drying conditions.

In the simplest case, the water content of absorbent blocks is determined by weighing. This requires the blocks to be set in access holes and, because repeated disturbance for weighing makes contact with the soil uncertain, this method has not proved attractive. Instead, the electrical conductivity of gypsum blocks was widely adopted as a routine method following its introduction by Bouyoucos and Mick (1940). Two electrodes are embedded in plaster of Paris which, after setting, forms a porous gypsum block (Fig. 3.6). Slight solubility of the gypsum masks the effect that other salts dissolved in the water have on the electrical conductivity of the block, provided the soil is not too saline. However, this solubility also reduces the life of the blocks, which deteriorate under prolonged wet conditions, so that the original calibration curve ceases to apply to the readings obtained. To avoid this, nylon as an absorbent web has been used instead of gypsum, under conditions where salts do not affect the calibration curve unduly. Gypsum blocks are an inexpensive guide for the timing of irrigation and other field control practices where an approximate measure of matric potential will suffice. They can be used under drier conditions than the tensiometer and are more sensitive for $\psi < -100$ kPa than for higher potentials. They are not able to measure potential accurately because of hysteresis and the variability occurring within a given batch of blocks (Tanner and Hanks, 1952). Gypsum blocks

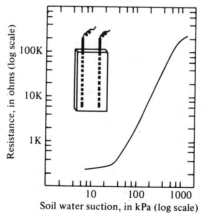

Fig. 3.6. Calibration curve of the gypsum block (after Aitchison and Butler, 1951). The inset shows a block cast in rectangular shape with bared copper wire electrodes: size of block is usually about $6 \times 3 \times 1$ cm.

and associated measuring equipment are readily assembled and are available commercially. Work done during the period of active development of the electrical conductivity method was reviewed by Marshall (1959a). An example of a calibration of the gypsum block, which can vary with type of block, in terms of its electrical resistance versus suction of the soil is shown in Fig. 3.6.

Another property that varies with the water content of a porous block is its thermal conductivity. Phene, Hoffman and Austin (1973) developed a technique to measure the rate of heat dissipation in a ceramic block as an indicator of the matric potential of the soil with which the block was in energy equilibrium. The device was incorporated as the sensor into a system for the automatic programming of irrigation. The way thermal properties of porous materials are affected by water content is discussed in Chapter 11 where it is shown in particular that thermal conductivity and diffusivity increase with water content (Fig. 11.5). The rate of dissipation of heat generated in a porous block is measured by means of a diode temperature sensor in the method developed by Phene, Hoffman and Rawlins (1971). References to earlier work on thermal methods are given by them.

3.3 Combined matric and osmotic potential

Relation to relative vapour pressure

The sum of matric and osmotic potential is important in plant-water-soil relations and in that context is often referred to as the water potential of the soil or of the root, leaf or plant tissue. Similarly, matric and osmotic suctions are together referred to as the total suction of a soil or other material. This combined retention potential can be estimated directly from the vapour pressure of the water in the material. The vapour pressure of water is lowered by interaction with the solid matrix and by the solutes it contains, and the combined retention potential can be calculated from the vapour pressure, e, expressed as a fraction of the vapour pressure, e_0, of pure free water at the same temperature. The fraction e/e_0 is called the relative vapour pressure.

The relation of matric potential to relative vapour pressure will be considered before dealing with the combination of matric and osmotic potentials. In Fig. 3.7 water under suction in a porous plate is in equilibrium with a hanging column of water. The whole system is in equilibrium at a uniform temperature. The water vapour pressure decreases with height, z, above the free water surface and at that surface it is e_0, the saturation vapour pressure of water. At the height of the porous plate it is the same as that of the water held in the pores of the plate under suction, otherwise there would be continuous circulation between the

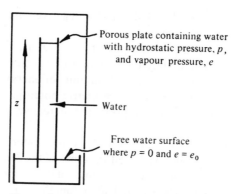

Fig. 3.7. Diagram illustrating equilibrium between hydrostatic pressure and vapour pressure of water.

two phases. The change, de, in the vapour pressure with height is equal to the weight of a column of water vapour of unit cross-sectional area and height, dz, so that

$$de = -g\sigma \, dz, \tag{3.18}$$

where σ is the vapour density and g is the acceleration due to gravity. Similarly within the water column we have

$$dp = -g\rho \, dz, \tag{3.19}$$

where p is the pressure and ρ the density of the water.

On combining Equations (3.18) and (3.19) we obtain

$$dp/\rho = de/\sigma. \tag{3.20}$$

From the gas law,

$$eV = (m/M)RT,$$

where m is mass and V volume of water vapour, M is the molecular weight of water, R is the gas constant per mole and T is the kelvin temperature. Hence since $\sigma = m/V$,

$$\sigma = eM/RT.$$

When this expression is substituted for σ, Equation (3.20) can be written as

$$dp = (\rho RT/M)(de/e).$$

Integration between the limits for p and e at the free water surface (where $p = 0$ and $e = e_0$) and a height, z (where the water pressure is p and the vapour pressure is e) leads to

$$p = (\rho RT/M) \ln (e/e_0). \tag{3.21}$$

Since $e/e_0 < 1$, the right hand side has a negative value as required for p when water is under suction.

The relation between pressure potential of water and relative vapour pressure is given immediately by Equation (3.21) since p can be taken to represent the pressure potential, P, in units of energy per unit volume of water (Section 2.2). Further, since matric potential is the only component of pressure potential represented here, we can replace p in Equation (3.21) by ψ.

It may be noted that the pressure, p, in the water at the top of the column in Fig. 3.7 can be expressed in terms of the radius of curvature, r, of the water-air interfaces in the porous plate by means of Equation (2.8). Hence r can be represented as a function of e/e_0. Also, since $p = -\rho g z$, the height, z, of the column required for equilibrium with any given value of e/e_0 can be calculated from Equation (3.21). From the example worked in Section 2.4, it can be seen that for $e/e_0 = 0.85$ a height of the order of 2 km is required so that the hypothetical column is impossible to realize experimentally.

If the porous plate is now taken to be a semipermeable membrane through which solutes cannot pass, we can consider the relation of osmotic pressure to e/e_0. The plate is assumed to be covered by an aqueous solution with the proper concentration to give it an osmotic pressure, π, equal to the suction $|p|$ of the water at the top of the column. Equilibrium between the vapour and liquid columns remains as before and we may rewrite Equation (3.21) with a change of sign as

$$\pi = -(\rho RT/M) \ln (e/e_0). \tag{3.22}$$

In this case the vapour pressure is reduced according to Raoult's law as follows

$$e = e_0 n_1 / (n_1 + n_2),$$

where n_1 and n_2 are the number of moles of solvent and solute respectively in the solution. Since the osmotic potential, Π, when expressed on a volume basis is equal to $-\pi$, it can also be calculated from Equation (3.22). Finally, the water potential, $\psi + \Pi$, can be determined from the contribution these together make to the lowering of the vapour pressure. Hence

$$\psi + \Pi = (\rho RT/M) \ln (e/e_0), \tag{3.23}$$

where ψ and Π are expressed per unit volume of water. The osmotic component in this equation can be measured for water in soil from the relative vapour pressure of the extracted soil solution. Other methods for doing this are discussed more fully in Section 10.6 dealing with soil salinity. With Π determined in one or other of these ways, ψ can then be found by difference in Equation (3.23).

Vapour sorption methods

Soil exposed to an atmosphere that is in vapour equilibrium with an aqueous solution at the same temperature will absorb or lose water vapour until its liquid is also in equilibrium with the vapour pressure, e, of the solution. Its water potential can then be calculated from e/e_0 by means of Equation (3.23), as in the example given in Section 2.4. The relationship is shown graphically in Fig. 3.8. The relative vapour pressures of aqueous solutions which can be prepared for this purpose at known concentrations are tabulated by Robinson and Stokes (1959), O'Brien (1948), and Kaye and Laby (1973, p. 28). The vapour sorption method has been widely used to obtain the relation between water content and matric potential in relatively dry soils and clays of negligible salt content, as in Fig. 2.5. Because the vapour pressure of water increases greatly with temperature, adequate control of temperature is required and this becomes a limiting aspect of the method as e/e_0 approaches 1. It is unsuited to the range of e/e_0 from 0.98 to 1. Other methods described below have been developed for use in this range which as shown in Section 2.4, covers

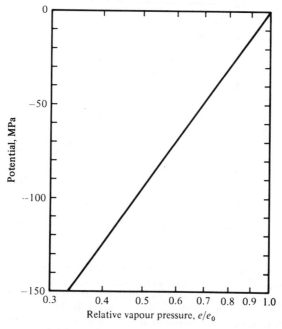

Fig. 3.8. Water potential ($\psi + \Pi$) in soil in equilibrium with water vapour at 20°C.

the moisture conditions suitable for plant growth. The relationship between water potential and e/e_0 within this range is illustrated in Fig. 3.9.

The relative vapour pressure or relative humidity of air is commonly measured for meteorological purposes by comparing the temperatures given by a pair of thermometers, one with a dry bulb and the other with a bulb that is kept wet from a supply of water. The wet bulb has its temperature lowered in accordance with the rate at which water evaporates from it under standard conditions of ventilation. This principle is followed on a micro scale in psychrometers for measuring water potential of soils and plants using thermocouples to measure temperature. In the method originated by Spanner (1951), a film of water is deposited at the sensing junction by passing a current through the thermocouple in the proper direction for cooling by the Peltier effect until water condenses on the junction. The temperature of the wet 'bulb' is then allowed to adjust itself according to the rate of evaporation into air in vapour pressure equilibrium with soil or plant material. The water potential corresponding to a pair of wet and dry bulb temperatures can then be

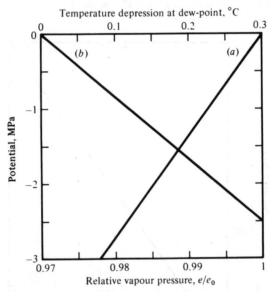

Fig. 3.9. Water potential ($\psi + \Pi$) in soil in equilibrium with water vapour at 20°C. (a) Relation with relative vapour pressure for $e/e_0 > 0.978$. Over this narrow range the curve can be represented as a straight line within the limited accuracy of the diagram, although the abscissa is not logarithmic as it is in Fig. 3.8. (b) Relation with temperature depression at the dew-point (Neumann and Thurtell, 1972).

determined using a thermocouple that has been calibrated over aqueous solutions of known concentrations at the same temperature. Tables showing the relation between water potential and the concentration of solutions suitable for calibrating thermocouples have been prepared by Lang (1967) for sodium chloride and Campbell and Gardner (1971) for potassium chloride solutions. Fig. 3.10 illustrates one such relation. Thermocouples sealed into small ceramic chambers are capable of determining soil water potential in the field within about 50 kPa, according to Merrill and Rawlins (1972). Although temperature control is not possible under field conditions, temperature changes occur slowly enough below the soil surface to allow the method to be used successfully. Variations in the method include the wetting of the sensing junction by adding a drop of water to a loop attached to the junction instead of creating a film by Peltier cooling (L. A. Richards and Ogata, 1958). However, this is not suited to *in situ* measurements of soil water potential.

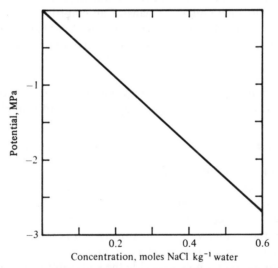

Fig. 3.10. Water potential $(\psi + \Pi)$ in soil in equilibrium with water vapour at 20°C. Relation to concentration of an aqueous solution as used in the calibration of thermocouple hygrometers (from data of Lang, 1967).

Neumann and Thurtell (1972) have developed a method for measuring water potential from the dew-point temperature depression instead of the wet bulb depression. By adjustment of the temperature of a droplet formed by Peltier cooling, its vapour pressure is brought to that of the water vapour in the chamber of the instrument. It has the advantage of causing less disturbance to the equilibrium of the water vapour with the soil or plant material than is caused by continuing wet bulb evaporation.

In this method the temperature depression at the dew-point is determined using a four-terminal thermocouple in which both temperature depression and the cooling current producing it can be measured simultaneously. The determination depends on the principle that, at the dew-point, the temperature produced in a junction by a given cooling current will be the same whether or not it is in a dry or humid atmosphere. The reason for this is that, at the dew-point, a water droplet on the junction is in equilibrium with the humid atmosphere and neither gains nor loses heat by condensation or evaporation. If, however, the temperature is lowered below or raised above the dew-point by increasing or decreasing the cooling current, the latent heat of condensation in the first case or evaporation in the second will cause its temperature to be respectively higher or lower than that produced by the same current in a dry atmosphere. When temperature depression is plotted against cooling current, the point of intersection of curves for dry and moist air thus fixes the dew-point. Water potential is then obtained from the temperature depression at the dew-point by means of an empirical relation used by Neumann and Thurtell and represented in Fig. 3.9. Measurements made in the laboratory over calibrating solutions gave the potential to within 20 kPa of the correct value. Neumann and Thurtell called their instrument a thermocouple dew-point hygrometer.

A full account of the development and use of thermocouple psychrometers and hygrometers is given by Rawlins and Campbell (1986).

3.4 Suitability of methods for use in the field

The choice of a method for measuring wetness and dryness of a soil depends on whether water content or potential is the better criterion for the purpose, whether the measurements are to be made in the laboratory or in the field, the precision required, the labour involved and the cost, reliability and availability of the measuring equipment.

Potential measured by tensiometers (in moist soil) and psychrometers provides information on how available the soil water is to plants and on the gradients causing movement of water in the soil. On the other hand, water content measured gravimetrically or by neutron scattering shows how much water there is in the soil and enables the change with time to be checked for various hydrological purposes. These four methods are widely used for soils in the field. The gamma ray method can be used rather than the neutron scattering method for experiments that require close definition of water content profiles and the automatic recording of changes with time. It has been particularly successful in laboratory experiments concerning the rate of infiltration into columns of soil, when accurate monitoring of the advance of the wetting front is needed.

The precision of all these diverse observations varies widely as their purposes differ. The resistance offered to a penetrometer by a clay soil, as it is forced vertically downwards may be an adequate measure of the soil wetness, in order to judge whether seeding can commence (Fawcett, 1972). But greater precision than this is needed in most laboratory and field measurements. Generally, the variability of soil water content, as measured gravimetrically, proscribes its direct use in field experimentation. The poor accuracy of the method is unacceptable and an example of this was given at the beginning of this chapter.

The neutron moisture meter technique, by repeatedly measuring the same soil sample, non-destructively, provides much improved accuracy in water content changes. So does the time-domain-reflectometry technique, although it has not yet been fully proven for field work. Nevertheless, there are sources of variance other than variability of the soil, that contribute to observed differences in results from site to site. For soil water content these undoubtedly include variable plant cover, root density and depth, root and plant pathogens that affect transpiration rates and variable micrometeorological conditions, all of which contribute to variation in the rate of drying of the soil. Few experiments have been conducted specifically to determine their effects.

The variability of measurements of matric potential by the tensiometer has been scarcely studied, though that is the instrument most often used in the field. But a notable set of observations by Greminger *et al.* (1985) has helped to redress that deficiency. Part of their field experiment provided data of the matric potential of the soil (Yolo loam), when it was allowed to drain to the water table following flood irrigation of 4 days duration. Table 3.1 extracts some of their data. It may be seen that the percentage standard deviation is large. That is consistent with every-day experience that tensiometers do show a lot of variation in replicate readings. The data allow a calculation of the vertical hydraulic potential gradients, $d\phi/dz$, between 0.3 and 0.6 m depths. These results, shown plotted in Fig. 3.11 *versus* time of drainage, reveal quite a smooth transition through several values greater than unity for drainage times less than about 500 hrs, to a rapid approach (on the log scale) to zero gradient at about 800 hours, i.e. the soil tended to become an equipotential volume for hydraulic potential in the upper part of its profile before the gradient reversed. The smoothness of the curve strongly suggests that the differences ($\psi_{0.3} - \psi_{0.6}$) are less variable than individual readings.

The whole question of the spatial variability of soil properties began to receive much attention from statisticians and soil scientists following the appearance of a book by Matheron (1971): *The theory of regionalised variables and its applications.* Webster (1985) has reviewed the application of this new geostatistics to soil science.

80   *Measurement of water content and potential*

Table 3.1. *Matric potential of Yolo loam during drainage to the water table, following 4 days of flood irrigation (after Greminger et al, 1985)*

Time	Soil depth, m	[1]ψ, kPa	[2]S.D	% S.D
4 hr	0.3	−1.9	1.2	63
	0.6	−3.4	0.5	15
28 hr	0.3	−4.5	2.1	47
	0.6	−6.9	0.7	10
72 hr	0.3	−6.3	2.5	40
	0.6	−8.9	1.1	12
720 hr	0.3	−18.3	2.7	15
	0.6	−16.5	1.1	7

[1] Means of 100 observations each at 0.3 and 0.6 m
[2] Standard deviations

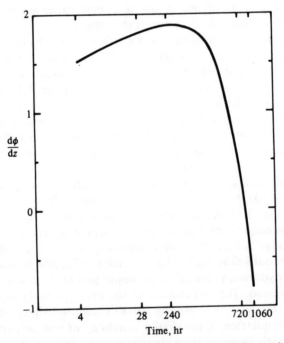

Fig. 3.11. The hydraulic potential gradient over the depth interval 0.3 to 0.6 m, during drainage after flooding, as a function of time. The flow is downwards when $d\phi/dz > 0$. (Calculated from data of Greminger *et al.*, 1985, as shown in Table 3.1).

As a final comment on the variability of soil properties, it should be emphasised again that uncertainty about the 'best' mean value can be diminished by making the measurements of interest upon the largest possible sample volume. For water content, the neutron moisture meter samples about 3×10^{-2} m^3, the T.D.R. method is influenced by a volume of about 2×10^{-4} m^3, and gravimetric samples have a volume of about 5×10^{-5} m^3, so the precisions are likely to be in that order of relative magnitude.

For chemical analyses and some physical measurements large volumes of soil can be reduced by sub-sampling, using good mixing techniques. Physical properties often depend upon the structural arrangement of the soil (see Chapter 8), so such recourse is usually denied to the soil physicist. Field measurements are often essential and the size-scale of the field set-up is important. This is particularly apparent in field drainage studies when the hydraulic conductivity, as measured, may have to serve as a design criterion for drains that traverse a field of several hectares in area (see Chapter 6).

Cost, reliability and availability of instruments are covered fairly well in the literature referred to including especially the book on methods edited by Klute (1986). Labour costs vary widely in different methods. For measuring water content at intervals of time, the neutron moisture meter does not require the great amount of labour used to obtain samples for gravimetric measurement, in boring holes in sufficient number and to a sufficient depth. One operator can make measurements covering a small catchment area to a depth of 6 m in 15 permanent access holes in one day with a neutron moisture meter (Australian Water Resources Council, 1974), while for gravimetric measurements numerous fresh holes have to be sunk on each occasion.

The influence of temperature upon the rate and direction of flow of soil water will be considered in Chapter 4, but it should be noted here that a controlled-temperature laboratory offers the best environment for any experimental work that requires freedom from temperature-induced effects and the best possible precision. In field observations measurement errors are usually less serious than sampling errors provided temperature variations can be tolerated or corrected for. Mistakes such as those due to site disturbance or a drift in instrument calibration or response must always be guarded against.

4

Principles of water movement in soil

4.1 Darcy's law

The total potential, Φ, of soil water was stated in Chapter 2 to be the sum of the gravitational, pressure and osmotic potentials, $Z + P + \Pi$. Wherever the potential energy of water differs from point to point, it experiences a force, F, that tends to make it move from positions of greater energy to those of lesser magnitude. The influence of Π on such movement is confined to cases in which salt sieving occurs, as for example in uptake of water by plant roots. In this chapter, Π will be neglected and we shall consider only the sum, $P + Z$, called the hydraulic potential, ϕ. We shall on occasion, in dealing with unsaturated soils, replace P by its component, the matric potential ψ.

The earliest quantitative description of water flow through a porous medium was that given by Henri Darcy (1856), who reported on the filtration of water flowing through a sand bed for an improved supply for Dijon. The modern statement of Darcy's law, for flow in a saturated medium, is

$$v = Q/A = K(\phi_2 - \phi_1)/l_{12}, \tag{4.1}$$

where Q is the rate of discharge of water through a cross-sectional area A, which is taken normal to the direction of flow, K is a proportionality constant discussed below, ϕ is the hydraulic potential given by $P + Z$, and v is the rate of discharge per unit area. The potentials ϕ_1 and ϕ_2, with $\phi_2 > \phi_1$, are measured at two positions separated by a distance l_{12} along a straight line parallel to the direction of flow.

It is convenient to express potentials on the basis of unit weight of water in dealing with water movement. ϕ then has the dimensions of length and represents the hydraulic head, H, which is obtained, as in Fig. 4.1, from

$$\phi_w = H = h + z$$

where h is the pressure head (equal to P_w or ψ_w) and z is the elevation

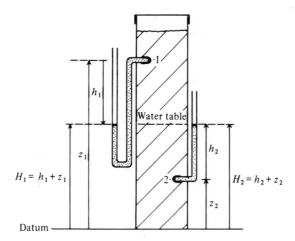

Fig. 4.1. Hydraulic head, *H*, of water at two points in a soil column at static equilibrium, for both saturated and unsaturated conditions. Note that the pressure head, h_1, in the tensiometer at point 1 has a negative value as represented by the hanging water column. At point 2, the head at the piezometer opening is h_2 and it represents a positive pressure. *z* is the elevation above the height datum.

above a suitable datum (equal to Z_w). The subscript, referring to the weight basis, will not be further used.

With potential thus given the dimensions of length, the constant of proportionality, *K*, in Equation (4.1) has the dimensions LT^{-1} and can be expressed in m s^{-1} or some convenient multiple of these units. It is called the hydraulic conductivity of the soil. As can be seen from Equation (4.1) it is the rate of discharge per unit area when the potential gradient, $d\phi/dl$, is unity.

For greater generality it would be preferable to have a proportionality constant that is independent of viscosity, which varies with the temperature and type of fluid. To do this, Darcy's law may be written as

$$Q/A = (k/\eta)(\phi_2 - \phi_1)/l_{12}, \qquad (4.2)$$

where η is the dynamic viscosity and *k* the permeability or intrinsic permeability of the porous material. This form is used in many applications, including petroleum technology, with ϕ in units of energy per unit volume of water, so that *k* then has the dimensions of area and can be expressed in units of m^2. The relation between hydraulic conductivity and intrinsic permeability is

$$K = \rho g k / \eta, \qquad (4.3)$$

where ρ is the density of water and *g* is the acceleration due to gravity. It may be remarked that, in a soil whose structure is affected by water,

k itself is not really independent of the fluid and K is more generally used in hydrology and soil science. The word 'permeability' has a broad descriptive use as well as being a specific term for k.

The relationship between a potential and its associated force field is a fundamental concept in physics (Macmillan, 1958). It can be expressed as

$$F = -\nabla \phi, \tag{4.4}$$

where the operator ∇ (*read as del*) is defined as performing the operations $[i(\partial/\partial x) + j(\partial/\partial y) + k(\partial/\partial z)]$ upon a scalar quantity. Here i, j and k are, respectively, the unit vectors in the (positive) x, y and z directions of the Cartesian coordinate system. Written in full Equation (4.4) is

$$F = -\left(i\frac{\partial \phi}{\partial x} + j\frac{\partial \phi}{\partial y} + k\frac{\partial \phi}{\partial z} \right).$$

Darcy's law, Equation (4.1), written in vector notation, is

$$v = -K\nabla \phi. \tag{4.5}$$

We assume for a start that the soil is homogeneous and isotropic so that the hydraulic conductivity is uniform and has no dependence upon the direction of water flow. The gradient of the potential $\nabla \phi$, being a vector, has its direction determined in three-dimensional space by the vector addition of its components resolved along the x-, y- and z- axes. If, in the x- direction, for example, the resolved part of ϕ increases so that $\partial \phi_x/\partial x$ is positive, the direction of the component of the water flow, v_x, must be $-i$, because the flow is from regions of higher to regions of lower potential. A negative sign is therefore needed in Equations (4.5) and (4.4) to ensure consistency in the algebraic signs. The magnitude of v is the rate of discharge of water through the soil, expressed as volume per second per unit cross-sectional area normal to the direction of flow.

If we compare Equations (4.4) and (4.5), we can see that the water flow through the porous medium has a steady velocity conferred on it by the force specified by Equation (4.4). The energy dissipated in its travel from a place of higher potential to another of lower potential disappears as heat generated in overcoming the viscous resistance to flow associated with the laminar planes of shear in the water as its velocity increases with distance from the solid phase boundaries. The water experiences no net residual force that would cause acceleration.

The discharge rate, v, when it is referred to the unit cross-sectional area of the whole soil body is sometimes spoken of as the Darcy velocity. The mean speed through the soil pore space is given by

$$v' = v/\theta, \tag{4.6}$$

where θ is the water content per unit volume of the soil.

We assume that Darcy's law is valid for the whole range of flow velocities observed in soil. It applies to cases for which the Reynolds number, $\rho r v / \eta$, is <1. Velocity, v, and pore channel radius, r, are always small enough to ensure that this requirement is met for water flowing through soil. The flow is then laminar and accelerations are unimportant, although clearly there can be to and fro changes in velocity as the water traverses pores of differing sizes. It is thus assumed that the energy that is dissipated in flow is expended in overcoming the shearing resistance of water and is not used in overcoming its inertia nor in creating turbulences.

It has been claimed by numerous investigators that Darcy's law does not hold exactly when the flow velocity is very small, because the water is not set in motion until the potential gradient becomes larger than a small threshold value. Swartzendruber (1969) reviewed many experimental results, including his own, that are sufficiently numerous to require the possibility to be entertained seriously, although it is difficult to eliminate significant experimental error from measurements of very small velocity of flow. If water possesses a shear strength in the close proximity of clay-mineral surfaces that confer on it non-Newtonian viscosity, as suggested by Low (1961), then a threshold force would have to be exceeded before the water moved. The evidence for non-Darcy behaviour does come principally from experiments about flow through clay. Such flow is very sensitive to changes in electrolyte concentration, as discussed later in this chapter, and the proposition that Darcy's law does not hold is probably unproven to date. Childs and Tzimas (1971) undertook a careful experiment to check upon the effect of a finite threshold to flow and could find no evidence of any departure from strict proportionality as potential gradients tended to zero.

4.2 Effect of porosity and pore size on the conduction of fluids

A number of models have been proposed for setting up relations between the structure of a porous material and its permeability. These make use of the equation of Hagen–Poiseuille for the laminar flow of a fluid through a capillary tube of radius r, namely

$$Q = (\pi r^4 / 8 \eta) \{ \rho g (H_2 - H_1) / l \}, \qquad (4.7)$$

where Q is the rate of volume flow, and H_2 and H_1 are the hydraulic heads at the ends of the tube of length l. (For a derivation of Poiseuille's equation, see for example, Childs, 1969, p. 194.) For comparison with Darcy's law as given in Equation (4.2) this may be written for a porous material as follows. The rate of flow through the tube per unit cross-sectional area is Q/A, where $A = \pi r^2$. The rate of flow, v, through a material containing the tube may be referred to the cross-sectional area of the material. Then the rate of flow, v, is $\varepsilon Q/A$ and the porosity ε here represents the cross-sectional area of the tube per unit cross-sectional area of the material. Hence

$$v = \varepsilon Q/A = (\varepsilon r^2/8\eta)\{\rho g(H_2 - H_1)/l\}.$$

Since Darcy's law as given in Equation (4.2), may be written as

$$v = (k/\eta)\{\rho g(H_2 - H_1)/l\},$$

this simple model leads to

$$k = \varepsilon r^2/8, \qquad (4.8)$$

where k is the intrinsic permeability. Equation (4.8) shows the fundamental significance of water-filled porosity, ε, and the square of a characteristic 'radius' of the pore spaces, r^2, in determining the magnitude of k. The range of pore sizes actually present in any porous medium can be derived from the moisture characteristic, as will be shown in Section 8.6, by an application of Equation (2.10), already discussed in Chapter 2.

In the conduction of water through soil, the whole of the pore space is involved only when the soil is saturated. As the degree of saturation decreases, the larger pores empty of water and become non-conducting. With flow confined to the smaller pores and also to less of the pore space, conductivity decreases. This is illustrated in Fig. 4.3 where the effect of water content on hydraulic conductivity is shown for two soils whose moisture characteristics are given in Fig. 4.2.

Another aspect of the hydraulic conductivity of unsaturated soils is shown in Fig. 4.4 where the effect of matric suction is illustrated. The

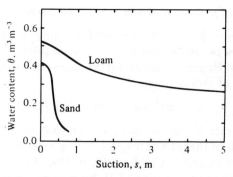

Fig. 4.2. Moisture characteristics (drying curves) for the two soils represented in Figs. 4.3 and 4.4 (Day and Luthin, 1956, for the sand; Elrick and Bowman, 1964, for the loam).

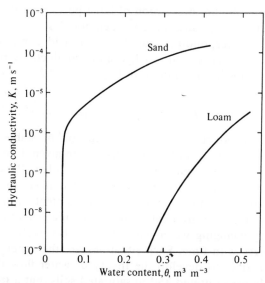

Fig. 4.3. Hydraulic conductivity as measured by Day and Luthin (1956) for a sand and by Elrick and Bowman (1964) for a loam soil. The effect of water content is shown.

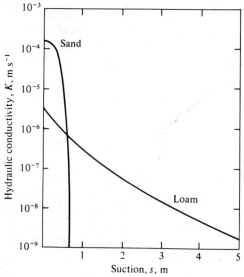

Fig. 4.4. Hydraulic conductivity as measured by Day and Luthin (1956) for a sand and by Elrick and Bowman (1964) for a loam. The effect of suction is shown.

sandy soil with a higher proportion of large pores than the loam is the better conductor at zero suction. But when both have dried to a suction of about 1 m, corresponding roughly under field conditions to field capacity, the loam has the higher hydraulic conductivity because little water then remains in the sand according to Fig. 4.2.

4.3 Flow through saturated soils

The hydraulic conductivity of a saturated soil of stable structure is constant, as is evident from the brief discussion given it above, because the whole of the pore space is always water-filled. By contrast, the hydraulic conductivity of an unsaturated soil is likely to change continuously in response to the changes of matric potential, the gradient of which is a part of the gradient of the hydraulic potential, and those changes imply changing water content. It is also likely that the water content changes with time. Such situations are the general rule in the field. It is possible to develop the theory of water flow through soils independently for saturated and unsaturated soils, but a treatment that brings out the underlying unity of the steady and the unsteady states is more pleasing.

When the water content changes with time we have to consider the law of conservation of matter, together with Darcy's law, in the same way as is required in developing diffusion theory from Fick's first law (see, for example, Bird, Stewart and Lightfoot, 1960). The usual descriptive statement of the law of conservation as applied to water flow through soil is that the rate of change of water content of an infinitesimal volume element is equal to the difference between the flux of water that enters the test volume and that which exits from the volume plus any sources of liquid water located in the element of volume and minus any sinks (see Appendix C for a derivation of the continuity equation). The mathematical statement is

$$\frac{\partial \theta}{\partial t} = -\nabla \cdot v + \text{sources} - \text{sinks}. \tag{4.9}$$

The quantity $\nabla \cdot v$ is the divergence of v (often abbreviated as div v) and it is defined as $\partial v_x/\partial x + \partial v_y/\partial y + \partial v_z/\partial z$ (see Appendix C).

If we neglect any change of state of the water in the system, and assume that plant roots are absent, sources and sinks must be zero. We also assume isothermal conditions. Combining Equations (4.9) and (4.5), we have, with these assumptions,

$$\nabla \cdot K \nabla \phi = \frac{\partial \theta}{\partial t}, \tag{4.10}$$

or

$$\frac{\partial}{\partial x}\left(K_x\frac{\partial\phi}{\partial x}\right)+\frac{\partial}{\partial y}\left(K_y\frac{\partial\phi}{\partial y}\right)+\frac{\partial}{\partial z}\left(K_z\frac{\partial\phi}{\partial z}\right)=\frac{\partial\theta}{\partial t},$$

where K_x, K_y and K_z are respectively the hydraulic conductivity in the x, y and z directions.

If the soil is saturated, the water content remains constant so that $\partial\theta/\partial t = 0$. Furthermore if it is isotropic, $K_x = K_y = K_z$. We assume that it is homogeneous so that K does not possess any dependence upon the space variables. With these assumptions, water flow in the saturated soil, in steady state conditions, is described by

$$\nabla^2\phi = 0, \tag{4.11}$$

where the operator ∇^2 (*read del squared*) is defined as performing the operation

$$\left(\frac{\partial^2}{\partial x^2}+\frac{\partial^2}{\partial y^2}+\frac{\partial^2}{\partial z^2}\right)$$

and it operates upon a scalar quantity. Equation (4.11) is Laplace's equation in the three space variables and is one of the intensively studied equations of mathematical physics (see, for example, Carslaw and Jaeger, 1959). Written out in full, Equation (4.11) becomes

$$\frac{\partial^2\phi}{\partial x^2}+\frac{\partial^2\phi}{\partial y^2}+\frac{\partial^2\phi}{\partial z^2}=0. \tag{4.12}$$

Note that the hydraulic conductivity does not appear explicitly in Equation (4.12), but it emerges as a constant of integration. Consider one-dimensional, saturated flow for which the equation

$$\frac{d^2\phi}{dx^2}=0 \tag{4.13}$$

is appropriate. Integrating Equation (4.13) once gives

$$\frac{d\phi}{dx}=C, \tag{4.14}$$

where C is a constant that can be identified with $-v/K$ of Equation (4.5), allowing that equation to refer to one-dimensional flow. Thus, Equation (4.14) is just Darcy's equation.

Now consider Laplace's equation in two dimensions. We have

$$\frac{\partial^2\phi}{\partial x^2}+\frac{\partial^2\phi}{\partial y^2}=0. \tag{4.15}$$

The solution of this equation could represent the hydraulic potential, ϕ, as a function of the space coordinates, x and y, in the horizontal plane. That solution would be appropriate to flow of water confined between an upper and a lower, horizontal, impermeable stratum. It would not be applicable to flow beneath the water table, except approximately, because that flow usually includes segments of directions sloping upwards or downwards with respect to the horizontal and Equation (4.15) does not include the (vertical) third spatial coordinate.

An approach to the solution of Equation (4.15) that is often used, is to take a trial function and examine what physical situation is described by it.

Suppose that the complex variable z (not to be confused with the z-coordinate) is defined by $z = x + iy$, in the horizontal plane, where $i = \sqrt{-1}$. In books on functions of a complex variable (for example, Carrier, Krook and Pearson, 1966) it is shown that a complex potential, $\Omega(z)$, exists in which the function ϕ, if it satisfies Equation (4.15), appears as a component, namely

$$\Omega(z) = \phi(x, y) + i\psi(x, y), \tag{4.16}$$

and ψ is another function of the real variables, x and y, related to ϕ in a particular way (and not to be confused with ψ for the matric potential). Together, ϕ and ψ are referred to as conjugate functions. As a trial solution, put

$$\Omega(z) = A \ln z = A \ln (x + iy). \tag{4.17}$$

Equation (4.17) can be written out, according to the theory of a complex variable, as

$$\begin{aligned}
\Omega(z) &= A \ln [|z|(\cos \theta + i \sin \theta)], \\
&= A \ln |z| \, e^{i\theta}, \\
&= A \ln |z| + A \ln e^{i\theta}, \\
&= A \ln |z| + iA\theta, \tag{4.18}
\end{aligned}$$

where $|z|$ is the modulus of z, equal to $(x^2 + y^2)^{1/2}$ and illustrated in Fig.

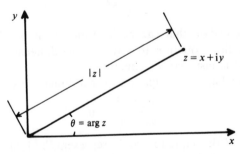

Fig. 4.5. The Argand diagram of the complex variable $z = x + iy$.

4.5. It is obviously, also, a radius vector r. Comparing Equations (4.16) and (4.18), we may assign the functions ϕ and ψ to be

$$\phi = A \ln r \qquad (4.19)$$

and

$$\psi = A\theta. \qquad (4.20)$$

The constant A can be evaluated from a boundary condition after the shape of the function ϕ has been settled. From the way $|z|$, i.e. r, has been defined, ϕ possesses circular symmetry and its representation in any (ϕ, r) plane is shown in Fig. 4.6. By Darcy's law we have

$$v = -K\frac{\mathrm{d}\phi}{\mathrm{d}r},$$

and the total quantity flowing through the full vertical, cylindrical section of the confined aquifer is

$$Q = -v2\pi rb, \qquad (4.21)$$

where b is the thickness of the aquifer and the negative sign is to preserve consistency with the definition of v, the vector discharge per unit cross-sectional area. From Equation (4.19), we have $\mathrm{d}\phi/\mathrm{d}r = A/r$, so with substitutions and re-arrangement of Equation (4.21), we derive

$$A = Q/2\pi Kb. \qquad (4.22)$$

Therefore, Equations (4.19) and (4.20) become

$$\phi = (Q/2\pi Kb) \ln r \qquad (4.23)$$

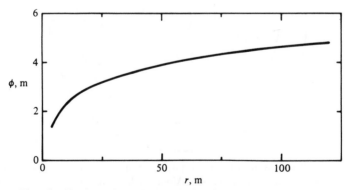

Fig. 4.6. The distribution of the hydraulic head of water in a confined aquifer with increase of distance from a pumped well. (See the text for the details of the parameters used for calculation of ϕ.)

and

$$\psi = (Q/2\pi Kb)\theta. \tag{4.24}$$

Equation (4.23) shows the distribution of hydraulic head, ϕ, in a confined aquifer as a function of distance, r, from a fully penetrating well that is being pumped at the steady rate $Q(\mathrm{m}^3\,\mathrm{s}^{-1})$. A vertical section of symmetry through the aquifer, its pumped well and the hydraulic head distribution produced by the pumping is shown in Fig. 4.6.

The representation of ϕ in the x, y plane shows an incremented set of equipotential lines, in Fig. 4.7. ψ is the corresponding stream-line function. We can now assign realistic values to the parameters. Figs. 4.6 and 4.7 have, in fact, been drawn using $Q = 0.785 \times 10^{-3}\,\mathrm{m}^3\,\mathrm{s}^{-1}$ and $Kb = 5 \times 10^{-5}\,\mathrm{m}^2\,\mathrm{s}^{-1}$. The product, Kb, of the hydraulic conductivity and depth of the permeable stratum is referred to as the transmissivity.

The stream function may be written out as

$$\psi = (Q/Kb)(\theta/2\pi). \tag{4.25}$$

Its maximum value is

$$\psi_{\max} = Q/Kb,$$

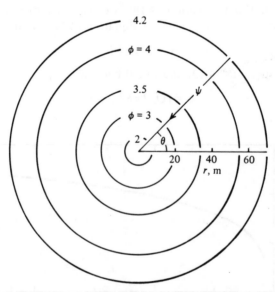

Fig. 4.7. The distribution of the hydraulic head, in plan view, corresponding to the vertical section shown in Fig. 4.6. Along each circle the hydraulic head has the value $\phi = (Q/2\pi Kb)\ln r$. The direction of the stream function, ψ, crosses the equipotentials at right angles. The magnitude of the stream function is $\psi = (Q/2\pi Kb)\theta$.

when $\theta = 2\pi$. ψ_{max} is therefore the normalized pumping rate of the well, with dimensions L, as it should have, because ψ is the conjugate function of ϕ.

The foregoing discussion of a particular solution of Laplace's equation in two dimensions is meant to be a demonstration of how the theory of functions of a complex variable is applied to derive the stream-function conjugate to the potential function. A properly rigorous account can be found in the book by Childs (1969), where the hydraulic conditions of a water table aquifer are also considered.

It is pertinent to remark that only one constant of integration has emerged from the solution, though Equation (4.15) is a second-order, partial differential equation. This is because the chosen, trial solution possesses cylindrical symmetry and the boundary condition close to the pumping well is effectively specified by the steady pumping rate $Q = 2\pi rbK \, d\phi/dr$, where r may be taken as the radius of the well to the outer surface of its screen. An alternative treatment of the single, pumped well will be given in Section 6.3.

4.4 Flow through unsaturated soils

The pore spaces of an unsaturated soil contain both water and air, so its pore water pressure is negative with respect to the pressure in the air phase, which is customarily assumed to be atmospheric. Although the gradient of the hydraulic potential may be produced by a gradient of one or other of its components alone, generally flow is caused by gradients of both. A matric potential gradient implies a change of water content from place to place. Changes with time as well, caused by wetting or drying out, require the equations that describe unsaturated flow to be written explicitly with space and time variables.

The simplifications of Equation (4.10) that allowed Equation (4.11) are not valid for flow through unsaturated soils. The hydraulic potential is given by

$$\phi = \psi + Z.$$

When expressed in the units of hydraulic head, $Z = z$ and Equation (4.10) takes the form

$$\frac{\partial}{\partial x}\left(K\frac{\partial \psi}{\partial x}\right) + \frac{\partial}{\partial y}\left(K\frac{\partial \psi}{\partial y}\right) + \frac{\partial}{\partial z}\left(K\frac{\partial \psi}{\partial z}\right) + \frac{\partial K}{\partial z} = \frac{\partial \theta}{\partial t}. \tag{4.26}$$

If flow is constrained to take place only in the horizontal direction, the appropriate governing equation is

$$\frac{\partial}{\partial x}\left(K\frac{\partial \psi}{\partial x}\right) = \frac{\partial \theta}{\partial t}. \tag{4.27}$$

If in the vertical direction only, it is

$$\frac{\partial}{\partial z}\left(K\frac{\partial \psi}{\partial z}\right)+\frac{\partial K}{\partial z}=\frac{\partial \theta}{\partial t}.$$ (4.28)

The hydraulic conductivity of the unsaturated soil must be regarded as depending strongly upon the water content (or matric potential) as has been explained in Section 4.2 above.

Flow that is entirely horizontal can be set up by experimental artifice in the laboratory, but it is hardly realistic for natural, field conditions. By contrast, flow that is vertical and possesses no significant component in the horizontal is a common feature of unsaturated soils in the field. Equations (4.27) and (4.28) have both water content and matric potential explicit in them, two properties of soil that are linked functionally by the soil moisture characteristic illustrated in Fig. 2.2. By the chain rule of differentiation

$$\frac{\partial \psi}{\partial x}=\frac{\mathrm{d}\psi}{\mathrm{d}\theta}\frac{\partial \theta}{\partial x},$$

so Equations (4.27) and (4.28) may be written as

$$\frac{\partial}{\partial x}\left(D\frac{\partial \theta}{\partial x}\right)=\frac{\partial \theta}{\partial t},$$ (4.29)

and

$$\frac{\partial}{\partial z}\left(D\frac{\partial \theta}{\partial z}\right)+\frac{\partial K}{\partial z}=\frac{\partial \theta}{\partial t},$$ (4.30)

where D is the soil water diffusivity defined by the relation

$$D=K(\theta)\bigg/\left(\frac{\mathrm{d}\theta}{\mathrm{d}\psi}\right).$$ (4.31)

It is the ratio of the hydraulic conductivity to the water capacity of the soil, $\mathrm{d}\theta/\mathrm{d}\psi$.

If the soil water diffusivity were constant, Equation (4.29) would reduce to

$$D\frac{\partial^2 \theta}{\partial x^2}=\frac{\partial \theta}{\partial t},$$ (4.32)

the well-studied diffusion equation. The diffusion equation with the diffusivity dependent upon the concentration of the diffusing substance

(Equation 4.29) has also received intensive study in other contexts. The treatment of Equation (4.30) is more difficult. Solutions of this equation are deferred for consideration to Chapter 5, when the phenomena of soil water infiltration, upward capillary rise, and steady state drainage are discussed. The absorption of water from the soil by plant roots is another, important pathway for water flow in the hydrologic cycle. It is considered in Chapter 12.

4.5 Hydraulic conductivity and soil water diffusivity

The practical use of any of the differential equations describing the flow of soil water requires reliable measurements of the water content, the conductivity and the diffusivity. Methods for measuring water content have been described in Chapter 3. The saturated hydraulic conductivity may be measured, in the laboratory, by means of a conductivity cell, referred to sometimes as a permeameter or oedometer cell in the soil mechanics literature, the principle of which is illustrated by Fig. 4.8.

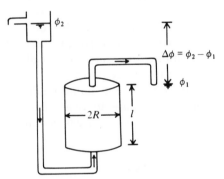

Fig. 4.8. Schematic of permeameter cell for the measurement of saturated hydraulic conductivity. For symbols, see text.

The hydraulic conductivity by this arrangement is given from Equation (4.2) by

$$K = \frac{Ql}{t\pi R^2 \, \Delta\phi},$$

(4.33)

where Q is the volume of water collected in time t, $\Delta\phi$ is the hydraulic head difference between the entry and exit planes of the test sample of length l, and R is the radius of the cell.

Klute and Dirksen (1986) describe a falling head permeameter convenient for soils with small hydraulic conductivities. The diameter of the
tube (see Fig. 4.9) serving as both reservoir for water flow and variable
head device may be chosen to suit the expected conductivity of the soil.
The rate of flow through the falling head permeameter at any instant of
time is given by

$$\frac{dQ}{dt} = \frac{K\pi R^2 \phi}{l},$$
(4.34)

with $\phi = h + l$, and $dQ = \pi r^2 \, dh$. Then, it is easily verified that

$$K = \frac{r^2 l \ln\{(h_1 + l)/(h_2 + l)\}}{R^2(t_2 - t_1)},$$
(4.35)

in which the symbols are as defined on Fig. 4.9.

A number of precautions must be adopted if reliable results are to be
obtained from laboratory permeameters. Disturbance of the soil sample
can be minimized best by taking the sample in the field in a coring device
which, when transported to the laboratory, serves also as the permeameter
cell. Even if this is done, the disturbance of the pore space at the
permeameter walls can create large pores and gaps which give a preferred
path for water flow that is not present in the soil *in situ*. To overcome
this disadvantage a flexible rubber sleeve can be made to serve as the
cell boundary. Slight suction of the water in the cell may also be applied
to drain the artificially created large pores.

The water used for the experiment should be de-gassed by boiling it
and cooling under reduced pressure, to avoid effects of gaseous entrap-
ment in the pore spaces. Any dissolved gas coming out of solution in

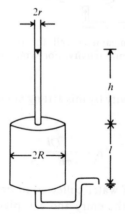

Fig. 4.9. Schematic of the falling-head permeameter.

the flow path migrates at once to the largest available pore, by the laws of surface tension, and thereby causes the greatest interference to the flow. The permeating fluid should therefore be kept as much undersaturated with respect to dissolved gases as possible.

The solution used for the experiment should be approximately isotonic with the naturally occurring soil water, to ensure normal behaviour of the clay colloids. Abnormal swelling of the clay domains could so alter the pore size distribution that the conductivity could change very markedly. The effect is discussed fully in Chapter 8, where Fig. 8.9 summarizes the mutually interacting influences of exchangeable sodium percentage and electrolyte concentration upon the equilibrium swelling of the colloidal fraction of the soil, which so significantly alters the permeability. Hydraulic conductivity is stable when the pore size distribution is stable. If swelling, and in the extreme case dispersion of clay platelets, is promoted by too low an electrolyte concentration or too high exchangeable sodium the hydraulic conductivity declines. It is hardly possible to be precise about the extent of the decline, but it can be two orders of magnitude if the cation is sodium. The effect is moderated by the divalent cations of common occurrence, calcium and magnesium.

Permeability is sometimes observed to decrease markedly when the soil is subjected to prolonged flooding. Field experience of this kind, gained in the context of water spreading for water conservation, led Allison (1947) to investigate the influence of microbial growths and associated metabolic products such as 'gums' and 'slimes' upon permeability of soils as measured in laboratory permeameters. Sterilization of the soil by the use of ethylene oxide and of the permeating solution by 10 mg l^{-1} mercuric chloride enabled the hydraulic conductivity to be maintained constant for up to 50 days, whereas untreated soil in control experiments showed a decline to less than one-tenth of the initial conductivity. If the soil was not completely sterilized, considerable improvement could be achieved by using such substances as phenol $\sim 500 \text{ mg l}^{-1}$ or formaldehyde ($\sim 200 \text{ mg l}^{-1}$), which would disinfect the soil to a useful degree when incorporated into the percolating water.

The measurement of unsaturated hydraulic conductivity needs a contrived control of the pore water pressure of the soil, unlike the test on a saturated sample. Gardner (1956) proposed to use the pressure membrane apparatus in an unsteady flow mode. A suitable geometrical arrangement of the sample of soil in that apparatus (see Chapter 3) enables the unsaturated hydraulic conductivity to be determined from observations of the rate of draining of the sample after the gas pressure has been increased. The instantaneous flow of water in a cylindrical sample of soil on the membrane of the pressure chamber can be described by Equation (4.28), modified to neglect the relatively small effect of gravity, namely

$$\frac{\partial}{\partial z}\left(K\frac{\partial \psi}{\partial z}\right) = \frac{\partial \theta}{\partial t}, \tag{4.36}$$

so that we may use ψ instead of ϕ.

If the experiment is performed over such a small range of ψ, from ψ to $\psi + \Delta\psi$, that the hydraulic conductivity remains approximately constant and represents a mean value in this range, Equation (4.37) then describes the flow, viz.

$$K\frac{\partial^2 \psi}{\partial z^2} = \frac{\partial \theta}{\partial t}. \tag{4.37}$$

We assume a functional relationship

$$\theta = a + b\psi \tag{4.38}$$

which again is approximately satisfied when the increments $\Delta\theta$ and $\Delta\psi$ of the experiment are kept small. Equation (4.37) becomes

$$D\frac{\partial^2 \psi}{\partial z^2} = \frac{\partial \psi}{\partial t}, \tag{4.39}$$

in which D, the soil water diffusivity, first defined in Equation (4.31) is here given by $D = K/b$, and the parameter b, introduced by Gardner, is again the water capacity of the soil. It may be seen by inspection of Equations (4.32) and (4.39) that either water content or matric potential may be chosen as the parameter to specify the moisture status of the soil. Gardner chose to work with ψ, but subsequently, as outlined in Chapter 5, the theory of infiltration was elaborated with θ as the dependent variable.

The initial and boundary conditions on ψ that must be satisfied by any solution of Equation (4.39) are determined by the nature of the experiment. In the pressure membrane apparatus (see Fig. 4.10) the soil sample of height L and volume V is assumed initially to be at moisture equilibrium at the gas pressure p_1. When the gas pressure is raised to p_2, water begins to flow out through the membrane, at $z = 0$, from the bottom of the sample until eventually the water content of the whole sample again becomes steady and at equilibrium with the gas pressure p_2. The pressure of the pore fluid in the soil sample relative to the chamber atmosphere is initially $-p_1$ and this is also the relative pressure of the water in the pores of the membrane boundary. When the chamber gas pressure is raised to p_2, the relative pressure of the pore fluid of the soil remains $-p_1$ because the gas pressure increment $(p_2 - p_1)$ is transmitted at once to the pore fluid. This relative pressure continues to be supported across menisci that have the same radii of curvature that they had initially.

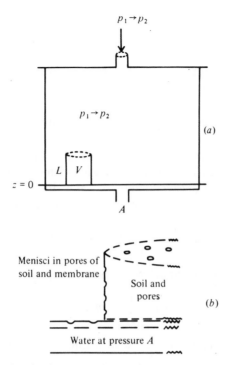

Fig. 4.10. Schematic of soil sample in the pressure membrane apparatus (a), and microscopic view (b), of continuity of water from membrane pores to soil pores, to illustrate the outflow method of Gardner (1956).

The pressure of the pore fluid in the membrane boundary, by contrast, becomes $-p_2$ relative to the chamber atmosphere when the gas pressure is raised to p_2. This relative pressure change is supposed to occur instantaneously and to be supported across menisci of the membrane pores whose radii of curvature decrease without any significant time lag. Since the soil sample rests upon the membrane boundary and the soil pores at $z = 0$ are hydraulically continuous with the membrane pores, the pressure in the soil pore fluid at $z = 0$ also becomes $-p_2$ relative to the chamber atmosphere instantaneously when the gas pressure is increased.

At the top of the soil sample there can be no flow of water across that boundary; a specification that requires the gradient of the matric potential to vanish at $z = L$. The boundary and initial conditions may therefore be written as

$$z = L, \qquad t > 0, \qquad \frac{\partial \psi}{\partial z} = 0 \qquad (4.40)$$

$$z = 0, \qquad t > 0, \qquad \psi = 0 \qquad (4.41)$$

$$0 < z < L, \qquad t = 0, \qquad \psi = \Delta\psi. \qquad (4.42)$$

The solution of Equation (4.39) subject to (4.40), (4.41) and (4.42) is (see Carslaw and Jaeger (1959), p. 96)

$$\psi = \frac{4\,\Delta\psi}{\pi} \sum_{(n=1,3,5,\ldots)}^{n} \frac{1}{n}\,e^{-\alpha^2 Dt}\,\sin\alpha z, \qquad (4.43)$$

where $\alpha = n\pi/2L$ and $\Delta\psi$ is now understood to be the change of matric potential in the soil sample that can have a maximum value of $\Delta\psi$ initially $(=(p_1 - p_2)/\rho g)$ and decays to zero as water drains out to equilibrium. The water content at any time t, from Equations (4.43) and (4.38) is

$$\theta = a + \frac{4b\,\Delta\psi}{\pi} \sum_{(n=1,3,5,\ldots)}^{n} \frac{1}{n}\,e^{-\alpha^2 Dt}\,\sin\alpha z, \qquad (4.44)$$

and the integrated water outflow from the sample at time t is

$$w(t) = \int_0^L (V\theta\,dz/L). \qquad (4.45)$$

Performing the integration indicated in Equation (4.45)

$$w(t) = V\left\{ a + \frac{8b\,\Delta\psi}{\pi^2} \sum_{(n=1,3,5,\ldots)}^{n} \frac{1}{n^2}\,e^{-\alpha^2 Dt} \right\}. \qquad (4.46)$$

The initial water content (at $t = 0$) and the final water content $(t \to \infty)$ of the soil sample are respectively

$$w_i = (a + b\,\Delta\psi)V, \qquad (4.47)$$

and

$$w_f = aV, \qquad (4.48)$$

as may be seen by summing the series of Equation (4.46). The total outflow from the sample is then

$$w_i - w_f = b\Delta\psi V. \qquad (4.49)$$

The outflow at any time t is $w_i - w(t)$ and may conveniently be designated

$$Q(t) = w_i - w(t) = (w_i - w_f)\left(1 - \frac{8}{\pi^2} \sum_{(n=1,3,5,\ldots)}^{n} e^{-\alpha^2 Dt} \right). \qquad (4.50)$$

For all $t > 0.3/\alpha^2 D$, the second term of the series is less than 1 per cent of the first. Therefore if the results of the measurement of outflow after a few hours from the start of the experiment are used for plotting, the equation

$$Q(t) = (w_i - w_f)\left(1 - \frac{8}{\pi^2}e^{-(\pi^2 Dt/4L^2)}\right), \qquad (4.51)$$

should be satisfied. It may be verified that

$$\ln\{Q_\infty - Q(t)\} = \ln\frac{8Q_\infty}{\pi^2} - \frac{\pi^2 Dt}{4L^2} \qquad (4.52)$$

where Q_∞ has been written for $(w_i - w_f)$, and is the total outflow from the sample, either measured, or calculated from the independently determined moisture characteristic of the soil.

Equation (4.52) allows the diffusivity to be obtained from a straight-line plot of $\{Q_\infty - Q(t)\}$ versus t, on semi-log paper when the slope is equal to $\pi^2 D/4L^2$, for t sufficiently large. The unsaturated hydraulic conductivity can then be obtained by calculation using Equation (4.31).

Gardner (1956) tested his theory with three experimentally determined unsaturated hydraulic conductivities, of a sand, a sandy loam and a silty clay loam. The procedure seemed to be satisfactory. Gardner's method necessarily assumes that the conductivity of the membrane is much greater than the conductivity of the soil and therefore presents negligible impedance to the outflow rate. Subsequently Miller and Elrick (1958) and Rijtema (1959) extended the treatment to cases when the impedance of the membrane must be taken into account. It is often found (for example, Bruce and Klute, 1963) that there is an inexplicable, long-continued flow of water out of the sample although the theory requires that the flow declines sooner. Despite this blemish on the method its utility in enabling estimates of diffusivity, $D(\theta)$, and $K(\theta)$ is unquestioned. More sophisticated methods of analysing the experimental data have been reported by Doering (1965), Gupta, Farrell and Larson (1974), and Passioura (1977).

The outflow method of determining the unsaturated hydraulic conductivity, first introduced by Gardner, is versatile because it enables a large range of water contents of the soil and the corresponding conductivities to be attained. The pressure applied to the pressure membrane cell is the only controlled pressure required, because atmospheric pressure provides the control at the second boundary to the soil specimen.

Steady state methods are also available. One of these, introduced by L. A. Richards (1931), was designed so that controlled suctions could be applied through two porous plates, one at each end of the test soil enclosed in the permeameter cell. An increment of suction between the ends then creates a gradient, to which may be added the gravity gradient if the flow is vertically downwards, to provide the hydraulic potential gradient needed to sustain the measured flow through the cell (see Klute and Dirksen, 1986, for some experimental details). The range of suctions for which conductivity can be measured is limited to a maximum of about 80 kPa, if the pore-water pressure control depends upon partial vacuums created by a vacuum pump. Positive pressure control is possible, but a limit to the operation of the cell is set by the air-entry value, characteristic of the porous plates.

The crust or throttle method, described by Green *et al.* (1986), can be used both in the field and the laboratory. The principle is explained in Section 5.5, where the discussion concerns a two-layered soil possessing a less permeable upper layer. Flow through that layer, downwards at its limiting rate, necessarily creates a condition of unsaturation in the lower-lying layer, in which there is a zone of unvarying matric potential. The flow rate then occurs in response to the gravity gradient only and it is

Table 4.1. *Hydraulic conductivity of soil materials when saturated*

Range m s^{-1}	Soils associated with range	Effects on land use under agriculture
$<10^{-7}$	Soils of fine texture and poorly aggregated soils of medium texture. Mottling of subsoils common in humid areas.	Drainage may be so poor as to restrict use of land to shallow-rooted plants. Restricted drainage is required for flooded rice.
10^{-7} to 10^{-4}	Wide range of soils including those of fine texture with stable aggregates.	Land used for crops, including those under irrigation, is largely in this range.
$>10^{-4}$	Soils of coarse texture.	This high conductivity is usually associated with poor retention of water. Use may be restricted to deep-rooted plants.

Table 4.2. *The unsaturated hydraulic conductivity and soil moisture diffusivity of a sandy loam soil, as a function of water content and matric suction (from Gardner, 1956)*

Matric suction	Water content m^3 m^{-3}	K m s^{-1}	D m^2 s^{-1}
25 kPa	0.170	1.3×10^{-10}	1.5×10^{-8}
70 kPa	0.133	7.0×10^{-11}	1.4×10^{-8}
300 kPa	0.110	5.4×10^{-12}	1.3×10^{-8}
1.1 MPa	?	5.2×10^{-13}	?

equal to the hydraulic conductivity at the steady matric potential as measured by an inserted tensiometer.

Unsaturated hydraulic conductivity as a function of the water content shows very little hysteresis in the relationship. By contrast, its relation to matric potential is strongly hysteretic (Mualem, 1986), just as matric potential versus water content is strongly hysteretic (see Section 2.8). Experiments on unsaturated hydraulic conductivity should be so conducted that the direction of change of the mean matric potential of the test soil should not be reversed until the end of the run.

We have considered in this chapter laboratory methods of measuring hydraulic conductivity both saturated and unsaturated, and soil water diffusivity. Field methods will be considered in Chapter 6. It is appropriate to conclude a discussion of these parameters with a review of their magnitudes. Table 4.1 shows the range of hydraulic conductivities of soil materials when they are saturated. The effect of unsaturation is shown in Table 4.2 for an extended range of matric suctions as obtained by Gardner (1956) for Pachappa sandy loam. The data of Table 4.2 show the same dependence upon matric suction as is shown in Fig. 4.4.

We may calculate the soil water diffusivity corresponding to the unsaturated hydraulic conductivity values given in Table 4.2 if the appropriate water capacities at these points are known. It appears that, approximately, the diffusivities are as shown in column 4. It should be noted that the diffusivity decreases much less rapidly than the hydraulic conductivity, as the soil dries.

4.6 Flow through anisotropic soil

If the soil is anisotropic for hydraulic conductivity, the conductivities resolved in the principal axes, K_x, K_y and K_z may be unequal. The simplifications that led us to Equation (4.12) are then not valid and Equation (4.10) remains for further discussion. For the treatment of flow only in the saturated zone we have as before $d\theta/dt = 0$, and we assume that the soil is, at least, homogeneous, though anisotropic, so that $dK_x/dx = dK_y/dy = dK_z/dz = 0$. Then we can obtain Equation (4.53),

$$K_x \frac{\partial^2 \phi}{\partial x^2} + K_y \frac{\partial^2 \phi}{\partial y^2} + K_z \frac{\partial^2 \phi}{\partial z^2} = 0, \qquad (4.53)$$

as the equation describing the flow.

A change in scale, by putting

$$\lambda = (K_x/K_y)^{1/2} y = \alpha y, \qquad (4.54)$$

and

$$\mu = (K_x/K_z)^{1/2} z = \beta z \qquad (4.55)$$

transforms Equation (4.53) into Laplace's equation,

$$\frac{\partial^2 \phi}{\partial x^2} + \frac{\partial^2 \phi}{\partial \lambda^2} + \frac{\partial^2 \phi}{\partial \mu^2} = 0. \qquad (4.56)$$

The same solutions as are available for Equation (4.12), and which will be presented in Chapter 6, are available for analysis of flow in anisotropic soils, when we proceed to the analysis in the isotropic transformed space.

There have been only few investigations of anisotropy of soils in the field, probably because experimental variation has seemed to be associated more with the inhomogeneous nature of the soil profile than with genuine anisotropy. An early demonstration of anisotropy (Childs, Collis-George and Holmes, 1957) of soils on Romney Marsh, Kent, England, drew attention to their very finely laminar structure. The soil at a depth of about 1 m was laminated with alternating bands of clay and fine sand each not thicker than 5 mm and usually only about 1 mm thick. The horizontal hydraulic conductivity was 50 to 100 times larger than the vertical. The same authors reported on anisotropy of a peaty soil in valley bottom alluvium. This soil, in Bedfordshire, possessed fibrous, peaty laminates of swamp vegetation that, lying flat as it had fallen, made the soil much more permeable in horizontal directions than vertical.

From what has just been written it should be apparent that a decision is required as to whether a particular soil is anisotropic or layered, each layer within itself being homogeneous and isotropic. Hydrogeologists, who deal with saturated sediments that are often hundreds of metres deep have conventionally taken the view that variability is conferred by discrete, relatively thick layers. Suppose we have a number of horizontal layers with conductivities K_1, K_2, K_3 etc. Parallel to the layers the overall conductance of a cross-sectional area A is

$$KA = K_1 A_1 + K_2 A_2 + K_3 A_3 + \cdots. \qquad (4.57)$$

If the respective thicknesses are z_1, z_2, z_3, etc. and the total thickness is z it can be easily shown that

$$K = \frac{K_1 z_1 + K_2 z_2 + K_3 z_3 + \cdots}{z}$$

or

$$K = \frac{\sum_1^i K_i z_i}{\sum_1^i z_i}. \qquad (4.58)$$

Perpendicular to the layers the overall conductivity must be derived by an application of Darcy's law to each layer in turn. The magnitude of flow through layer 1 (see Fig. 4.11) is given by

$$v = K_1 \frac{\phi_1 - \phi_{12}}{z_1}, \qquad (4.59)$$

where ϕ_{12} is the potential at the boundary between layers 1 and 2. Similar equations give the flow, which must be the same, in all other layers. The flow is also given by

$$v = K \frac{\phi_1 - \phi_3}{z_1 + z_2 + z_3}, \qquad (4.60)$$

if there are three layers, where K is the mean hydraulic conductivity in the vertical direction. The three equations like Equation (4.59), lead to

$$\frac{z_1 v}{K_1} = \phi_1 - \phi_{12},$$

$$\frac{z_2 v}{K_2} = \phi_{12} - \phi_{23},$$

$$\frac{z_3 v}{K_3} = \phi_{23} - \phi_3.$$

Adding these equations, we have

$$v \left(\frac{z_1}{K_1} + \frac{z_2}{K_2} + \frac{z_3}{K_3} \right) = \phi_1 - \phi_3. \qquad (4.61)$$

Combining Equations (4.60) and (4.61) gives

$$\frac{1}{K} = \frac{z_1/K_1 + z_2/K_2 + z_3/K_3}{z_1 + z_2 + z_3},$$

or in general

$$\frac{1}{K} = \frac{\sum_1^i z_i/K_i}{\sum_1^i z_i}. \qquad (4.62)$$

Equations (4.58) and (4.62) are analogous to summing resistances for

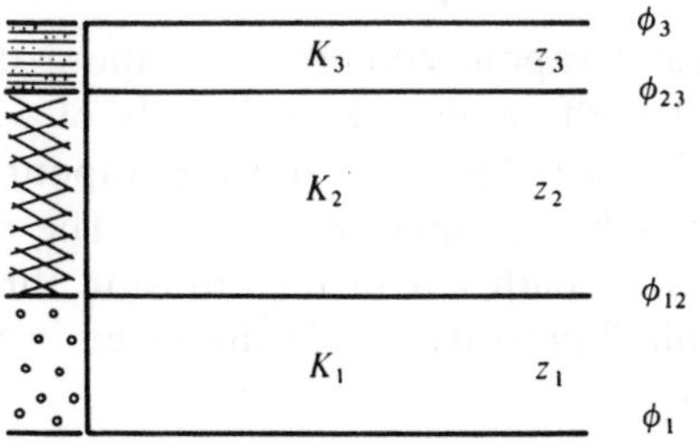

Fig. 4.11. Schematic of layered soil for illustration of method of calculating mean vertical and horizontal conductivities.

series connection and reciprocals of resistances for parallel connection in electric circuits, bearing in mind that conductance is the reciprocal of resistance. This statement suggests that hydrologists should use the terminology hydraulic conductance for quantities of the form $K_i z_i$. The traditional nomenclature for such a quantity is transmissivity, and it will be discussed further in Chapter 6.

4.7 Movement of water under a temperature gradient

Temperature affects soil water movement both in the liquid phase, by its effect on surface tension, and in the vapour phase by its large effect on the vapour pressure of water. The relation between surface tension, γ, and matric potential is given (from Equation 2.8) by

$$\psi = 2\gamma/r, \tag{4.63}$$

where r, the radius of curvature of the water–air interface, has negative values and ψ is expressed per unit volume of water (with dimensions of pressure). Since γ decreases with temperature as shown in Table 2.2, ψ will increase (it will have smaller negative values) as temperature increases when r is constant. Hence, a gradient in temperature in soil of uniform water content will give rise to a gradient in ψ favouring movement in the liquid phase in the direction of decreasing temperature.

A gradient in temperature can cause water to move in the vapour phase, also in the direction of decreasing temperature, in response to the vapour pressure gradient. Vapour pressure of water is strongly affected by temperature and large gradients are quite commonly caused in this way. The vapour pressure, e_0, of water in bulk changes, for example, from 2.3384 kPa at 20 °C to 4.2451 kPa at 30 °C. A similar large effect of temperature occurs in the vapour pressure of water in moist soil because as shown in Section 2.4 this must lie between e_0 and $0.989\,e_0$ for solute-free soil between saturation and the permanent wilting point. The vapour flux, q is given by Fick's law,

$$q = -D\nabla\rho_v, \tag{4.64}$$

where ρ_v is the water vapour content (i.e. vapour density) of the soil atmosphere and D is the molecular diffusivity of water vapour in the porous medium. The partial pressure of water vapour and its density are ideally related through the equation of state. The diffusivity depends upon the tortuosity of path length for diffusion through the air-filled pores and the air-filled porosity, ε_a. As discussed in Section 11.3, it can be approximated as

$$D = 0.66\,\varepsilon_a D_0, \tag{4.65}$$

where D_0 is the molecular diffusivity in free air.

The overall response of water in soil to a temperature gradient is then for water to tend to move from hot to cold regions according to the laws expressed by Equation (4.5) and (4.64). Gurr, Marshall and Hutton (1952) experimented with an enclosed 10 cm long, horizontal soil column, the two ends of which were kept at about 10 and 25 °C, in an attempt to assign relative magnitudes to the fluxes of water in the liquid and vapour phases. The experiment was repeated numerous times with increments of initial water content so that the range of water content from very dry to nearly saturated pore space could be investigated. They found that a tracer of chloride, initially uniformly distributed throughout the column, had accumulated at the hot end at the termination of the experiment, for several initial water contents in the middle of the water content range. The dissolved chloride could only be transported by a liquid water flux. The interpretation of this result is that although both vapour and liquid flux would initially go in the direction from hot end to cold end, the vapour flux would be associated with condensation of water at sites close to the cold end. This soon caused a matric potential gradient in response to the water content gradient that overcame the matric potential gradient initiated by the temperature gradient. Liquid water therefore moved from cold to hot against the temperature gradient, bearing with it the dissolved chloride, which accumulated at sites near the hot end where vaporization occurred.

Some of the results of Gurr *et al.* are shown in Fig. 4.12. It can be seen that the maximum transfer of water, as shown by the final water contents, occurred in columns *C* and *D* at a degree of saturation of 0.2 to 0.3. Because return flow occurred in the liquid phase, as shown by

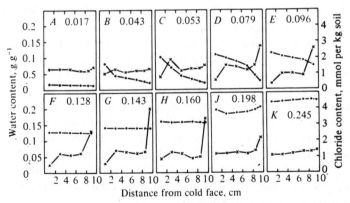

Fig. 4.12. Distribution of water (●) and chloride (×) in enclosed columns of a loam soil after being subject to a temperature gradient of 1.6 °C per cm for 5 days. The initial water content (in $g\,g^{-1}$ dry soil) is shown for each column. (After Gurr *et al.*, 1952.)

the salt migration, this gives an underestimate of the amount of vapour movement. Salt migration in the return flow of wetter columns shows that substantial amounts of water moved in the vapour phase in them also. Calculations showed that the vapour flux was up to three times the rate that could be accounted for by Fick's law (Equation 4.64), in some cases.

Philip and de Vries (1957) suggested that, in general, the flux of water in response to a temperature gradient in the soil would proceed by vapour movement in air-filled pores, subsequent condensation of the vapour at a liquid water meniscus and re-evaporation from the part of the water-filled pore exposed to the next air-filled pore in series. Because of the differing thermal conductivities of the three phases present in moist soil viz. solid, water and gas, the temperature gradient across an air-filled pore should be larger than the mean temperature gradient. These two factors can be shown, by semiquantitative calculation, to give a plausible explanation of the enhanced rate of vapour transfer observed by other experimenters (Rollins, Spangler and Kirkham, 1954, and others reviewed by Marshall, 1959a) in addition to Gurr and his co-workers.

The general equation to describe water flux under a temperature and water content gradient can be set down as

$$q = -D_T \nabla T - D_\theta \nabla \theta - Kk, \qquad (4.66)$$

in which k is the unit vector in the positive upwards direction. Such a phenomenological equation, expressing, by coefficients of suitable dimensional balance, the influence of the temperature gradient and water content gradient in addition to the gravitational gradient is not amenable to application as it stands.

Philip and de Vries (1957) showed how the diffusivity for thermal moisture transfer, D_T, can be separated into two components appropriate to describe the movement of liquid water and water vapour caused by the effects of a temperature gradient. Similarly the diffusivity for water content gradient transfer, D_θ, can be separated into two components appropriate to the liquid and vapour phase movements. The thermodynamics of irreversible processes was applied, in a formal statement of these linked flows, by Cary (1965, 1966). Practical application of such theories, simple and formal as they appear, depends upon suitable measurements of the linked and unlinked coefficients such as the components of D_T, etc. in Equation (4.66) above. There has been little further development of solutions to the problem, due to the lack of experimental design to evaluate these coefficients.

There are numerous natural phenomena that can be qualitatively described from the understanding that has been so far gained about thermally induced moisture flow. Nocturnal cooling by radiation from the ground surface is often observed to cause an early morning accumulation of moisture in a rather thin layer near the surface of the soil. Rose (1968) and Jackson (1973) reported a marked diurnal variation in water content of the surface soil under field conditions. This variation was superimposed on the trend in water content as the soil dried out from day to day (see Fig. 4.13). At about 05:00 hours the maxima in water content occurred in the soil near the surface and minima occurred at about 16:00 hours each day. During the daytime, heating of the surface layers creates a vapour pressure gradient tending to make water vapour move not only upwards into the atmosphere at the soil surface but also downwards towards cooler soil. The surface then cools at night to a temperature lower than that of the deeper soil and the vapour moves upward into the surface soil. Rose (1968) concluded from his field experiments that the vapour flux due to the strong, daily temperature wave in arid Australia was similar in magnitude to the liquid flux that occurred independently of temperature gradients. Flow in the liquid phase in response to the temperature gradient was considered to be negligible.

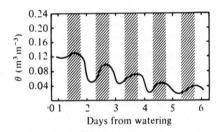

Fig. 4.13. History of water content of the soil at a depth of 0 to 1.3 cm, under the influence of a diurnal temperature fluctuation (after Rose, 1968). The time intervals run from noon to noon for each day and the night periods are shown by shading.

During a complete seasonal cycle, Abramova (1963) has measured movements of 13 mm of water downward in summer and upward in winter in soil in the northwestern Caspian lowland. This followed early work by Lebedeff in the USSR in 1927 that drew attention to the possible importance of water vapour movement. Generally, however, vapour movement in response to a temperature gradient in soils is overridden by the greater fluxes in the liquid phase during infiltration, drainage, evaporation and water uptake by roots.

5

Distribution of water in soil

The movement of water through the soil, as discussed in Chapter 4, can sometimes be regarded as being in the steady state. There can then be no time-dependent change in the water content of the soil and a study of any particular zone of the soil presupposes that the fluxes of water across the boundaries of that zone can be specified, and that inflow and outflow just balance. This approach has been particularly useful in developing drainage theory, as will be shown subsequently in Chapter 6. The accurate satisfaction of the steady state criteria may not be necessary if the time scale of interest is so long that short-term variations in water content of the soil, and lack of balance between inflow and outflow at the boundaries, can be smoothed over and eliminated on a larger scale of time. Rapid movements of water into and through the soil, however, cannot be described satisfactorily by the steady state approach. Infiltration of rainwater or irrigation water through the soil surface and the descent of a wetting front into relatively dry soil is one of the basic natural processes. The life-environment of terrestrial plants includes this zone of intermittent or cyclical wetting of the soil. In this chapter the theory of infiltration of water will be described, together with the subsequent redistribution of water from wet to dry soil. Some aspects of the transport and redistribution of solutes in the soil solution will also be discussed, so that such principles can be used to assist in an understanding of soil salinity, which can often develop in consequence of changed land use.

5.1 Infiltration from ponded water

It was shown in Chapter 4 how the general equation that describes water flow in soils can be derived, and Equations (4.29) and (4.30) demonstrated that the absorption of water by soil has features that are similar to diffusion processes. These equations rewritten here for convenience are

Equations (5.1) for horizontal flow and (5.2) for vertical flow:

$$\frac{\partial}{\partial x}\left(D\frac{\partial \theta}{\partial x}\right)=\frac{\partial \theta}{\partial t},$$ (5.1)

$$\frac{\partial}{\partial z}\left(D\frac{\partial \theta}{\partial z}\right)+\frac{\partial K}{\partial z}=\frac{\partial \theta}{\partial t}.$$ (5.2)

Here θ is the water content (by volume fraction), D is the soil water diffusivity and K is the hydraulic conductivity. Both these soil properties are functions of θ.

For horizontal flow Equation (5.1) can be solved analytically, if D can be regarded as constant. Suppose water is absorbed into dry soil that has been packed into a horizontal tube of small diameter, as illustrated by Fig. 5.1. If the tube is of finite length, l, the problem has its initial and boundary conditions described by

$$t=0, \qquad \theta=0, \quad \text{for} \quad 0<x<l,$$

$$t>0, \qquad \theta=\theta_0, \qquad x=l \quad \text{and} \quad \frac{d\theta}{dx}=0, \qquad x=0. \tag{5.3}$$

It has its direct analogy in the theory of the non-steady flow of heat in a rod of finite length, for which the solution given by Carslaw and Jaeger (1959, p. 309) can be taken over for our problem as

$$\frac{\theta}{\theta_0}=\sum_{n=0}^{\infty}(-1)^n\,\text{erfc}\,\frac{(2n+1)-xl^{-1}}{(4Dtl^{-2})^{1/2}}$$

$$+\sum_{n=0}^{\infty}(-1)^n\,\text{erfc}\,\frac{(2n+1)+xl^{-1}}{(4Dtl^{-2})^{1/2}} \tag{5.4}$$

where the error function complement is defined to be

$$\text{erfc}\,z=1-\text{erf}\,z,$$

and

$$\text{erf}\,z=(2/\pi^{1/2})\int_{0}^{z}\exp(-\xi^2)\,d\xi. \tag{5.5}$$

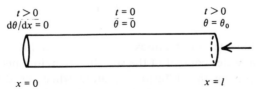

Fig. 5.1. Schematic for experiment to measure the infiltration of water horizontally into dry soil.

Here ξ is a dummy variable, the independent variable z being the upper limit of the integration. Tables of numerical values usually are for the error function only (see Jahnke and Emde, 1945, and Carslaw and Jaeger, 1959).

Suppose we select a diffusivity of 1×10^{-8} m^2 s^{-1} for a soil, and calculate the propagation of change of water content in a horizontal direction, corresponding to the experimental arrangement depicted in Fig. 5.1, for the absorption of water at one end of the column. Fig. 5.2 shows a graphical representation of the solution given as Equation (5.4) for several values of the dimensionless parameter $(4Dtl^{-2})^{1/2}$. If our column were 1 m long the water content half way along the column would be 0.16 of its final equilibrium value after about 72 days had elapsed. The example is meant to demonstrate the relatively long time scale that is involved in absorption of water into soil, when it is unassisted by gravity. However, this treatment is far from precise because it has been assumed that the diffusivity remains constant.

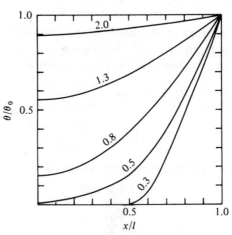

Fig. 5.2. Diffusion of water horizontally into dry soil, with diffusivity constant, according to the problem sketched in Fig. 5.1, the solution of which is given by Equation (5.4). The curves are distinguished by values of the dimensionless parameter $(4Dt/l^2)^{1/2}$.

Many experimental observations have shown that diffusivity increases with increasing water content of the soil. It is therefore necessary to use methods for solving the diffusion equation when the diffusivity is a function of concentration, if solutions are desired that are better than just a qualitative guide to the time scale for absorption. Klute (1952) used a method of numerical solution due to Crank and Henry (1949*a*, *b*) and subsequently Philip began to publish his well-known series of papers on the theory of infiltration. Horizontal infiltration is of course highly artificial, although the soil water diffusivity can be determined by laboratory experiments using horizontal columns and a suitable application of the theory (Bruce and Klute, 1956).

For infiltration vertically downwards, the equation to be solved is Equation (5.2). Following Philip (1957a), Equation (5.2) is written as

$$\frac{\partial \theta}{\partial t} = \frac{\partial}{\partial y}\left(D\frac{\partial \theta}{\partial y}\right) - \frac{\partial K}{\partial y}, \qquad (5.6)$$

where y is the vertical co-ordinate, positive downwards from the soil surface as datum. The infiltration process is also described by the following initial and boundary conditions appropriate to Equation (5.6).

$$
\begin{aligned}
t = 0, & \quad y \geqslant 0, & \quad \theta = \theta_n \\
t > 0, & \quad y = 0, & \quad \theta = \theta_0.
\end{aligned}
\qquad (5.7)
$$

Suppose as a first approximation that the effect of gravity upon the infiltration process can be neglected. Then

$$\frac{\partial \theta}{\partial t} = \frac{\partial}{\partial y_1}\left(D\frac{\partial \theta}{\partial y_1}\right) \qquad (5.8)$$

is the equation that describes the process, subject to

$$
\begin{aligned}
t = 0, & \quad y_1 \geqslant 0, & \quad \theta = \theta_n \\
t > 0, & \quad y_1 = 0, & \quad \theta = \theta_0.
\end{aligned}
\qquad (5.9)
$$

The co-ordinate y_1 can be thought of as a distance of propagation for any particular water content change caused by the gradient of the matric potential. It must be less than the distance y at which the same change is propagated when infiltration is further assisted by gravity.

Unlike the problem described mathematically by Equations (5.3) and illustrated by Fig. 5.1 as horizontal diffusion into a tube of finite length, the problem described by Equations (5.9) is that of diffusion into a soil column of infinite linear extent. The solution, when D is constant, looks rather like Equation (5.4) in that the error function complement is required. It is again readily obtained from the theory of conduction of heat in solids (see Carslaw and Jaeger, 1959, p. 59) and is

$$(\theta - \theta_n)/(\theta_0 - \theta_n) = \mathrm{erfc}\; x/(4Dt)^{1/2}, \qquad (5.10)$$

where, for generality we write, for the moment, x for y_1. An inspection of Equation (5.10) reveals that, for diffusion into a medium of semi-infinite extent, the solution is a function of $xt^{-1/2}$. This characteristic combination of the two independent variables, well known in diffusion theory, has the following consequences.

(i) The distance of penetration of the diffusing substance specified by a given concentration is proportional to $\sqrt{t}$.

(ii) The time taken for the concentration to reach a given concentration at any distance x from the boundary is inversely proportional to D and proportional to x^2.

An early demonstration of these rules was given by Bell and Cameron (1906), for a single capillary tube on the basis of the Hagen-Poiseuille law. Following Green and Ampt (1911) we can do this in terms of Darcy's law, thus. For horizontal flow into a soil column arranged similarly to Fig. 5.1, we have,

$$v = -K\frac{\Delta\psi}{x}, \tag{5.11}$$

where $\Delta\psi$ is the matric potential difference from the end of the column where water is allowed to enter (effectively there $\psi = 0$) to the position of the wetting front as a distance x. The rate of water entry is also given by

$$v = \varepsilon\frac{dx}{dt}, \tag{5.12}$$

where ε is the porosity that is being filled with water as the wetting front advances. From Equations (5.11) and (5.12) we have

$$x\,dx = -(K/\varepsilon)\Delta\psi\,dt,$$

$$\int x\,dx = -(K/\varepsilon)\Delta\psi\int dt + \text{constant},$$

$$\text{or } x^2 = -\frac{2K}{\varepsilon}\Delta\psi t. \tag{5.13}$$

The constant of integration is zero because $x = 0$ at $t = 0$. Therefore

$$xt^{-1/2} = \text{constant}, \tag{5.14}$$

remembering that $\Delta\psi$ is a negative quantity and that, therefore $(-2K\,\Delta\psi/\varepsilon)^{1/2}$ is a real number.

The significance of Equation (5.14) is that the constant may be regarded as specifying a numerical value of the matric potential at the wetting front. Therefore, however long it takes (t) for the wetting front to advance to a position x from the entry end, the product $xt^{-1/2}$ is invariant. Experimental data do not follow Equation (5.13) perfectly, but Kirkham and Feng (1949) showed that the relation

$$x = Nt^{1/2} + a \tag{5.15}$$

agrees closely with the experimentally observed rate of horizontal advance of a wetting front into a dry soil, when N corresponds to $(-2K\,\Delta\psi/\varepsilon)^{1/2}$ and a is a small constant.

Finally, in discussion of the manner of derivation of Equation (5.14), we should note that it requires that the value of $\Delta\psi$ between the point of entry and the wetting front should always be the same, so that the integration that led to Equation (5.13) can proceed. If the wetting front should reach the other end of the horizontal column the theory is invalidated. Physically, a qualitative explanation of the $xt^{-1/2}$ rule is that the same potential difference is spread over a longer and longer path from entry so that the potential gradient becomes less and less. The rule applies equally whether the diffusivity coefficient, D, is constant or depends upon water content.

This discussion has anticipated the more precise account of the wetting front developed by Philip, to which we now return. Mathematically the combination of variables as $xt^{-1/2}$ may be formalized by a transformation used by Boltzmann (1894) to convert the partial differential equation to an ordinary differential equation. Using

$$\lambda = y_1 t^{-1/2}, \tag{5.16}$$

Equation (5.8) may be converted to

$$-\frac{\mathrm{d}\theta}{\mathrm{d}\lambda} = \frac{2}{\lambda}\frac{\mathrm{d}}{\mathrm{d}\lambda}\left(D\frac{\mathrm{d}\theta}{\mathrm{d}\lambda}\right). \tag{5.17}$$

If the initial and boundary conditions of the problem can be accommodated by the same transformation, it is possible to proceed to analytical and numerical solutions of Equations (5.8) and (5.1). In general the transformation can be made for diffusion into infinite or semi-infinite media when the initial concentration in the medium is uniform or zero.

If Equations (5.8) and (5.6) are compared, the difference, $y - y_1$, of the space variables is to be understood as the further distance of diffusion of a given water content change caused by the influence of the term $-\partial K/\partial y$ in Equation (5.6). Philip derived a power series solution of Equation (5.6) as

$$y = \lambda t^{1/2} + \chi t + \psi t^{3/2} + \cdots. \tag{5.18}$$

The coefficients λ, χ, ψ etc. are functions of the water content θ only. They arise as transformations of the kind $\lambda = y_1 t^{-1/2}$, $\chi = (y - y_1)t^{-1}$, $\psi = y_2 - (y - y_1)t^{-3/2}$ etc., where $y_2 - (y - y_1)$ etc. are new residual errors, subject to processes of solution similar to that used for Equation (5.17).

The physical explanation of the terms on the right hand side of Equation (5.18) is that infiltration unassisted by the gravity field is described by the first term. All the other terms should be regarded as correction terms to add the effect of gravity to absorption caused by the matric potential gradient. The mathematical development of the solutions, which are numerous for the variety of situations that can be analysed, is beyond the intended scope of this book.

Philip (1957b) remarked that it is often customary to assume a series solution of a partial differential equation, such as Equation (5.18), and then to proceed to evaluate numerically or analytically the coefficients of the terms. Because the water content θ at y and t does not appear explicitly in Equation (5.18), the presentation of that equation as a solution has to be graphical for a particular problem. Fig. 5.3 shows the development of the water content profile in the soil (Yolo light clay) calculated by Philip (1957a), for continuous infiltration. Equations (5.7) give initial and boundary conditions. The soil hydrological data were taken from work by Moore (1939), as used by Philip (1957a) in his calculations. The actual values of θ_0 and θ_n were 0.4950 and 0.2376, respectively. The water content is presented on a normalized scale in Fig. 5.3, with $\theta_* = (\theta - \theta_n)/0.2574$.

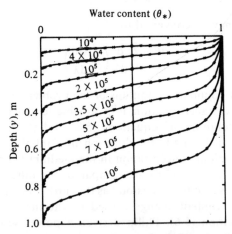

Fig. 5.3. Computed water content profiles during infiltration into Yolo light clay (numbers on each profile represent time in seconds). (After Philip, 1957a.)

The cumulative infiltration of water into the soil possessing the sequence of water content profiles shown in Fig. 5.3 could be given approximately by the product of depth of the wetting front and the mean water content, $\bar{\theta}$, behind the wetting front. If the coefficients λ and χ of the first two terms of Equation (5.18) could have a water content dependence written explicitly as $\lambda = \lambda'/\bar{\theta}$ and $\chi = \chi'/\bar{\theta}$, then Equation (5.18) could be approximated as

$$y\bar{\theta} = \lambda' t^{1/2} + \chi' t, \tag{5.18a}$$

where y is the depth of the wetting front. In Section 5.7 we give an account of the measurement of infiltration in the field, and it will be seen that Equation (5.18a) is equivalent to the equation that Philip suggests should be used for field purposes.

A simple test of the influence of absorption of water in response to the matric potential gradient upon the total infiltration process can be applied to the data of Fig. 5.3. If a diffusion-type process dominates, the ratio $yt^{-1/2}$ should be invariant for a given water content. Choosing $\theta_* = 0.5$ we find that the dependence of $yt^{-1/2}$ upon the time of infiltration is as shown in Table 5.1. In the early stages of infiltration under the influence of matric potential gradient, absorption predominates. At later stages of infiltration the gravitational effect becomes dominant.

Table 5.1. *Test of the influence of diffusion-type absorption upon the infiltration process illustrated in Fig. 5.3*

t, s	$yt^{-1/2}$, m s$^{-1/2}$
10^4	0.053
4×10^4	0.054
10^5	0.057
10^6	0.073

It is instructive to contrast Figs. 5.2 and 5.3, and to consider the differences between the physical systems that they depict. The profile of water content, using a diffusivity function calculated by Philip from Moore's data has a very steep wetting front. Its rate of descent obeys approximately the rule that $yt^{-1/2} = $ constant, in the early stages of infiltration. By contrast, there is no wetting front at all shown by the analytical solution (Fig. 5.2) using the (unreal) assumption of a constant D. The propagation of a chosen water content up to the time that water first reached the further end of the tube would, nevertheless, also obey the $xt^{-1/2}$ rule.

Turning our attention now away from the water content profiles as the descriptive feature of infiltration, we consider the flux of water through the soil surface. The amount of water contained in the soil from the surface to any depth y is

$$\int_0^{\eta=y} \theta \, d\eta = \bar{\theta}y, \tag{5.19}$$

which serves to define a mean content $\bar{\theta}$. If y is the distance to the wetting front then $\bar{\theta}$ is the mean water content of the soil from the surface to the wetting front. The lengthening zone of wet soil behind the advancing wetting front, in which the water content is essentially constant, was called the transmission zone by Bodman and Colman (1944) and it is illustrated by the data of Fig. 5.3.

Let us assume that the infiltration can be ideally depicted by a very steep wetting front behind which the water content has the value $\bar{\theta}_0$, as shown in Fig. 5.4. Suppose that the depth of water ponded upon the surface is h and that at time t the wetting front has descended to $-y$ in the soil. The hydraulic potential difference between the soil surface and the wetting front is $(h + y - \psi)$, reckoning $y = 0$ at the surface and positive upwards. The matric potential at the wetting front is ψ and $-\psi$ is the capillary suction so designated by Green and Ampt (1911), who originated the following discussion of the phenomenon. The hydraulic potential difference between the surface and a depth $-y$ can also be obtained from Darcy's law rewritten as

$$d\phi = -(v/K_0)\, dy,$$

and following integration,

$$h + y - \psi = -v/K_0 \int_0^{-y} dy = vy/K_0. \tag{5.20}$$

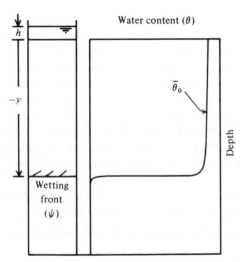

Fig. 5.4. The downward movement of water into the soil according to the concept of a capillary 'pull' at the wetting front. (After Green and Ampt, 1911.)

Rearranging Equation (5.20) we have

$$v/K_0 = 1 + (h - \psi)/y. \tag{5.21}$$

The water flux through the soil surface is the infiltration rate, i.e.

$$v = \mathrm{d}i/\mathrm{d}t \tag{5.22}$$

where i is the cumulative infiltration, which is also given by

$$i = (\bar{\theta}_0 - \theta_n)y. \tag{5.23}$$

Combining Equations 5.21, 5.22 and 5.23, we obtain

$$\frac{\mathrm{d}i}{\mathrm{d}t} = K_0 \left\{ 1 + \frac{(h - \psi)(\bar{\theta}_0 - \theta_n)}{i} \right\}. \tag{5.24}$$

The solution of Equation (5.24) that satisfies $i = 0$, $t = 0$ is

$$t = \frac{1}{K_0} \left\{ i - (h - \psi)(\bar{\theta}_0 - \theta_n) \ln \left(1 - \frac{i}{(h - \psi)(\bar{\theta}_0 - \theta_n)} \right) \right\}. \tag{5.25}$$

Equation (5.25), derived also by Philip (1954), is one example of numerous infiltration equations that attempt to represent the infiltration phenomenon. Unless a quantitative estimate of ψ, the matric potential of the wetting front can be inserted, Equation (5.25) cannot be applied with precision. Values of -0.5 to -1.5 m seem to fit the data for many soils over a range of water contents. The theory of Green and Ampt, who found that a value of -0.9 m fitted their data best, is gaining in favour, because good data about ψ at the wetting front are accumulating (see for example, van Schilfgaarde, 1974b).

Infiltration rate in the field usually shows a steep decline with time from the start of the application of water, particularly if water is ponded upon the soil surface. Horton (1940) suggested that the equation

$$v = \frac{\mathrm{d}i}{\mathrm{d}t} = v_f + (v_i - v_f) \, \mathrm{e}^{-\beta t} \tag{5.26}$$

represents the infiltration rate reasonably well, where v_f, the final infiltration rate as $t \to \infty$ can usually be identified with K_0, the saturated hydraulic conductivity. The initial infiltration rate v_i can be very large at the moment of first application of water, either as rain or in the infiltration test. The actual value used in an equation such as Equation (5.26) depends upon the time interval that is used to get a measurement of v_i, and rarely can it be less than one minute. The actual rate of fall of infiltration is usually faster than exponential. Difficulties may arise in giving a suitable value to the constant β, which has no physical meaning.

An example of an infiltration curve is shown in Fig. 5.5, together with a sketch of a Horton-type curve. In this example the observed infiltration rate declines faster than exponential in the early stages.

Distribution of water in soil

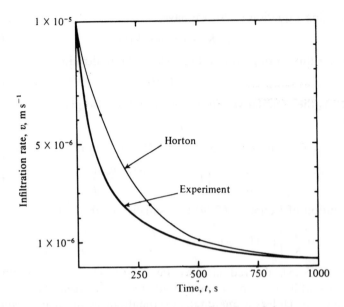

Fig. 5.5. Infiltration rate as a function of time.

The final infiltration rate for ponded infiltration should approach the saturated hydraulic conductivity when the water table remains very deep, because the potential gradient for downwards flow approaches unity in the transmission zone, i.e. the gravitational gradient. But, in the context of irrigation agriculture for which infiltration experiments were developed initially, the water table may be quite shallow at the start of the infiltration. The development of a ponded water table condition would invalidate the theory and expectations given above. In that case the final infiltration rate may be very much less than the saturated hydraulic conductivity. It would depend upon the disposition of the drains, as will be discussed in Chapter 6, where the flow net for underground flow will be presented. Water entry into the soil at a location remote from a drain, under ponded conditions, is likely to be much slower than into the soil immediately above a tile-drain or in the near vicinity of a ditch-drain.

Conditions that affect the rate of infiltration of water into a given soil are discussed in other chapters, but two of general significance may be mentioned here. The first of these is the initial water content of the soil. The drier the soil the greater is the initial rate of entry because the gradient of the matric potential is then of greater magnitude. This persists because the wetting front advances more slowly when more water is required to bring the larger air-filled porosity nearly to saturation. This generalization may not hold for soil that develops shrinkage cracks or is difficult to wet. Secondly, there may be a limitation on the rate of supply of water, as in rain or in sprinkler irrigation. The actual infiltration rate, v, may then be less than the potential rate for water ponded upon the surface. If the limited supply rate, for example, a constant precipitation P, is maintained, v equals P until the transmission zone has lengthened sufficiently for v to fall below P. The curve $v(t)$ for this common situation thus has a horizontal portion before v begins to decrease towards the minimum infiltration rate for the soil and ponding occurs (see Rubin and Steinhardt, 1964).

Infiltration rate has been intensively studied since 1960, not only for its undoubted importance in irrigation layout and design, but also because it is an important parameter to be given the best possible estimate in catchment run-off models (see, for example IAHS, 1974).

Further consideration of rain-fed infiltration, at an approximately constant flux across the soil surface, will be given in Section 5.6.

5.2 Redistribution following infiltration

When the infiltration of water through the soil surface ceases there ensues a quite prolonged period of redistribution of water in the soil profile from the parts that have been wetted to the dry soil beyond the wetting front. An example of such redistribution, chosen from many experimental demonstrations, is that given by Gardner, Hillel and Benyamini (1970) and illustrated by Fig. 5.6. At day 0 the water content profile at the cessation of infiltration shows a zone of near saturation where matric potential is approximately zero, above a steep wetting front and a dry zone where matric potential is much less and is characteristic of the initial dryness of the soil.

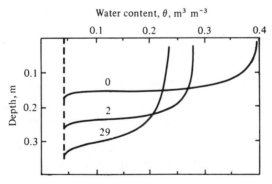

Fig. 5.6. Successive water content profiles in a soil column, following application of 50 mm of water. 0, 2 and 29 are the times in days after irrigation. (After Gardner *et al.*, 1970.)

The reduction of water content of the soil in the zone above the position of the wetting front at the start of redistribution of the soil water defines the draining zone. In this zone the relationship between θ and ψ must be the draining characteristic. Hysteresis in the $\theta(\psi)$ function, discussed in Chapter 2, is an important factor in quantitative descriptions of redistribution.

Sometimes redistribution profiles exhibit the shape shown in Fig. 5.7, not that shown in Gardner's experiment (Fig. 5.6). As can be seen from Fig. 5.7, rather than exhibiting a region whence water drains downwards from all depths, the draining zone is located near to the soil surface, then there is a region of insignificant drainage near the wetting front that acts effectively still only as a transmitting zone. These contrasting redistribution profiles seemed to indicate to many contributors to the development of infiltration theory that there was an unresolvable conflict in experimental evidence. Youngs and Poulovassilis (1976) have shown that, in theory, the existence of the two different kinds of redistribution profiles presents no inconsistency.

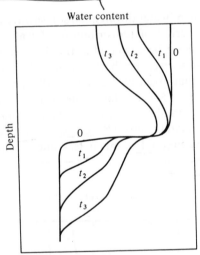

Fig. 5.7. Successive water content profiles in a soil column, following the onset of redistribution, with drainage from near the soil surface, a persistent transmission layer showing little water content decline, and water penetration into the dry zone below the initial wetting front.

Following the arguments of Youngs and Poulovassilis (1976), the water flux in the soil at any depth z, positive upwards, is obtained from Darcy's law as

$$v = -K\left(\frac{\mathrm{d}\psi}{\mathrm{d}z} + 1\right). \tag{5.27}$$

There is a maximum water content, θ_L, at any depth, that the soil can attain on wetting up. With further increase in time the soil begins to drain at that plane. We have, for the total differential,

$$\frac{d\psi}{dz} = \left(\frac{\partial\psi}{\partial\theta}\right)_{\theta_L} \left(\frac{d\theta}{dz}\right) + \left(\frac{\partial\psi}{\partial\theta_L}\right)_{\theta} \left(\frac{d\theta_L}{dz}\right). \tag{5.28}$$

By a rule for partial differentiation

$$-\left(\frac{\partial\theta}{\partial\theta_L}\right)_{\psi} = \left(\frac{\partial\psi}{\partial\theta_L}\right)\left(\frac{\partial\theta}{\partial\psi}\right),$$

we can derive from Equations (5.27) and (5.28) that

$$\frac{d\theta}{dz} = -\left(\frac{v}{K}+1\right)\left(\frac{\partial\theta}{\partial\psi}\right)_{\theta_L} + \left(\frac{\partial\theta}{\partial\theta_L}\right)_{\psi}\left(\frac{d\theta_L}{dz}\right). \tag{5.29}$$

The quantity $d\theta/dz$ is the gradient of the water content at any depth, and whether it is zero, positive or negative appears to be determined by the magnitude of the individual terms on the right hand side of Equation (5.29). The situation physically is that

$$\left(\frac{\partial\theta}{\partial\psi}\right)_{\theta_L} > 0, \qquad \left(\frac{\partial\theta}{\partial\theta_L}\right) > 0 \quad \text{and} \quad \frac{d\theta_L}{dz} \geqslant 0.$$

Therefore

$$\frac{d\theta}{dz} = 0 \quad \text{when} \quad -\frac{v}{K} = 1 + \left(\frac{\partial\psi}{\partial\theta_L}\right)_{\theta}\left(\frac{d\theta_L}{dz}\right) \tag{5.30}$$

$$\frac{d\theta}{dz} > 0 \quad \text{when} \quad -\frac{v}{K} > 1 + \left(\frac{\partial\psi}{\partial\theta_L}\right)_{\theta}\left(\frac{d\theta_L}{dz}\right), \tag{5.31}$$

and

$$\frac{d\theta}{dz} < 0 \quad \text{when} \quad -\frac{v}{K} < 1 + \left(\frac{\partial\psi}{\partial\theta_L}\right)_{\theta}\left(\frac{d\theta_L}{dz}\right). \tag{5.32}$$

Youngs and Poulovassilis have developed this useful general analysis to show that water content gradients may be zero, positive or negative in the draining zone, according to classical flow theory.

In the wetting zone $-v > K$ (remember that here z is positive upwards and therefore v, the water flux in the downwards direction, has a negative sign associated with it), and the flux must have a magnitude greater than K at the plane of transition from wetting to draining. From considerations of continuity of flux $|v| > K$ for a small distance above the transition plane also. Equation (5.29) then requires that the water content gradient $d\theta/dz$ is always positive just above the transition plane.

The hydraulic conductivity K increases with increasing θ, so that K also increases with height above the transition plane as long as $d\theta/dz > 0$. Since this is the draining zone, the flux $|v|$ must decrease with height. Equation (5.29) then requires that $d\theta/dz$ decreases with height above the transition plane. It may remain positive right to the surface, or it may change through $d\theta/dz = 0$ to negative values within the draining zone. The requirement that $d\theta/dz = 0$ at the surface for no flux across the surface, imposes conditions upon the magnitude of the terms in Equation (5.29). However, the analysis is general and can predict the trend of $d\theta/dz$ when v is positive at the surface, i.e. evaporation is allowed to occur, or is negative, i.e. a reduced infiltration is allowed during the redistribution phase. It is easy to see by inspection of the shapes of the profiles exhibited in Figs. 5.6 and 5.7 that profiles of water content that have negative values of $d\theta/dz$ are more likely to occur for lesser values of $d\theta_L/dz$, and this tends to occur for deeper infiltration. Shapes like the profiles of Fig. 5.7 are readily produced of course by evaporation, as is well known to any worker experienced in field observation, but it is now possible to accept the validity of the observations on laboratory columns of profiles of this shape for redistribution when the flux of water through the surface is zero or when it is less than the maximum possible infiltration rate.

The quantity $(\partial\psi/\partial\theta_L)_\theta$ is the slope of the moisture characteristic curve and for different soils this slope, evaluated at the water content of the transition plane, shows a change from smaller values for coarser textured soils to larger values for finer textured soils. Therefore, according to Equations (5.31) and (5.32), the redistribution profile shapes tend to be those of Fig. 5.6 as the soil texture becomes finer.

The discussion on redistribution following infiltration has proceeded on the assumption that the soil is uniform, although of course its soil moisture characteristic displays the well-known hysteresis loop of the draining and wetting situation. There remains to obtain, if we can, some validation of the guiding theory from observed soil water change in the field.

Fig. 5.8 shows water content profiles obtained during an experiment on water use of pine trees (Holmes and Colville, 1970a) that were unpublished at the time but are convenient for the present purpose. The progress of the wetting front under the influence of rainfall can be seen, but the heterogeneous textural composition of the soil tends to distract the attention, particularly near the two layers at 4.2 m and 6.3 m where the soil profile has a heavy clay accumulation. Fig. 5.9 shows only the increase of water content in the profile. It is much more readily perceived that indeed the descent of the wetting front is consistent with the theory and the idealized experimental observations on laboratory columns of

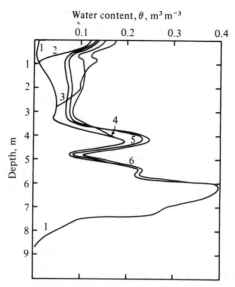

Fig. 5.8. Water content profiles measured in a deep sand profile (overlying two layers of clay accumulation at 4.2 m and 6.3 m) during rain-fed infiltration in the sequence 1 to 6 (Holmes and Colville, unpublished).

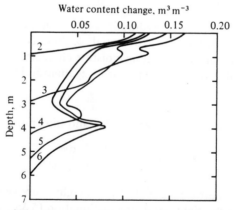

Fig. 5.9. The data of Fig. 5.8 replotted to show water content increment independent of absolute magnitude of the water content, for the purpose of comparison of field data with infiltration theory and prediction.

homogeneous soil. The data from which Figs. 5.8 and 5.9 were constructed are representative of innumerable experiments on water budgeting of vegetation done in many parts of the world since the introduction of the neutron moisture meter about 1956. The very variable nature of soil profiles probably forbids forever any close theoretical analysis of water content in the field despite the attention paid to infiltration as a challenging problem in applied mathematics. By contrast, the validation of the solutions by experimental results on idealized laboratory columns is very satisfactory.

5.3 Steady state infiltration through a soil draining to a stationary water table

There are several steady state situations that are instructive and, when understood, help us to interpret field observations. Suppose that a soil and its associated soil water profile is at equilibrium with the water table. If water is then made to infiltrate through the soil profile at a steady rate, v, that is less than the saturated hydraulic conductivity, K_0, a new profile of water content will eventually be established. We have to assume that the water table can be held at a stationary level by suitable underdrainage of an extensive, very permeable, deep horizon. Under the steady state criterion there is no change of water content at any given level with time, and the unsaturated flow regime is described by an application of Darcy's law,

$$-v = K \frac{d\phi}{dz},$$ (5.33)

where K is the unsaturated hydraulic conductivity, dependent upon the water content, θ, or the matric potential, ψ, these two parameters being related to each other by the wetting moisture characteristic. The potential, ϕ, has the usual two components for its evaluation in soil, namely

$$\phi = \psi + Z.$$ (5.34)

Combining Equations (5.33) and (5.34) and integrating yields

$$z = \int_0^z dz = -\int_0^\psi \frac{d\psi}{1 + v/K},$$ (5.35)

where ψ is the matric potential at any height z above the water table. The integral on the right hand side of Equation (5.35) can be evaluated if the function $K(\psi)$ is known, usually necessarily by graphical means.

Suppose we have a soil that has the $K(\psi)$ function given in Table 5.2. Using these data the hydraulic potential as a function of height above the water table was derived for the case of $v = -1.0 \times 10^{-8}$ m s^{-1} infiltration rate. Note that a negative sign has to be associated with v, because it is a downwards water flux. The result is shown in Fig. 5.10, for which the individual plotting points were obtained by graphical integration of Equation (5.35).

Table 5.2. *Soil hydrological data for a well-struc-tured clay loam (composite, hypothetical data), for purposes of calculation of Equation (5.35) and Equation (5.37)*

θ $m^3 m^{-3}$	ψ m wetting	drying	K $m\ s^{-1}$
0.400	0	0	3.0×10^{-7}
0.395	−0.2	−0.22	2.9×10^{-7}
0.340	−0.4	−0.65	4.0×10^{-8}
0.316	−0.6	−1.04	1.4×10^{-8}
0.300	−0.8	−1.60	5.0×10^{-9}
0.260	−2.5	−4.80	5.0×10^{-10}
0.220	−6.6	−10.7	4.0×10^{-11}
0.180	−17.4	−29.5	3.0×10^{-12}
0.140	−60	−89	1.5×10^{-13}
0.100	−220	−355	7.0×10^{-15}

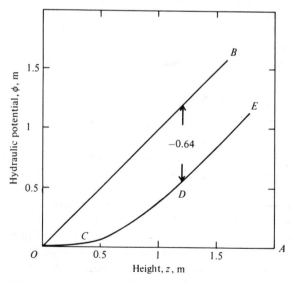

Fig. 5.10. Hydraulic potential, ϕ, as a function of height above the water table in a uniform soil through which water steadily percolates at a rate 0.033 times the saturated hydraulic conductivity, K_0. (See text for details.)

Any result derived from Equation (5.35) must plot within the region *AOB* of Fig. 5.10. When there is no downward flux at all, the whole of the soil above the water table is an equipotential volume, $\phi = 0$ and plots along the axis *OA*. If the downwards flux is just equal to the saturated hydraulic conductivity, by definition the soil is saturated, possesses its maximum water content at all depths, the matric potential $\psi = 0$ and the hydraulic potential $\phi = z$. This is represented by the line *OB*. When $0 < |v| < K_0$ the hydraulic potential profile has some shape like that of *OCDE*, which can be exactly determined. As can be seen from the data in Table 5.2, the hydraulic conductivity is only a little less than K_0 at $\psi = -0.2$ m. That matric potential is meant to represent the approximate extent of the capillary fringe above the water table. The soil pore spaces only begin to drain significantly for $\psi < -0.2$ m with the consequent decrease in K shown in Table 5.2. The part of the line *OC* represents this region of slowly changing K within the capillary fringe. The part of the line *CD* represents heights above the water table with decreasing ψ (going more negative), and decreasing K. Eventually as $K \rightarrow |v|$, the function $1/(1 + v/K) \rightarrow \infty$ and the integral cannot be evaluated precisely. Physically it represents the height above the water table where the matric potential has the value appropriate to $K = |v|$. This is given by the point *D* on Fig. 5.10. The soil at all heights above *D* has a uniform matric potential, in the example -0.64 m, and the hydraulic potential gradient becomes the gravitational gradient. The region *DE* is therefore also a region of uniform water content. Again using the data of Table 5.2 we can calculate the water content profile corresponding to *OCDE* of Fig. 5.10. It is shown in Fig. 5.11.

The steady downward percolation of water used in this example has a magnitude of 1.0×10^{-8} m s^{-1}, or 315 mm yr^{-1}. Such an annual recharge to the groundwater is consistent with the water harvest from catchments in relatively humid climatic zones. That is not to imply that there is always a downwards flux throughout the year. In the Gambier Plain region of southern Australia, for example, where the rainfall is about 750 mm yr^{-1} and the potential evaporation from grass 850 mm yr^{-1}, the annual water harvest of about 80 mm of water drains through the soil to an aquifer during the months June to November, i.e. the actual evaporation is 670 mm yr^{-1}. In the south of Britain the annual water harvest is about 100 mm where the rainfall is about 550 mm yr^{-1}, actual evaporation 450 mm yr^{-1} and potential evaporation 500 mm yr^{-1}. In fact so much of the populated part of the world lies in climatic regions where rainfall exceeds actual evaporation by 50 to 300 mm yr^{-1} that the example has been chosen to suggest a characteristic soil moisture status for those parts. Further attention will be paid to this matter in Chapter 12, but in concluding here it may be remarked that Fig. 5.10 is a more useful general

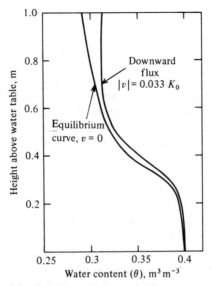

Fig. 5.11. The water content profiles of the soil corresponding to lines *OA* and *OCDE* of Fig. 5.10, namely equilibrium states with $v = 0$, and $|v| = 0.033 K_0$.

specification of the common soil moisture regime than is Fig. 5.11. Soil variability makes it preferable for us to use the matric potential as an unambiguous measure of soil water status. Water content profiles are useful only when the equilibrium water content profile can be specified, as has been done on Fig. 5.11.

5.4 Steady state flow of water upwards during evaporation from the soil surface

If, instead of downwards flow of water to a water table, water flows through the soil vertically upwards from the water table to the soil surface where it evaporates, an equation similar to Equation (5.35) can be derived on the same considerations. Into the equation that describes Darcy's law

$$-v = K\frac{d\phi}{dz},$$

we substitute $E = v$, where E is the evaporation rate at the soil surface, a positive quantity, assuming steady state conditions and no change in water content anywhere with time, i.e. no continued drying out of the soil. The substituted Equation (5.36)

$$-E = K\frac{d\phi}{dz}, \qquad (5.36)$$

together with Equation (5.34) gives, on integration,

$$z = \int_0^z dz = -\int_0^{\psi_z} \frac{d\psi}{1 + E/K}. \qquad (5.37)$$

The integral on the right hand side can be evaluated graphically if the relationship between K and ψ is known to sufficiently small values of these parameters. It gives the depth to the water table, z, for a chosen E. The data of Table 5.2 were used to derive the results shown in Fig. 5.12 which describes the relationship between the steady state evaporation rate and the equilibrium depth to the water table. The soil hydrological data of Table 5.2 were chosen carefully to give a representative set of values for a well structured clay loam soil. As may be seen from Fig. 5.12 the evaporation rate is 10^{-8} m s^{-1} or about 1 mm day^{-1} when the water table is 1.7 m deep. When the water table is at a steady depth of 0.7 m below the surface, the evaporation rate from the soil surface could be 5 mm day^{-1}, if meteorological conditions allowed it. Such a rate is possible for summer evaporation in the humid, middle latitudes. Further consideration is given to evaporation rates in Chapter 12.

The data given in Fig. 5.12 accord well with field observations of the rate of salt accumulation under irrigated and partially waterlogged conditions, when the salt is transported to the soil surface by upward capillary

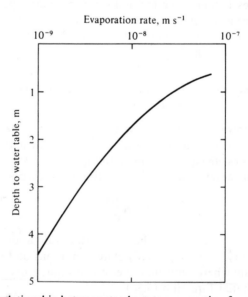

Fig. 5.12. The relationship between steady state evaporation from the soil surface and depth to the water table that can be in equilibrium with the assumed evaporation rate.

rise from the water table. The salinity problem that develops so frequently in irrigation agriculture has prompted many studies of the minimum permissible water table depth. Talsma (1963) found that the water table should be kept deeper than 1.5 m unless heavy leaching could be employed from time to time. Scientists in the USSR have given general specifications that drainage should keep the water table deeper than 1.5 m in the cotton districts of Central Asia (Kovda, van den Berg and Hagan, 1973). The significance of 1 mm day^{-1} upwards flow under equilibrium conditions is that this rate maintained, let us say, for several months when land is spelled or in fallow would cause to precipitate at the evaporation sites near the soil surface an amount of salt transported with about 100 mm of water flow. This salt could subsequently be dissolved and leached downwards by a similar amount of water, readily achievable on an annual cycle either by irrigation or wet-season rainfall.

Some caution is required in applying results suggested by Fig. 5.12 and similar data for other soils. Evaporation rate varies greatly through the day with a maximum usually shortly after noon. The soil at the surface actually dries far more during the day, and during the night it tends to a water content that is larger than the equilibrium appropriate to a steady flux for 24 h. Furthermore, strong heating at the surface during the day and cooling at night initiates thermo-osmotic flow of water. For these reasons a more precise prediction of evaporation rate than the general guide to be had from Fig. 5.12 would need to employ a more sophisticated analysis.

5.5 Steady percolation of water into a layered soil

The principles outlined in Section 5.3 that enabled the steady state profiles of potential and water content to be constructed for a uniform soil can be used just as successfully for a layered soil. Suppose that we have a two-layered soil such as that depicted in Fig. 5.13, in which the hydraulic conductivities are K_1 and K_2. The more permeable soil horizon is the lower and it drains into a water table that is held stationary. The equation that describes the relationship between height above the water table and the matric potential is again Equation (5.35). If water is infiltrating through the surface of the upper layer at a steady rate that is sufficient just to saturate the surface of layer 2 to $\psi = 0$, the infiltration rate through layer 1, being the same, will be equal to its hydraulic conductivity at some matric potential less than zero. Since the matric potential must be continuous across the interface between the two soils, layer 2 will also have a matric potential less than zero at its lower boundary. The infiltration rate therefore must be greater than the saturated hydraulic conductivity of layer 2.

Distribution of water in soil

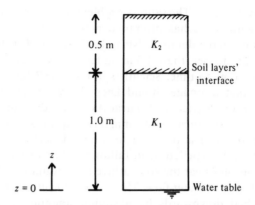

Fig. 5.13. Schematic for steady state infiltration into a two-layered soil. The lower layer drains to a water table at $z = 0$, and its hydraulic conductivity function, K_1, is characteristically larger in magnitude than that of the upper layer, K_2.

The matric potential profile of layer 2 is given by

$$z = -\int_{\psi_{12}}^{\psi} \frac{d\psi}{1 + v/K_2},$$ (5.38)

where ψ_{12} is the matric potential at the interface between layer 1 and layer 2. Similarly, the matric potential profile of layer 1 is given by

$$z = -\int_{0}^{\psi_{12}} \frac{d\psi}{1 + v/K_1}.$$ (5.39)

The computation has to be completed in a number of iterative steps until the correct value of ψ_{12} emerges. Using the data of Table 5.2 for layer 2 and a set of hydraulic conductivities ten times larger for layer 1, the result for the geometry of the two layers depicted in Fig. 5.13 is shown in Fig. 5.14. The matric potential is zero at the water table and at the soil surface, as it should be. At all depths between, the soil is unsaturated in both layers.

Fig. 5.15 shows the dependence of hydraulic potential upon height above the water table. In layer 1 there is a region where ϕ has a constant slope corresponding to the part DE of Fig. 5.10. In this interval $\psi = -0.38$ m. At the interface at $z = 1$ m, there is an abrupt change in the slope of ϕ, such that from $d\phi_1/dz = 1$, $v = K_1$, the slope $d\phi_2/dz$ must increase to make $K_2 \, d\phi_2/dz = K_1$. The actual magnitude of v is -4.42×10^{-7} m s^{-1} and the interested reader may verify that the tangent to the curve in the region of K_2 is everywhere consistent with the requirement that $4.42 \times 10^{-7} = K_2 \, d\phi_2/dz$.

An analysis of this kind was presented by Takagi (1960) in rather more detail, with the remark that the problem is of great interest in soils that

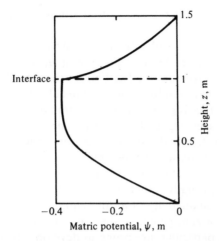

Fig. 5.14. The matric potential profile for a two-layered soil with the geometry depicted in Fig. 5.13, when the steady state infiltration is sufficient just to saturate to $\psi = 0$ the surface of the upper, less permeable layer. (See text for details.)

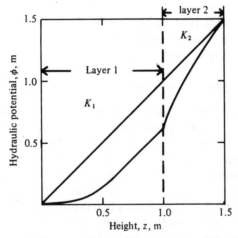

Fig. 5.15. The dependence of hydraulic potential, ϕ, upon height, z, above the water table for the conditions depicted by Figs. 5.13 and 5.14, i.e. infiltration rate sufficient to keep layer 2 just saturated at its surface, and $K_2 < K_1$. (See text for details.)

are puddled for paddy-rice cultivation. The relatively impermeable upper layer has to keep the water in the irrigation bay, and a more permeable zone below can have the effect of increasing the infiltration rate. In the example as worked out here, that increase is about 1.5 times the saturated conductivity of the upper layer, caused simply by an increase of the potential gradient.

If the layering of the soil is such that the less permeable layer lies beneath the more permeable layer, the consequent effect upon the infiltration rate is not as pronounced. The limit to the rate of water flow is, of course, as before, imposed by the less permeable layer when the application rate at the surface is greater than its saturated hydraulic conductivity. Then the upper layer begins to have a perched water table established within it. The positive hydraulic head at the upper surface of the less permeable layer increases the potential gradient through it and so increases the flux to a value greater than the hydraulic conductivity. The effect is, however, usually rather small.

The hydraulic conductivity of a differentiated soil profile, possessing an A- and a B-horizon, generally decreases markedly with depth. The solonetz soils, in particular, are prone to develop a perched water table in the A-horizon. This situation is usually accompanied by water flow in a predominantly horizontal direction, below the perched water table towards the natural drainage lines. Engineering hydrologists have referred to this kind of run-off as inter-flow. The arrival of this water in the stream discharge and its resulting influence upon the stream hydrograph occurs more quickly than could be possible if it traversed a flow path along the deeper aquifer stream lines and entered as base flow (see Chapter 6 for other details).

5.6 Infiltration from rainfall or sprinkler applications

The theory of ponded infiltration, with some experimental tests of the success of the theory, was presented in Section 5.1. The boundary condition at the soil surface, in that case, is that the water content there is θ_0, the saturation value. The subscript 0 may help to remind the reader that the matric potential is zero, or slightly positive and equal to the depth of the ponded water, in addition to representing the water content where $y = 0$ (see Equation 5.7). Considering that the saturated hydraulic conductivity of the surface horizon of many soils lies in the range of 5 to 50 mm hr^{-1}, and that the initial infiltration rate that a relatively dry soil can sustain is at least an order of magnitude greater again, there is obviously a need to investigate the infiltration process when the flux of water across the soil boundary is similar to rainfall rates. These latter might be taken, for present purposes, as up to 30 mm hr^{-1} averaged over an hour.

The process is described by Equation (5.6), written here again as

$$\frac{\partial \theta}{\partial t} = \frac{\partial}{\partial y}\left(D\frac{\partial \theta}{\partial y}\right) - \frac{\partial K}{\partial y}. \qquad (5.40)$$

The initial and boundary conditions are

$$t = 0, \qquad y \geqslant 0, \qquad \theta = \theta_n,$$

and (5.41)

$$t > 0, \qquad y = 0, \qquad -D(\partial\theta/\partial y) + K = V_0.$$

Philip (1973) initiated numerous attempts to find solutions appropriate to the boundary conditions (5.41). As was found with the theory of ponded infiltration, here too there is advantage in neglecting gravity for a start, so that the equation to be solved reduces to

$$\frac{\partial\theta}{\partial t} = \frac{\partial}{\partial y}\left(D\frac{\partial\theta}{\partial y}\right), \tag{5.42}$$

and the solutions are often referred to as absorption solutions. White *et al.* (1979) showed that the Equations (5.43) and (5.44) then proceed from Equation (5.42), subject to (5.41), namely

$$V_0 y = -\int_{\theta_0}^{\theta}\frac{D\,\mathrm{d}\theta}{F} \tag{5.43}$$

and

$$V_0^2 t = \int_{\theta_n}^{\theta_0}\frac{(\theta - \theta_n)D\,\mathrm{d}\theta}{F}, \tag{5.44}$$

where F is a flux-concentration relation due to Philip (1973), which he defined to be $F(\Theta, t) = V(\theta, t)/V_0$, for the absorption solution, and Θ represents any water content increment in the profile from the initially uniform value θ_n, normalised by dividing by $\theta_0 - \theta_n$, that is, $\Theta = (\theta - \theta_n)/(\theta_0 - \theta_n)$. Note that $D = D(\theta)$ is the soil water diffusivity and $\theta_0 = \theta_0(t)$ is the water content at the soil surface, which increases with time from the initial value, θ_n, to a final steady value as $t \to \infty$. The flux V_0 is the steady rainfall rate or sprinkler application rate crossing the boundary, and it is not restricted in magnitude. It is to be noted too that if $V_0 < K_s$, the hydraulic conductivity of the soil at saturation, ponded conditions cannot develop. Then, as $t \to \infty$, the water flow proceeds through a lengthy transmission zone, which is unsaturated and possesses a water content θ such that $K_\theta = V_0$. In this zone the hydraulic potential gradient is unity. The flux $V(\theta, t)$, in the definition of $F(\Theta, t)$ above, is the flux across any horizontal plane in which the water content is θ.

Inspection of Equation (5.43) reveals that the magnitude of the integral depends only weakly upon the value of the water flux at the boundary, V_0, and not at all if the approximation $F(\Theta, t) = (\theta - \theta_n)/(\theta_0 - \theta_n)$ suggested by White (1979) is used. With that assumption $V_0 y$ is invariant for all values of V_0. By the same argument $V_0^2 t$ is also invariant for all

Distribution of water in soil

values of V_0. Equation (5.43) therefore gives an expression for the distance of penetration of a wetting front specified by a chosen water content $\theta = \theta_{wf}$, or by repeated integrations the θ, y relationship can be established. Equation (5.44) gives an expression for the time taken for the water content at the soil surface to develop to any (chosen) new value θ_0.

Perroux *et al.* (1981) reported the results of a laboratory experiment designed to test the implications of the theory summarized above. Two soil materials were used, one a fine sand with no secondary aggregation, the other a silty clay loam possessing very stable aggregates. Water was allowed to infiltrate vertically. Fig. 5.16 shows some of their results obtained with the silty clay loam, in which the water content of the loam

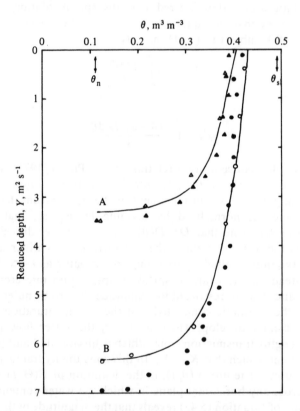

Fig. 5.16. Water content *v.* depth curves for infiltration into a dry soil, by water controlled to a steady flux at the surface, V_0, simulating mean rainfall rates. Parameters are ▲ ● $V_0 = 4.6 \times 10^{-6}\,\mathrm{m\,s^{-1}}$; △ ○ $V_0 = 1.1 \times 10^{-5}\,\mathrm{m\,s^{-1}}$; $Y = V_0 y$; For Curve *A*, $T(= V_0^2 t) = 8.0 \times 10^{-8}\,\mathrm{m^2\,s^{-1}}$; For Curve *B*, $T = 1.7 \times 10^{-7}\,\mathrm{m^2\,s^{-1}}$. (After Perroux *et al*, 1981.)

is displayed as a function of depth in the column. Time is expressed as the reduced time, $T = V_0^2 t$; and depth is the reduced depth $Y = V_0 y$. By that device data from numerous experiments with differing values of V_0 could be plotted together. Actual depths of penetration of the wetting front for the few data selected and shown here, ranged between about 140 mm and 30 mm. Actual times of infiltration to cessation and dismantling of the columns for sampling ranged from $2\frac{1}{4}$ hours to 10 minutes. The infiltration profiles given by the experimental points are similar in shape to the full curves, which are the predicted profiles for absorption (gravity neglected), calculated according to the theory. Note that, with increasing time and depth of penetration, the water content at the surface θ_0 increases towards the fully saturated value but cannot reach it. Since the application rates were less than the saturated hydraulic conductivity $(3.3 \times 10^{-5} \text{ m s}^{-1})$ the water content at the surface adjusts to a value appropriate for $K_\theta (< K_0)$ to equal V_0, as $t \to \infty$, but then the absorption theory becomes no longer valid.

A test of the theory against field infiltration measurements was reported by Clothier *et al.* (1981). When the application rate is greater than K_0 at the surface, ponding eventuates and the time to ponding, t_p, may be calculated approximately from the relation

$$t_p = S^2 / 2 V_0 (V_0 - K_0),$$
(5.45)

where S is the sorptivity, to be defined in Section 5.7 following. The measured and predicted times to ponding, for one of their experiments, were in acceptable agreement. The ability to calculate the time to ponding, from a knowledge of the sorptivity, a hydraulic property of the soil in its unsaturated condition, obviously confers enhanced skill in the management of catchments.

In our discussion of infiltration processes, nothing has been written about the contemporaneous loss of water by transpiration and evaporation from the soil. The theories of entry of water, followed by its redistribution within the soil, when the source of water has stopped, have served as conceptual guides. In many climatic zones of the world during the growing season, the evapotranspiration roughly balances the rainfall, at a time scale of 10 to 20 days. We might suppose that the water loss over 20 days in summer could be 80 mm, so that redistribution such as shown in Fig. 5.6 of 50 mm of added water, could not be observed out-of-doors except beneath a bare, covered soil. Notwithstanding that generality, Figs. 5.8 and 5.9 were constructed from field data, gathered sequentially at 28 days intervals, so the deepest water content profile shown in Fig. 5.9 reached its location after 140 days. The contemporary evapotranspiration was about 250 mm. Wetting fronts were preserved over the period of those observations, during which the soil absorbed and stored about

200 mm. The season was April to August inclusive, in southern Australia, during which the winter rains replenish the soil moisture that becomes greatly depleted during the summer drought.

5.7 Measurement of infiltration and associated parameters

The rate at which water can enter soil when not limited by the rate of supply is measured in the field with water either ponded on the surface or falling on it as artificial or natural rain at a rate sufficient to cause some to run off. It is expressed in m s^{-1} or some convenient multiple of these units and is called infiltration capacity (Horton, 1940) or infiltrability (Hillel, 1971). It is a potential rate that is characteristic of the soil under specified conditions. In particular, it varies with time during a test (Fig. 5.5) and with initial water content.

Ring infiltrometers used to measure this rate consist of a metal ring, preferably 0.3 m or larger in diameter, that is pressed a few centimetres into the soil. Water is ponded within this, and the cumulative amount required to maintain a constant shallow depth is recorded at successively longer times of application. Because water spreads laterally below the ring, measured rates of entry are greater than those obtained for a large area in which flow is one-dimensional. A guard ring inside which there is a smaller concentric ring, may help to limit the lateral spread of water from the smaller ring, if the water levels in the annulus and the central circle are kept at the same head. However Bouwer (1986) states that electric analogue studies deny that such is the case.

The infiltration characteristics of a soil can be represented by parameters of an infiltration equation. Kostiakov (1932) and Lewis (1937), independently of each other, suggested that the equation

$$i = at^b \qquad (5.46)$$

fitted the data of many experiments, with the arbitrary constants, a and b, to be determined by curve fitting. Equation (5.26) given above in Section 5.1, also empirical, has more physical significance because Horton's 'minimum infiltration capacity', v_f, which is the final, quasi-steady rate of infiltration, must be the hydraulic conductivity in the transmission zone, K_1. The hydraulic conductivity of the saturated soil, K_0, is usually reported to be larger than K_1, but any measurements of K near to $\psi = 0$, in the field, are prone to error due to cracks and fissures that may empty or fill capriciously. Therefore, the association of v_f with either K_1 or K_0 is acceptable in concept.

The rate of infiltration at the start of the experiment is often so large that the pressure potential gradient completely dominates the effect of gravity. Philip (1957c) proposed an infiltration equation that approximates to his series solution, noted as Equation (5.18), by truncation after the second term, namely

$$i = St^{1/2} + At, \tag{5.47}$$

in which S is the sorptivity of the soil. A, the other parameter, also has a physical meaning for it can be taken as equal to K_1, when t is large. In the early stages it is approximately $K_1/3$. The instantaneous infiltration rate, v, is

$$v = \mathrm{d}i/\mathrm{d}t = \tfrac{1}{2}St^{-1/2} + A, \tag{5.48}$$

from which it can be seen that v becomes steady and equal to A, at times long enough for the first term to become negligible.

Sorptivity is an important parameter in the description of both cumulative infiltration and the instantaneous infiltration rate. It is not a constant but becomes smaller as the initial water content (presumed uniform) increases. Talsma (1969) worked out the optimum conditions for deriving S from data gained in the field from the standard ring-infiltrometer. As Equation (5.47) shows, i should plot on a straight line against $\sqrt{t}$, soon after the start of the experiment and before the processes represented by the second term contribute significantly. For many soils the straight line plot of i v. $t^{1/2}$ is limited to three minutes, and up to 10 minutes sometimes, after the first application of the water. Clothier and White (1981) applied the water to the soil surface through a porous plate, with hydrostatic pressure regulation, so that the matric potential at the soil surface is about -0.04 m. That slight suction is sufficient to drain gross fissures and holes and markedly improves the estimate of S, particularly when it is to be used in calculations involving rain-fed infiltration. They isolated a short soil column by sculpturing it and enclosed it with a 'sorptivity tube', made of clear Perspex, so that the wetting front could be seen.

In sprinkler infiltrometers, excess water is sprayed or dripped onto the soil at a measured rate and the runoff from a central test plot within the sprayed area is collected and measured. The surrounding guard area is effective in preventing error from lateral movement. Natural rainfall can be used as the source of supply when surface runoff and precipitation data are available for rainstorms of sufficient intensity and duration. Zegelin and White (1982) describe the design of a field sprinkler infiltrometer, intended for use in constant flux experiments for a range of values of V_0 both smaller and larger than K_0. They were able to verify that the expression for the time to incipient ponding, if $V_0 > K_0$, was calculated acceptably by the relation

$$t_p = S^2/2V_0(V_0 - K_0). \tag{5.49}$$

Lower infiltration rates are obtained by sprinkling than by ponding because the falling drops and running water affect the structure of the exposed soil. Sprinkler infiltrometers are usually preferred for catchment studies. However, the simplicity of the unbuffered ring method makes it useful for obtaining comparative values for different treatments when a high degree of replication is necessary because of variability in soil conditions. It has been widely used on land already under irrigation or intended for that use.

Methods of measurement of infiltration and related parameters, in the field, have been described and critically evaluated by Bouwer (1986) and Green *et al.* (1986). Sharma *et al.* (1980) reported a study of the spatial variability of infiltration on a small watershed in Oklahoma. Sorptivity varied by one order of magnitude and the *A*-factor by nearly two orders. They used the theory of scale change to discover some improved homogeneity of their data. The principal feature of scaling theory can be illustrated by a comparison of saturated hydraulic conductivities of two different soils. Regarding them as ideal porous media, we suppose that they are geometrically similar but differ in size, so that if L_1 is the length of an indentifiable channel in medium 1, and L_2 is the length of the corresponding identifiable channel in medium 2, then

$$\frac{K_1}{L_1^2} = \frac{K_2}{L_2^2}.$$
(5.50)

If $L_1 = NL_2$, where $N > 1$, that is medium 1 possesses larger grains and pore spaces than medium 2, then

$$K_1 = N^2 K_2.$$
(5.51)

This relationship is proved rigorously by Childs (1969, p. 192). All soil hydraulic parameters scale in ways that can be discovered by dimensional analysis. It is unlikely that field heterogeneity is caused by simple scale of length differences, but variability of properties does seem to respond to analysis based upon that principle.

Clay soils have special features affecting their infiltration behaviour. Their hydraulic conductivity in the saturated condition is often low and it may decrease with time in initially dry soil of natural structure due to delayed swelling. A gross example of delay is the slow closure of shrinkage cracks which, when a dry clay soil is flooded, can greatly affect the infiltration rate if they are open to the surface. Also in a swelling soil, part of the overburden load is transmitted to the soil water, as shown in Equation (2.11), so that the envelope pressure component of matric potential increases with depth. This has the effect of reducing the gradient in hydraulic potential (Philip, 1969, 1971*a*), as will now be shown.

The hydraulic potential, $\phi = \psi + Z$, of a swelling clay soil can be written (see Section 2.7) as

$$\phi = \psi_u + P_e + Z, \tag{5.52}$$

where ψ is the matric potential of the loaded soil, ψ_u is the matric potential of the unloaded soil and P_e is the envelope pressure potential arising from an overburden load, L, of soil lying above. From Equation (2.12),

$$P_e = \beta L \simeq -\beta g \rho_{wb} z \qquad (z \leqslant 0), \tag{5.53}$$

where β is the compressibility factor of the soil (see Section 2.7), ρ_{wb} is the wet bulk density, g is the acceleration due to gravity, z is the height relative to the surface where $z = 0$, and potential is expressed per unit volume of water. Hence, on combining Equations (5.52) and (5.53) and replacing Z by $\rho g z$, where ρ is the density of water

$$\phi = \psi_u - \beta g \rho_{wb} z + \rho g z$$
$$= \psi_u + \rho g (1 - \beta \rho_{wb}/\rho) z. \tag{5.54}$$

Equation (5.54) shows that the gradient in hydraulic potential, $d\phi/dz$, suffers a reduction in the gravitational component due to the overburden. Philip concluded that the infiltration rate would be reduced accordingly in a swelling soil and conversely that the upward flux would be increased during the wetting of a swelling soil from below.

6

Groundwater in soils and aquifers

The movement of water through soil has been intensively studied in the laboratory, particularly with respect to flow under unsaturated conditions. These experiments and their results, which have been described in Chapters 4 and 5, have provided confirmation of the present understanding of the theory of water flow through soils. The scale of such experiments has, of necessity, been limited. Controlled experimentation at the field size has generally been daunting, principally because it has appeared that the financial cost could not be justified. Childs (1953) and his co-workers established a full-scale field drainage laboratory near Cambridge, in which numerous three-dimensional problems have been investigated by direct experiment. Their drainage tank was about 10 m by 10 m in horizontal dimensions and was filled with a sand of homogeneous and isotropic properties. Application of the results from this installation and others (for example, van Schilfgaarde, Frevert and Kirkham, 1954) was intended to provide design data for improving the economy and efficiency of artificial field drainage, the principles concerning which we now proceed to describe.

6.1 Artificial drainage

In low-lying coastal regions the water table can be found close to the soil surface at an elevation a little above sea level. The hydraulic potential of the reservoir, or sink, to which the flow of the groundwater must be directed is the level of the sea or its tidal inlets. Alluvial plains, reclaimed deltas and their swampy margins have often needed artificial drainage to improve agricultural or pastoral production.

As the sea determines the base level for groundwater in coastal plains and dunes, so the base level of inland regions is set by the water level of rivers and lakes, to which the groundwater beneath lake-marginal flats and river flood plains directs its flow. The distinction between the saturated zone of the soil and a lower-lying aquifer is arbitrary. It is often unnecessary. Only when a water-bearing formation is confined at an upper boundary by a less permeable bed is the distinction readily perceived and likely to be useful. In the latter case there is always an overlying, unconfined aquifer where a free water surface, the water table, may be found, and this may be within the soil zone if the local climate is sufficiently humid. The water table was defined in Section 2.1, by reference to the concept of a laboratory column demonstration. In the field the water table lies at that depth in an auger hole where free water just begins to flow in. It is accurately indicated by the static level of the water in the hole when the depth of the water is vanishingly small, a curious requirement that guards against hydraulic head gradients with a vertical component. On flat land, the depth to water in an open hole is also the water table depth, with little likelihood of error.

In the arid zones of the world, wherever irrigation has been developed extensively, the soil moisture regime often becomes too wet and subsurface drains are necessary. The extent of irrigation began to expand at the beginning of the nineteenth century, partly under pressure of population, partly with improving technology about water control, these two factors, of course, being not unlinked. Irrigation in ancient times is woven into the history of civilization. The Kingdom of Egypt, the dynasties upon the Mesopotamian plain and the little-known civilizations of the Indus were all totally dependent upon the irrigation of fields for their agricultural production. Since the dawn of the historical record of achievements and failures of mankind, healthy irrigation enterprise has depended upon the maintenance of efficient drainage works (see Stamp, 1961).

In the absence of artificial drains the soil may become water-logged in new irrigation districts, even though the water table before water application may have been very deep. On the northern plains of Victoria, Australia, the water table began to rise, following the beginning of extensive irrigation about 1880, and it had risen 20 m, to a position near the soil surface by about 1940. The hydrological processes of drainage that sufficed in naturally arid environments are usually not appropriate after irrigation has been established. In particular, groundwater flow cannot cope with the excess that escapes evaporation, fortuitously or by design, when water is applied artificially. Such an excess in humid regions would discharge into streams and become the well known base-flow of rivers. The surface drainage network itself is fashioned by the running water, which may erode or incise its channel into the rocks and sediments along its course and so promote drainage flow below the ground. But in the arid zones under irrigation, water-logging of flat-lying land often has to be corrected by drainage.

Artificial drainage in regions of the world that enjoy a high rainfall has been practised for several centuries. Open ditch drainage is now combined with sub-soil pipe drainage, that uses short lengths of ceramic or concrete 'tile', laid at the chosen depth in the trench which is then filled in. Perforated, rigid plastic pipe may also be used as an alternative. The interested reader may consult Singer *et al.* (1957) for an account of the history of drainage.

Consider the field drainage system illustrated in Fig. 6.1, which represents ditches of rectangular section, parallel to each other, and sufficiently long so that the problem may be considered two-dimensional. If the system is at a steady state, the water table is stationary and the discharge from the drains is just equal to the infiltration rate. It is easy to obtain a quantitative description that is accurate enough for many applications.

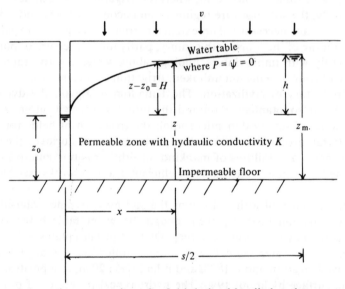

Fig. 6.1. Schematic cross-section of a field drained by ditches that penetrate to an impermeable layer. For symbols, see text.

Assuming a constant infiltration rate, v, the amount of water per unit time which has to cross the vertical plane at x, per unit width is

$$((s/2) - x)v,$$

where $s/2$ is the half-spacing of the drains. According to Darcy's law (Equation 4.1), the discharge through the vertical plane per unit width is alternatively given by $zK\,\mathrm{d}H/\mathrm{d}x$, where K is the hydraulic conductivity. Here we have assumed that the hydraulic gradient in the vertical section at x is equal to the slope of the water table, $\mathrm{d}H/\mathrm{d}x$, and is invariant with depth. This assumption is the well-known Dupuit-Forcheimer assumption that stream lines are horizontal and equipotential lines are vertical. The hydraulic potential or head, H, may conveniently be referred to the ditch water level as datum from which it follows that, if z is the elevation of the water table above the impermeable floor, $H = z - z_0$. We therefore have

$$zK\frac{\mathrm{d}z}{\mathrm{d}x} = \left(\frac{s}{2} - x\right)v. \tag{6.1}$$

Rearrangement of Equation (6.1) gives

$$z \, \mathrm{d}z = \frac{v}{K}\left(\frac{s}{2} - x\right) \mathrm{d}x,$$

which may be integrated for the corresponding limits of integration, from the base level of the water in the ditch, z_0, where $x = 0$, to the water table at the groundwater divide, z_m, where $x = s/2$. Therefore

$$\int_{z_0}^{z_m} z \, \mathrm{d}z = \frac{v}{K}\int_0^{s/2} \left(\frac{s}{2} - x\right) \mathrm{d}x, \quad \text{or} \quad s^2 = \frac{4K}{v}(z_m^2 - z_0^2). \tag{6.2}$$

Equation (6.2) specifies the drain spacing needed to maintain a maximum permissible water table height, z_m, when the ditch water level is controlled at z_0 for a given combination of infiltration (rainfall) rate and hydraulic conductivity, v/K, when the flow situation is in the steady state. The ratio v/K recurs in all steady state drainage formulae.

The geometrical arrangement shown in Fig. 6.1 is unreal because rarely could the bottoms of the drains be located on the impermeable floor. Usually, some water would flow along stream lines in the soil below the bottom of the drain. This situation is depicted schematically in Fig. 6.2, from which it can be seen that the average depth of saturated soil through which flow can occur is approximately $(D + z_0 + \frac{1}{2}h)$, whereas when the drains meet the impermeable floor it is $(z_0 + \frac{1}{2}h)$. The transmissivity, defined to be the product of hydraulic conductivity times the depth of the zone through which the water flows, is therefore either $K(D + z_0 + \frac{1}{2}h)$ or $K(z_0 + \frac{1}{2}h)$. Such an increased transmissivity for the flow situation depicted in Fig. 6.2 would tend to reduce the significance of h for a given v/K whereas the convergence of flow towards the bottom of the drain would tend to hold up the water table, i.e. to increase h when considered in relation to the geometry in Fig. 6.1. These opposing tendencies result in a closer correspondence of many observed drainage situations with the Dupuit-Forcheimer theory than its simplifying assumptions would lead us to expect. Many practical formulae for drain design have been proposed, guided by potential theory but eventually based upon empiricism. A useful review of drainage in theory and practice was given by Wesseling (1973).

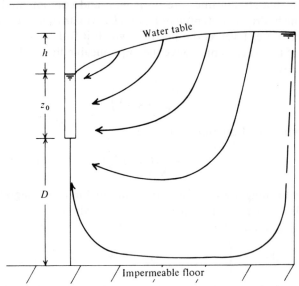

Fig. 6.2. Diagram of a field drained by a ditch to illustrate curved streamlines that represent the convergence of water flow towards the ditch. For symbols, see text.

If Equation (6.2) is rewritten as

$$s^2 = (8K/v)h\bar{d}, \qquad (6.3)$$

where $\bar{d}$ is the mean depth of saturated soil through which flow can occur ($= D + z_0 + \frac{1}{2}h$) a suitable adjustment of the magnitude of $\bar{d}$ to accommodate the opposing trends of convergence of stream lines and greater transmissivity has an understandable physical basis.

Where land has to be intensively drained, tile-drains may take the place of the short lateral ditches. The usual size of a tile is about 100 mm diameter by 600 mm. When laid in the trench (see Fig. 6.3) the individual pipes are butted together and the gap is protected by sheeting of plastic, rushes or coarse gravel against entry and clogging with earth. Although the tile may be porous, if fired at a low temperature, the route for entry of the water into the channel is through the gap. Continuous plastic tube into which slots have been cut is now also used.

When the trench in which the tile-drains have been laid is filled again with soil, the land has an unseen drainage system that does not interfere with the cultural work of ploughing, controlling weeds and pests, and harvesting. The main collecting drains, however, can be open ditches. Access to the tile-drains has to be provided in order that they can be

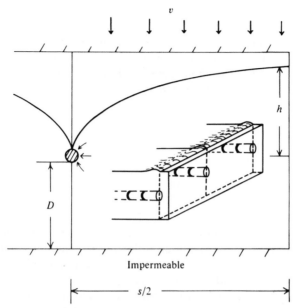

Fig. 6.3. Drainage of a field by tile-drains that are buried beneath the soil surface. The inset shows how tile-drains can empty to an open ditch. For symbols, see text.

cleaned mechanically from time to time as required. The tile-drains therefore often open into the side of a ditch as shown in Fig. 6.3.

Hooghoudt (1940) applied the concepts that led us to Equation (6.3), to tile-drained land, in which the drains are laid at a distance D above the impermeable horizon (see Fig. 6.3). The situation is then described approximately by the equation,

$$s^2 = 8(K/v)hd, \tag{6.4}$$

in which the equivalent depth, d, can be obtained from the tabulated values appropriate to s, D and r, the radius of the drains. Table 6.1 gives a selection of values from the more extensive published tables.

The theory of drainage given so far is not exact and leads to formulae that are approximate descriptions of the true situation. They are useful practically. It is hardly possible to measure the hydraulic conductivity of soil, in a drainage situation, to a precision that would warrant more exact calculations. Soil profiles and aquifers are often quite variable and the range of measured conductivities may span at least one order of magnitude in the same field.

Theoretical studies by van Deempter (1949) and Engelund (1951) of flow to tile-drains in homogeneous, infinitely deep soil yielded exact solutions. Watson (1959) presented some of these results in graphical

Table 6.1. *Values of the equivalent depth d, in m, for tile drainage design, following Hooghoudt (1940). (The radius of the tile-drain is taken to be 50 mm, s (drain spacing) and D (depth to impermeable layer) are in m, see Fig. 6.3 for details)*

s	10 m	20 m	40 m
D			
1 m	0.80	0.89	0.96
2 m	1.08	1.41	1.66
4 m	1.13	1.81	2.51
8 m	1.13	1.88	3.13

form which can be used to show the correspondence between the exact solution, the approximate solution following Hooghoudt's assumptions, and experimental solutions from Child's pioneering original work with electrical analogue models (Childs, 1943b).

Fig. 6.4 is reproduced from Watson's paper and shows the dependence of water table height, midway between tile-drains, upon the steady state infiltration rate for four values of the depth to impermeable lower boundary of the soil, which is the parameter distinguishing the four curves. The variables are all expressed in scaled, dimensionless form.

Of the four curves shown in Fig. 6.4, that for $D/s = \infty$ was calculated from the theoretical results of Engelund and van Deempter, and the other three were obtained from the electrical analogue experiments of Childs. Numerical examples can serve to show how closely the calculated design values correspond, when obtained by different methods. For example, suppose that the drain spacing has to be calculated according to the formula of Hooghoudt (Equation 6.4) for $v/K = 0.02$, $h = 1$ m and three values of D, namely zero, 3 m and infinitely deep. If the drain spacing so derived is then used to enter Fig. 6.4 and obtain values of h, the water table height mid-way between drains, the comparison of these different methods can be displayed as in Table 6.2. Equation (6.4) appears to predict the water table to be about 10 per cent higher than the height obtained by applying Fig. 6.4. Such a discrepancy is insignificant in actual drain design.

It should be noted that one of the assumptions underlying the ability to obtain the results of Fig. 6.4, either by Childs' analogue method or by mathematical methods is that the drain is just full of water, i.e. it represents an equipotential volume. The practical results are rather insensitive to this requirement, which is exactly met by a variable specification of drain diameter. The usual diameters of tile-drains lie in

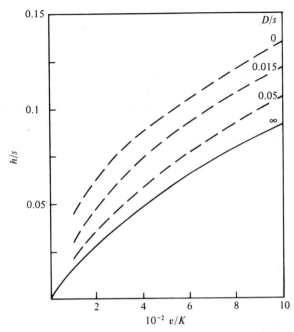

Fig. 6.4. The dependence of water table height above the tile-drain, h, as a function of rainfall rate, v, for steady state drainage. Parameters are presented as dimensionless ratios by dividing lengths by the drain spacing, s, and the rainfall rate is presented as the ratio v/K, K being the hydraulic conductivity. The family of curves are distinguished by different values of D/s where D is the depth to the impermeable layer beneath the tile-drain. See text for further details. (After Watson, 1959).

Table 6.2. *A comparison of design criteria for tile-drains obtained by using Equation (6.4) and from curves of Fig. 6.4, $v/K = 0.02$ in all cases*

	h	s
Depth to impermeable layer, D = 0		
By Equation (6.4)	1.0 m*	14.2 m†
From Fig. 6.4	0.89†	14.2*
Depth to impermeable layer, D = 3 m		
By Equation (6.4)	1.0 m*	27.2 m†
From Fig. 6.4	0.89†	27.2*
Depth to impermeable layer, D = ∞		
By Equation (6.4)	1.0 m*	32.6 m†
From Fig. 6.4	0.90†	32.6*

* By assumption, † calculated.

the range 100–150 mm and little error is to be expected to result from variations in drain diameter in practice, because of the larger hydraulic conductivity of the soil generally, in the disturbed zone of the trench in which the tiles have been laid.

One of the most arduous exact solutions of a drainage problem is that due to Kirkham (1958). Even this exact solution requires (see Fig. 6.3) that the flow above the horizontal plane through the drains is vertical and the hydraulic conductivity is so large that the hydraulic potential distribution at the water table is reproduced at this plane. The solution is then

$$h = \frac{sv}{K} \frac{1}{1 - v/K} \frac{1}{\pi}$$

$$\times \left\{ \ln \frac{s}{\pi r} + \sum_{m=1}^{\infty} \frac{1}{m} \left(\cos \frac{m 2 \pi r}{s} - \cos m\pi \right) \left(\coth \frac{m 2 \pi D}{s} - 1 \right) \right\}. \quad (6.5)$$

Toksöz and Kirkham (1961) give a nomograph, evaluated for a range of the parameters which determine Equation (6.5), that enables drainage design to be undertaken. It may be noted that Equation (6.5) is appropriately composed of Fourier series, as an analogue of steady state heat conduction should be.

6.2 Natural drainage

Artificial drainage situations allow the theoretician to use shapes of the boundaries that can be specified geometrically, and at least in principle could be amenable to the construction of exact solutions of Laplace's equation (4.11). In a natural catchment, water flows underground to the stream channel and augments the flow of the stream in just the same way as water in a waterlogged soil flows to the drains. But the shapes of natural boundaries preclude mathematical treatment except of an approximate or qualitative nature. The position of the water table in the field is usually detected, on level ground, by its observation in a shallow bore-hole. In hilly country it can be usefully indicated by the highest spring or start of seepage in the stream channel, below which flow is permanent. A first-order stream, defined to be the first drainage way from the divide that has no tributary but itself forms a tributary, may contain the spring line in a humid region. In more arid climatic zones the spring line may be much lower in the landscape, and in either case its location may shift a long way downstream during drought, indicating a dropping of the level of the water table. In the very arid zones there may not be any permanent spring line at all. The drainage from the land surface is then accommodated by underground flow, replenished by leakage from the stream beds when water does flow after a storm. The streams of such a region are entirely 'losing' streams, the term now recommended with its opposite 'gaining', in preference to the terms 'influent' and 'effluent' formerly in the vocabulary of hydrologists.

The Bremer River in South Australia is an example of a river which is a gaining stream in its hills tract and a losing stream in its tract across the plains. A diagrammatic representation is shown in Fig. 6.5. Above the fault line shown as point A, the effect of higher rainfall and lesser depth of permeable zone makes the river a gaining stream. At point A and below, the transmissivity of the underlying aquifer is sufficient to conduct most of the dry-weather flow of the river underground, to Lake Alexandrina, B. This little river loses its discharge at a rate of about 2.5×10^3 Ml yr^{-1} at and below A by leakage from the stream bed.

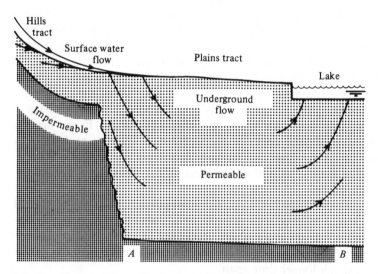

Fig. 6.5. The Bremer River, South Australia (schematic) as an example of a gaining stream in its hills tract and a losing stream in its plains tract.

Blyth and Rodda (1973) surveyed the channel of the River Ray in the Midlands of England, by systematically walking the catchment for one year. They noted on each occasion the position farthest upstream on it and its tributaries where flow began. As expected, the principal controls were found to be the effective rainfall during the preceding week and the length of stream flowing at the time of the previous week's inspection. The catchment of the River Ray is on the impermeable Oxford Clay. A study of general relevance to improving our understanding of natural drainage would be of a catchment with considerable groundwater flow.

Such studies, yet to be undertaken in a systematic and quantitative manner, are likely to prove instructive.

The quantity of water flowing underground, from wherever it first infiltrated into the soil to its discharge as base flow into the receiving stream must, in general, be equal to the stream discharge, by the law of continuity. In assessment of water resources it is not correct to add underground water resources to stream discharge. The temptation is strong to count our water resources twice over because of the large volume of water that is held in the underground reservoir. Yet all that water eventually issues as stream flow, unless it discharges directly to lakes or the sea. The ratio V_R/Q, where V_R is the volume of the underground reservoir and Q is the annual rate of flow through it, expresses a natural equilibrium, with Q identified as the annual replenishment which is just equal to the drainage rate. Note that this ratio has dimensions of years and approximately equals the mean residence time of the water underground. The equilibrium value of that ratio may shift one way or another in response to climatic variability that alters the recharge rate. Changing land use may also affect the recharge rate.

Natural hydrological phenomena possess several scales of time. There is a time of months, up to one year, during which the changes of the seasons bring the monsoon rains of the tropics, the winter rains of the low latitude temperate zones, or other wet seasons in the many other distinctive climatic zones of the world. Soil water is replenished then, water tables are high and streams and rivers are in spate. Steady state drainage theory, as outlined in Section 6.1 above, has proved to be a reliable guide.

Floods and droughts, as intermittent perturbations of the normal, are often stated to have recurrence intervals which grow longer as the perturbations become more severe. Reckoned in years or decades, nevertheless they have no lasting effect because a real change of climate occurs over a much longer time span.

A cause of hydrological change other than that due to climatic variation has come to prominence over the last three decades. The change from forest to arable or to pasture, and occasionally also the re-afforestation of denuded land, has caused unforeseen changes in the hydrology of the affected catchments. The evapotranspiration rate has been affected permanently. This influence upon stream discharge and soil and stream salinity will be discussed here, with some further discussion of the mechanisms of the process deferred to Chapter 12.

Large areas of forest, in which eucalypts dominated, were cleared in the southwest of Western Australia in the 1880s. The clearance is still continuing. The region, located between latitudes 30° and 35° S, receives its rain in winter but has a long summer drought. Subsequent to clearing, the land has been used for pastures wherever the rainfall exceeds about 700 mm yr^{-1} and for cereal cropping in the 700 to 300 mm yr^{-1} zones. The potential evapotranspiration is about 1200 mm yr^{-1} and is not strongly linked with the annual rainfall.

The actual evapotranspiration from forest and bush land was greater than it is from the cleared land. The surplus of rain over evaporation has therefore generally increased. For a start, such a surplus went unnoticed and undetected by stream-gauging, which at any rate did not exist in the early years of European occupation. The ephemeral streams probably flowed a little stronger during the winter and dried up a little later in the spring, but these changes were not noticed. Then the streams began to run salty and there was an accompanying widespread saliniz-ation of the valley soils.

The first measured record of the increasing salinity of stream water came from an engineer's responsibility to seek out good water for the steam locomotives of the railways. Table 6.3 shows data of Wood (1924) for the salinity of the Blackwood River. A large part of its catchment had been cleared by about 1900. The level of salinity at the 1904 sampling may be taken as the primeval bench-mark. An increase of salinity was noted by 1910 and the water was not suitable for locomotive supply by 1915. The salinity increased further and had reached about 15 g l^{-1} by 1940.

Table 6.3. *Increase in salinity of the Blackwood River, at Bridgetown, Western Australia, caused by changing hydrology after clearing of native vegetation. (Data from Wood, 1924)*

Date	Salinity mg l^{-1}	Remarks
26 Jan. 1904	270	First test of supply for steam locomotive use.
28 Dec. 1910	660	
23 Dec. 1912	1290	Water no longer suitable.
Feb. 1915	1650	Further testing discontinued.

Fig. 6.6 shows a simplified representation of the hydrology in the surface and subsurface of this region. Under the natural vegetation the valley slopes and valley bottoms were wooded, but not densely wherever the rainfall was less than about 500 mm yr^{-1}. The water table was below the ground surface along the valley bottoms AB as well as on the valley slopes. When the land was cleared there was a larger soil water input, with deep infiltration to the water table along ABC. Potential gradients

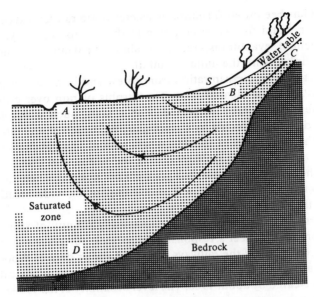

Fig. 6.6. Groundwater flow in a valley in a semi-arid environment, to illustrate the potentiality for salinization of soil between the ephemeral stream, A, and the break of slope, S, where a surface of seepage may develop.

generally had to steepen. The water table at locations AB often rose nearly to the surface. Seepages developed at the break of slope such as at B. Furthermore, since the pore water in a vertical section AD through the saturated zone was always exceedingly saline $(20 \text{ g } 1^{-1})$ the stream began to receive a larger contribution from the saline groundwater seepage relative to the surface run-off than it did formerly.

The time scale for the appearance of increased amounts of salts in the stream water and spread of soil salinity on valley bottoms has been as short as 5 years in high rainfall zones $(\sim 1000 \text{ mm yr}^{-1})$ or up to 40 years in the semi-arid margins of the arable land $(\sim 300 \text{ mm yr}^{-1})$. Peck and Hurle (1973) calculated a time scale for the recovery of the salinity to lower levels appropriate to a new hydrological equilibrium. Following their arguments suppose that, before clearing, the annual surplus of precipitation over actual evaporation from a catchment $(P - E)$ was v $(\text{m}^3 \text{ m}^{-2} \text{ yr}^{-1})$, and that a mean storage volume of the catchment could be assigned as $V (\text{m}^3 \text{ m}^{-2})$. The time required for complete replacement of one storage volume by an average annual increment of v would then be V/v, and this quantity is denoted by Peck and Hurle as the 'characteristic time' for equilibration following a disturbance to the natural hydrological regime. They suggest that characteristic times of about 100 to 400 years may be appropriate in many parts of the region studied by them, where rainfall is about 600 mm yr^{-1}, although it should be clearly understood that a general rule cannot be arrived at.

In the context of the salinity change associated with the hydrological change, the concept of characteristic time is appropriate, as affording a scale of time over which replacement of one pore volume of fluid can be expected, with an appropriate change in groundwater salinity that should then begin to approach a new steady state. But the replacement of one pore volume of groundwater by a fresher recharge does not imply that all the salts have been leached out or pushed through to the streams, as will be further discussed in Section 10.7. Peck and Hurle used the concept only to derive a lower bound to the time to recovery.

The place of storage of the soil salts in the profile, as indicated by Cl$^-$ concentration, is shown in Fig. 6.7, using data from other work of Peck and his colleagues (Peck *et al.* 1981). Curve *A* was obtained from bore-hole soil sampling in a region where rainfall is about 800 mm yr^{-1} and Curve *B* was from a region of 1150 mm yr^{-1}. The land was still in virgin forest at the places of sampling. It may be observed that the peak of the salt accumulation in the soil was, in each case, high above the water table and in the unsaturated zone, where the matric potential would

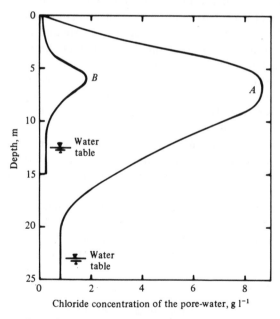

Fig. 6.7. The profiles of chloride concentration of the pore-water in the unsaturated zones of deeply weathered soils in Western Australia. Site *A*, rainfall 800 mm yr^{-1}; Site *B*, rainfall 1150 mm yr^{-1}; both afforested. (After Peck *et al.*, 1981.)

be −7 m for the soil of Curve *B*, when at equilibrium with the water table and about −15 m for the soil of Curve *A*.

There is ample experimental evidence from other sites to predict that the water table would rise to a position only 2 or 3 m below the surface, if these sites were to be cleared. The zone of salt accumulation would then lie beneath the water table. It would become mobile to the extent that groundwater flow would transport the saline pore water, laterally, towards the stream channels. After a lapse of some years it would begin to discharge as base flow. It can be seen that the intensity of salt accumulation in the soils was greater at the drier site than at the wetter site (Curve *B*), where natural leaching would be more effective, as expected.

The way in which the vertically directed, seasonal recharge to the water table appears to by-pass the salt accumulation in the forest soils is understood only qualitatively. These soils are deeply weathered, lateritic remnants of weathered granites, of which the main constituent is kaolinitic clay. The clay must possess fissure planes and cracks that are relatively isolated from the much smaller pore spaces of the dense clay peds. Rapid infiltration must occur through preferred pathways direct to the water table, if the rainfall is sufficiently intense. The theory of displacement of a resident solution by infiltrating new water, as discussed in Section 10.7, is not particularly apt in this case and it cannot explain the extraordinary freshness of the pore water at the water table. The saline water bulge, however it is created, is thus preserved, because sufficient time is not available for infiltration to pass through the fine pores where the saline water accumulates. It is thought that the saline pore water is a result of the summer-time transpiration of the forest vegetation as the soils become excessively dry.

Finally in this discussion, it is interesting to ponder upon why such a problem has developed on the continent of Australia, but apparently not to the same extent on other continents. Evergreen forest, kaolinitic clay profiles of great depth to bed-rock that possess a store of soluble salts and a highly seasonal incidence of rainfall in the winter are the ingredients of the phenomenon. Perhaps the problem has indeed occurred on some of the continents of the Old World before the age of scientific hydrology. It may have been a partial cause of the downfall of dynasties, but has gone unrecorded in scientific terms.

6.3 Groundwater drainage by pumped bore-holes

So far in this chapter, we have considered artificial drainage of fields where the drain spacing is quite close, using either open ditches or tile-drains. The effect upon the water table height is essentially limited to the zone between the level of the bottom of the drain and the soil surface. Such drainage is appropriate when land is intensively used in irrigated areas or in productive regions of high rainfall.

Extensive drainage, without the need for such a close control of the water table depth can be achieved by pumping from suitably located bore-holes or wells. Consider a tube-well that completely penetrates a confined aquifer, as depicted by Fig. 6.8. The flow rate, Q, through the aquifer at any distance r from the axis of the bore-hole is given by an application of Darcy's law, viz.

$$Q = 2\pi K b r \, \mathrm{d}H/\mathrm{d}r, \tag{6.6}$$

where b is the thickness of the conducting zone of the aquifer, K is its hydraulic conductivity, and H is the hydraulic head. If a steady state has been reached such that there is no withdrawal of water from storage in the aquifer between the steadily pumped well and the distance r, Q given by Equation (6.6) is also the pumping rate, which can be measured. Rearrangement and integration of Equation (6.6) gives

$$\int \mathrm{d}H = \frac{Q}{2\pi K b} \int \frac{\mathrm{d}r}{r}. \tag{6.7}$$

We meet a difficulty here in attempting to complete the integration. Thiem (1907) adopted the approximation that, at a sufficiently large distance from the pumped well it is permissible to neglect the effect of continued withdrawal of water from aquifer storage so that the hydraulic potential remains at its initial value, H_0. The hydraulic potential in the aquifer just at entry to the bore-hole, of radius r_w, should be nearly H_w, the

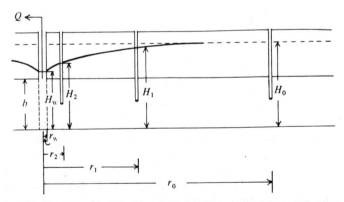

Fig. 6.8. Diagrammatic cross-section of a steadily pumped bore-hole that can be used for drainage purposes. Note that the aquifer is, in this case, confined to the depth b, although leakage would be required for effective drainage. For symbols, see text.

potential in the bore-hole itself. Therefore we can have

$$\int_{H_w}^{H_0} dH = \frac{Q}{2\pi Kb} \int_{r_w}^{r_0} \frac{dr}{r}. \tag{6.8}$$

This is equivalent to the statement that the flow Q is distributed across a boundary at r_0 where the hydraulic potential is H_0. It is the specification of the appropriate distance r_0 that is uncertain.

From Equation (6.8) we have

$$Q = \frac{2\pi Kb(H_0 - H_w)}{\ln r_0/r_w}. \tag{6.9}$$

Equation (6.9) is known as the Thiem steady state formula. It can give guidance upon the pumping rate to be expected for a variety of draw-down $(H_0 - H_w)$ situations. It is more widely employed in experiments to determine the transmissivity of an aquifer, $T(= Kb)$, or the hydraulic conductivity.

The reader will discern the equivalence of Equation (6.9) and Equation (4.23), which was derived by solving the Laplace equation by the application of complex variable theory. Equation (4.23), written out again here, is

$$\phi = (Q/2\pi Kb) \ln r. \tag{6.10}$$

It specifies the shape of the hydraulic head as a function of the distance r from the pumped well. At any two distances r_1 and r_2 $(r_2 > r_1)$ from the well, we would have

$$\phi_2 - \phi_1 = \frac{Q \ln r_2/r_1}{2\pi Kb},$$

which becomes on rearranging,

$$Q = \frac{2\pi Kb(\phi_2 - \phi_1)}{\ln r_2/r_1}. \tag{6.11}$$

The symbols H and ϕ denote the same quantity, the hydraulic head or potential. A convention often observed is that H is used in engineering applications such as the present drainage discussions. When more general flow considerations are described ϕ is the symbol to be used and that notation accords well with the extended solution of the partial differential equation, in which the family of equipotential curves are designated by ϕ_1, ϕ_2 etc. and the family of orthogonal stream lines are designated by ψ_1, ψ_2 etc.

Large areas of irrigated land have been drained by deep tube-wells, for example in the USSR (Kovda *et al.* 1973), Pakistan (FAO, 1971),

Australia (SRWSC, 1969), and in other countries. The requirements for economically successful drainage by wells seem to be principally that permeable sediments should occur at suitable depths below the surface and that pumped water can be sent to waste without doing harm to freshwater supplies in the environment. The Riverina of southeastern Australia is a region where there are numerous buried river channels, from which water can be pumped easily because those sedimentary deposits are very permeable. The location of the tube-wells is therefore determined by the prior and ancestral stream gravels and sands. The wells can be spaced about 1 km apart when sediments are permeable. The mutual interference of tube-wells introduces some difficulty in the design of the system, but guidance from the ideas of potential theory is helpful. The principle of superposition is valid, providing that any simplifying assumptions allow it. Therefore, the combined effect of all the pumping wells on the hydraulic head at a particular location can be obtained simply by summing the effect of each well calculated singly. This consequence is only invalidated if the total drawdown of the hydraulic head should be large enough to dewater the upper part of the aquifer, below the confining bed. In the case of a water table aquifer, the superposition principle requires that the total drawdown of the water table is small in comparison with the depth of the aquifer, through which flow can occur. Since water table aquifers are often shallow, this requirement is likely to be lacking.

The flow of groundwater from, and through a region is conveniently presented by a map that shows the equipotential lines and the streamline directions. Fig. 6.9 is such a map, showing the southeastern district of South Australia, chosen because it is an extensive plain and the flow-net is relatively simple to describe. The contours of the equipotentials were drawn from the data of water-levels in a system of bore-holes, established on a 5×5 km square grid. Correct interpolation procedures were followed, but the task was not computerized. The region is underlain by deep, very permeable limestones and the equipotentials depict the water table height.

The general direction of flow is towards the sea, where the groundwater discharges, much of it through diffuse submarine outlets. The streamlines are nearly horizontal, as can be judged from the water table slope, which lies between 1 : 1000 and 1 : 1500, except for some steep gradient parts. If a vertical plane, extending from the surface depiction of the streamlines to the lower boundary of the aquifer be imagined, the three-dimensional model of the region is made up of stream-tubes, of which there are nine along the coast, in this example. The streamlines were drawn carefully, not only to be everywhere normal to the equipotentials at their crossing points, but also to represent cross-sectional segments through which the

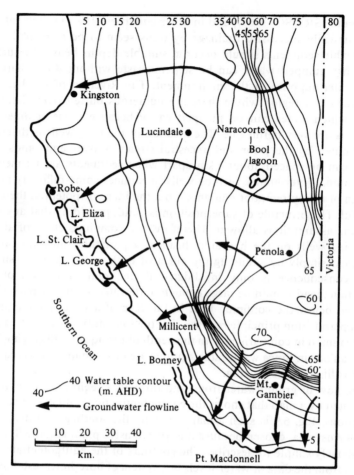

Fig. 6.9. Water table contours and flow directions (streamlines) for the Gambier limestone (unconfined) aquifer in South Australia (m. AHD - metres above Australian Height Datum).

underground flow is about $7 \times 10^7 \, \text{m}^3 \, \text{yr}^{-1}$. Of course, that kind of specification can derive only from rather detailed knowledge of a region's hydrology. A flow-net of this kind is a representation in two dimensions of a three-dimensional regime and is scarcely quantitative. The addition of rain water to the water table, as annual recharge, which can be visualized diagrammatically in Fig. 6.1 and Fig. 6.2 requires that the number of stream-tubes should increase in the downstream direction from the groundwater divide to the discharge region. Indeed there is an increase in their number shown in Fig. 6.9, rather arbitrarily placed to ensure that the discharge to the sea adds up to about $0.6 \times 10^9 \, \text{m}^3 \, \text{yr}^{-1}$.

Furthermore, it may be verified that the recharge to the water table is about 90 mm yr^{-1} in the most southerly part and about 15 mm yr^{-1} in the most northerly complete stream-tube. These data are consistent with the known hydrology. A regional map such as that displayed in Fig. 6.9 has a worth in visualization that is not lessened by its inherent inaccuracies, if these are properly understood.

6.4 Non steady state field drainage

In Chapter 4 the theory of water flow through soils was developed to show that Equation (4.10), rewritten here as

$$\nabla \cdot K\nabla H = \frac{\partial \theta}{\partial t}, \qquad (6.12)$$

is an equation with general relevance to the problem of discovering the hydraulic potential field, $H = H(x, y, z, t)$, its change with time, and the nature of the water content change, $\theta = \theta(x, y, z, t)$. Suppose we consider the time course of drainage of a field by ditches after the initial application of a large amount of irrigation water. The use of Equation (6.12) to develop a solution to this problem would require a knowledge of hydraulic conductivity as a function of the direction of flow, if the soil were anisotropic, expressed as K_x, K_y and K_z in the principal axial directions, the dependence of K upon water content of the soil, $K(\theta)$, for conditions in the unsaturated zone together with dependence of θ upon the matric potential, $\theta(\psi)$. The solution of Equation (6.12) would have to be valid for the part of the soil through which the water table had descended to a lower position at any time t, that is the zone of unsaturation. It would have to be valid for the zone below the water table and although solutions could, in principle, be obtained separately for the two zones they would have to correspond at the water table. The problem has, in fact, proved to be intractable for an exact solution in the absence of considerable simplifying assumptions.

Many degrees of simplification have been attempted. Let us make the following assumptions. Suppose that the variation of hydraulic potential with height z is negligible, i.e. $dH/dz = 0$. Furthermore, the zone through which flow can occur is restricted to an isotropic and homogeneous region below the water table whose thickness can be specified as a constant, b. The release of water from the unsaturated zone when the height of the water table falls can be simulated by the distribution of that same volume of water uniformly and instantaneously throughout a vertical section of the flow zone, the governing relationship being that

$$b \, d\theta = S \, dH, \qquad (6.13)$$

where S is the aquifer yield per unit change in water table elevation, a parameter widely known as the specific yield. The use of the adjective 'specific' is now recommended to be restricted to variation of a quantity per kg, in SI nomenclature. Therefore it is suggested that 'aquifer yield' may have to be adopted instead, or, at least, some similar phrase. Fig. 6.10 summarizes these assumptions geometrically. Equation (6.14) is consistent with the foregoing assumptions,

$$\frac{\partial^2 H}{\partial x^2} = \frac{1}{(Kb)/S}\frac{\partial H}{\partial t}. \qquad (6.14)$$

It is capable of describing horizontal one-dimensional non steady state flow to an acceptable accuracy.

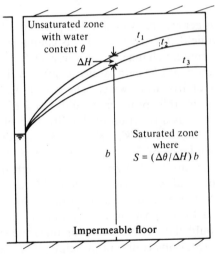

Fig. 6.10. Schematic of the draw-down of a water table aquifer with time, to illustrate the assumed distribution of the water released from storage uniformly through the saturated zone of depth b, to satisfy the assumption needed for solution of Equation (6.14). For symbols, see text.

The combination of constants $(Kb)/S$ is so written to facilitate direct comparison with Equations (4.32) and (4.39), where the soil water diffusivity is the important parameter. In the initial treatment of the flow problem in Chapter 4, the differential equations were developed rigorously by reference to an infinitesimal volume of the medium. In this chapter we treat the macroscopic problem of flow in the field and its simplification by considering that the whole of the flow depth, of thickness b, and horizontal increment dx, is equivalent to a multiplication by b of both the hydraulic conductivity and the water capacity, as defined in

Equation (4.31). We therefore have the equivalence (approximately) that

$$D = \frac{K}{d\theta/d\psi} \equiv \frac{Kb}{(d\theta/d\psi)b} = \frac{T}{S}, \tag{6.15}$$

where we have the transmissivity, T, and aquifer yield, S, stated in terms of the microscopic parameters K and $d\theta/d\psi$ and the thickness of the zone, b, through which the water is assumed to flow and to be distributed on its release from above the water table.

An equation similar to Equation (6.14) was first studied by J. Boussinesq in 1877 in the context of water flow. It is formally the same as the equation that describes flow in a confined aquifer. When recharge of the flow region or irrigation has to be considered, on a time-varying basis, another term has to be added to the left hand side of Equation (6.14) to describe that part of the water budget in order that the water continuity equation should be preserved. If we write out Equation (6.14) as

$$(Kb) \frac{\partial^2 H}{\partial x^2} = S \frac{\partial H}{\partial t}, \tag{6.16}$$

we note that the term $S\,\partial H/\partial t$ is the change in water content of the aquifer per unit time, and $(Kb)\partial^2 H/\partial x^2$ is the divergence of the flux. Surface recharge at a rate distributed uniformly across the water table is then equivalent mathematically to a source term v, and Equation (6.16) becomes

$$(Kb) \frac{\partial^2 H}{\partial x^2} + v = S \frac{\partial H}{\partial t}, \tag{6.17}$$

where v is to be regarded as a steady rainfall rate. Equation (6.17) may conveniently be written as

$$\frac{\partial^2 H}{\partial x^2} = \frac{1}{D} \frac{\partial H}{\partial t} - v', \tag{6.18}$$

where $D = Kb/S$ and may be termed the diffusivity coefficient and $v' = v/Kb$ is a recharge coefficient. First, it is interesting to note the steady state solution of Equation (6.18), which then reduces to

$$d^2 H/dx^2 = -v'. \tag{6.19}$$

Integration of Equation (6.19) gives

$$H = -\frac{v}{2Kb} x^2 + Ax + B, \tag{6.20}$$

where the constants A and B have yet to be evaluated and it is required to assume that the rainfall rate, v, is constant. For the flow situation depicted by Fig. 6.1 the constants A and B can be evaluated from the requirements that $H = 0$ at $x = 0$ and $x = S$. Also approximately the depth of conducting aquifer is given by $b = (z_m + z_0)/2$. Therefore Equation (6.20) becomes

$$H = \frac{v(Sx - x^2)}{K(z_m + z_0)}. \tag{6.21}$$

At the mid-point between the ditches, Equation (6.21) reduces to

$$S^2 = \frac{4K}{v}(z_m^2 - z_0^2),$$

the same equation as Equation (6.2).

Solutions of Equation (6.18) have been presented and discussed by Werner (1957), Maasland (1959) and van Schilfgaarde (1974*b*). The reader is referred to these papers and other references cited in them for the mathematical detail, which is formidable. We are here concerned with presenting an assessment of the time that may be required for the water table mid-way between drains to fall a prescribed distance after the cessation of a rainfall that had persisted long enough for a steady state drainage situation to have been established. It appears that the literature can give only imperfect guidance upon this point, from a theoretical point of view. Data presented by van Schilfgaarde (1974*b*), when interpreted, suggest that the rate of decline of the water table would be of order 0.1 m day^{-1} for a drainage system of adequate design.

6.5 Field measurement of hydraulic conductivity, transmissivity and storage coefficients

A rearrangement of Equation (6.9) shows that the transmissivity could be derived from suitable measurements of Q, H_1, H_2, r_1 and r_2, thus

$$T = Kb = \frac{Q \ln (r_1/r_2)}{2\pi(H_1 - H_2)}, \tag{6.22}$$

where H_1 and H_2 are the hydraulic heads measured in observation wells spaced at distances r_1 and r_2 from the steadily pumped bore hole (see Fig. 6.8).

The requirements for success are that the pumped well should penetrate the aquifer completely. If casing is needed it should be slotted over the whole interval of the aquifer. A sufficient time should be allowed to elapse from the start of pumping so that water levels, as read in the observation holes, are effectively steady. In addition, the pumping should be well regulated to a steady rate from the start, and the water should be discharged sufficiently remotely that there is no likelihood of it returning to the part of the aquifer under test.

The use of several observation wells, though expensive for installation, helps to ensure that the observations yield suitable values for calculation. The hydraulic conductivity derived from the direct calculation of transmissivity is necessarily a weighted mean for the whole depth of the aquifer. The method is suitable for confined aquifers and for thick water table aquifers.

There are other, well-tried methods for measurement of hydraulic conductivity of soils, under water table conditions. Such methods have been developed to meet the need for drainage design criteria.

For flow of water into an auger hole that penetrates beneath the water table, the essential geometrical features are summarized in Fig. 6.11. When the water level in the auger hole is perturbed from its equilibrium level, for example by rapid bailing, the subsequent rate of inflow into the auger hole should depend upon the hydraulic potential difference, z, the hydraulic conductivity, K, and geometrical features of the installation, including the depth, D, to the lower (supposed) impermeable boundary. We could write

$$\frac{d}{dt}\{\pi r^2 (H - z)\} = KAz, \qquad (6.23)$$

where A is a factor that specifies the effects of geometry, and the other symbols are as shown in Fig. 6.11. If we could regard A as a constant, Equation (6.23) could be integrated to give

$$z = \exp -(KA't), \qquad (6.24)$$

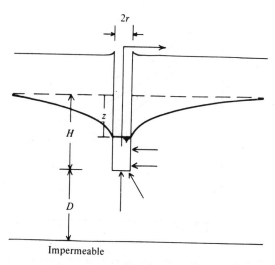

Fig. 6.11. The auger hole method of measuring soil hydraulic conductivity in the field.

where A' is another constant obtained by a combination of the constant terms of Equation (6.23). In fact, the recovery of the water level in a bailed auger hole does follow approximately an exponential relationship with time, as suggested by Equation (6.24). However, when a small change of water level in the auger hole is observed, after bailing, Equation (6.24) is approximated by

$$K = C \frac{\Delta z}{\Delta t},$$
(6.25)

where the constant C accommodates the effects of geometry of the hole and depth to an impermeable layer. The development of the auger hole method to its routine use was summarized by van Beers (1963). In his bulletin there are a number of graphical representations of C as a function of depth of penetration of the auger hole below the water table, H, which effectively determines the surface area of the hole through which water can enter. Fig. 6.12, taken from van Beers, shows C versus H, for an auger hole of diameter 0.1 m, where $D > H/2$. The parameter that distinguishes the individual curves of the family is mean depth of draw-down in the auger hole, z.

Kirkham (1945) and his co-workers (for example, Luthin and Kirkham, 1949) developed a field method for measuring the hydraulic conductivity

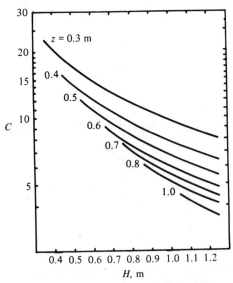

Fig. 6.12. The dependence of the parameter C upon penetration of the auger hole beneath the water table, H, for a variety of draw-downs, z. (After van Beers, 1963.)

of soil, in place, that was based upon an analysis of the rate of approach to the equilibrium level of water in a piezometer tube. If the tube is properly installed so that a tight seal is provided between the tube and the soil (see Fig. 6.13), then water can enter only through the exposed surface of the cavity at the bottom.

The hydraulic conductivity of the soil is given by (Luthin and Kirkham, 1949)

$$K = \frac{\pi r^2}{A} \frac{\ln (z_1/z_2)}{(t_2 - t_1)}, \tag{6.26}$$

where z_1 and z_2 are draw-downs of the water level in the piezometer tube measured at times t_1 and t_2. The A-factor obtained by electrical analogue experiment for a comprehensive set of conditions was reported by Youngs (1968), and Table 6.4 presents a selected set of values for the most used sizes of apparatus, as dimensionless ratios.

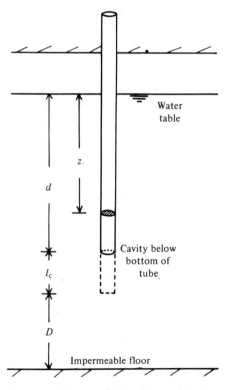

Fig. 6.13. The piezometer tube method of measuring soil hydraulic conductivity. (After Luthin and Kirkham, 1949.)

Table 6.4. *Values of the shape factor A expressed as the dimensionless ratio A/r for the piezometer-tube method (after Youngs, 1968): for a range of depths of the piezometer tube below the water table (d/r) and lengths of the cavity (l_c/r), both expressed in dimensionless form with respect to the radius (r) of the tube*

Depth D/r of impermeable floor				
l_c/r	d/r	∞	4.0	1.0
0	20	5.6	5.3	4.4
	12	5.6	5.4	4.5
	4	5.8	5.6	4.8
1.0	20	10.6	10.0	8.4
	12	10.8	10.2	8.6
	4	11.5	11.2	9.7
2.0	20	13.8	12.8	10.9
	12	14.0	13.2	11.2
	4	15.0	14.5	12.6

In contrast to the auger hole and piezometer-tube methods which have inexact solutions to a non steady flow, Childs' two-well method (Childs, Cole and Edwards, 1953) utilizes a steady state flow that provides an analytical solution. The principle of the method is illustrated by Fig. 6.14. The hydraulic conductivity is given by

$$K = \frac{\cosh^{-1}(s/2r)Q}{\pi(l+l_f)H},$$ (6.27)

where H is the hydraulic head difference between the two wells set up by the equilibrium pumping rate, Q. The distance apart of the wells, s, is conveniently about 1 m and a suitable diameter $(2r)$ would be 200 mm. The distance that the wells penetrate below the depth of the equilibrium water table, l, must have a correction added to it, l_f, that adds compensation for the flow beneath the bottom of the wells. This correction depends upon the depth to the lower impermeable boundary and may be up to about 50 per cent of l.

The methods for measuring the saturated hydraulic conductivity of soils, as outlined above, depend upon the presence of a water table. The flow to be measured is set up essentially by making small, controlled perturbations to the hydraulic head, so that the water for the experiment is provided by the water already present in the saturated zone. Another method, known as the well permeameter (or the auger-hole pump-in)

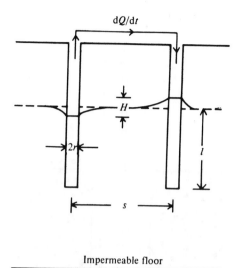

Fig. 6.14. The two-well method of measuring hydraulic conductivity. (After Childs *et al.*, 1953.)

method may be used for investigations above the water table. Fig. 6.15 illustrates its field arrangements.

The dry hole, of radius *r*, is augered to a depth sufficient to enable H/r to reach a value of 10, when water is added. The height H is maintained by a Marriotte-tube bleed of air as indicated. The initial rate of water seepage is fast, appropriate to infiltration into a dry soil. A steady rate of flow, Q, is approached as the soil surrounding the hole approaches saturation and the saturated hydraulic conductivity then controls the rate of infiltration rather than sorptivity and the unsaturated hydraulic conductivity (see Section 5.7).

A mathematical expression for K in terms of the parameters of the experimental set-up was obtained by Zangar as

$$K = Q[\sinh^{-1}(H/r) - 1]/2\pi H^2, \qquad (6.28)$$

and was quoted thus by Talsma and Hallam (1980) in their account of the use of the technique in rather inaccessible forest regions. Reynolds *et al.*, (1983) suggested the relationship

$$K = CQ/2\pi H^2, \qquad (6.29)$$

in which the parameter C replaces the expression $[\sinh^{-1}(H/r) - 1]$. They determined C by numerical modelling and gave the following values: for $H/r = 10$, $C = 3.3$ and for $H/r = 5$, $C = 2.2$. The apparently

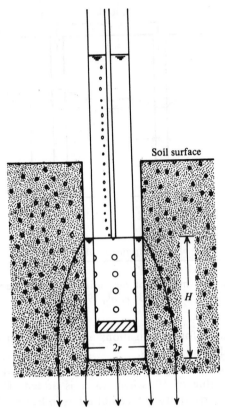

Fig. 6.15. The well permeameter (auger-hole pump-in) method for measuring saturated hydraulic conductivity above the water table. (After Talsma and Hallam, 1980.)

exact solution, Equation (6.28) would reduce to

$$K = Q/\pi H^2, \tag{6.30}$$

for $H/r = 10$, because $(\sinh^{-1} 10 - 1) = 3$ quite closely. The exact solution conceals its nature, that it has been derived with simplifying assumptions. The parameter C is a geometrical shape factor of the same kind as the A-factor in Equation (6.26).

Talsma and Hallam made good use of this method in their study of the likelihood of surface run-off from the forest soils of several mountainous catchments. The amount of water that can be carried to sites accessible only by walking obviously determines what technique can be used. One determination of K, using a hole of about 60 mm diameter would require one to two litres of water, depending upon how dry the soil is initially.

The volume of soil sampled is small, by this method, and it is to be expected that the results would be very variable so that many replications are necessary. Furthermore Reynolds *et al.* (1983) consider that this 'field saturated' K is about one-half of the truly saturated hydraulic conductivity.

The amount of water that can be released from a soil when the pore water pressure changes, under saturated conditions, is very small. It is related to the compressibility of water, $\beta(\mathrm{m}^2\,\mathrm{N}^{-1})$, and the vertical compressibility of the soil or aquifer grains, $\alpha(\mathrm{m}^2\,\mathrm{N}^{-1})$, by

$$\frac{\mathrm{d}\theta}{\mathrm{d}p} = \alpha(1-\theta)+\beta\theta. \qquad (6.31)$$

Here θ is the volumetric water content and p is the pore water pressure (de Wiest, 1966). Assigning the usual hydrostatic formula to the pore water pressure, namely $p = \rho g h$ and the symbol S_s for the storage coefficient per unit volume, we have

$$S_s = \frac{\mathrm{d}\theta}{\mathrm{d}h} = \rho g\{\alpha(1-\theta)+\beta\theta\}. \qquad (6.32)$$

In physical terms S_s is the volume of water that would be released from unit volume of the saturated soil or aquifer if the pressure head were reduced by 1 m. Therefore it has the dimensions L^{-1} and the total amount of water that would be released over the whole saturated thickness b of the soil or aquifer per unit change of piezometric head, known as the aquifer storage coefficient, is

$$S = bS_s, \qquad (6.33)$$

and is dimensionless. The magnitude of the storage coefficient per unit volume is typically about $10^{-4}\,\mathrm{m}^{-1}$ for sand. It depends principally upon the Young's modulus of the grains and appears to be larger for fine sand and clayey sand than for coarse sand. It is a property that merges into the shrinking or swelling property of fine textured soils, the nature of which forms part of the theory of consolidation (see, for example Taylor, 1948).

By contrast to release of water by elastic expansion from a saturated zone when the pore water pressure is lessened, water can drain out of pore spaces under unsaturated conditions. Then the pore water pressure is negative and any further reduction of it caused by a lowering of the water table will cause water to drain downwards to the water table, as has been already discussed in Section 6.4 above. Such additional water may be thought of as a flux across the water table vertically, and is additional to the elastic release of water. If the water table falls sufficiently slowly so that the soil water profile can remain essentially in equilibrium with it, unconfined aquifer yield may be calculated as follows.

A representation of the water content of the soil (aquifer) as a function of height above an arbitrary datum that is sufficiently deep to be always beneath the water table is given in Fig. 6.16. The mean water content from $z = 0$ to the soil surface, z_u is

$$\bar{\theta} = \frac{1}{z_u} \int_0^{z_u} \theta \, \mathrm{d}z. \tag{6.34}$$

At any particular height of the water table, say z_1, the mean water content is therefore $\bar{\theta}_1$ and at another height of the water table, z_2, it is $\bar{\theta}_2$. The difference in mean water content is

$$\bar{\theta}_1 - \bar{\theta}_2 = \frac{1}{z_u} \int_0^{z_u} (\theta_1 - \theta_2) \, \mathrm{d}z. \tag{6.35}$$

If the water table at positions z_1 and z_2 has been stationary there for sufficient time for equilibrium water content profiles to be established, then θ_1 and θ_2 will have the same shapes, with the exception of a portion of the profiles near the soil surface. If the water table is deep and the water content near the soil surface is the rather elusive field capacity, we can regard the water content at the surface as approximately the same, namely θ_u. A simple geometrical representation is shown in Fig. 6.17. It can be proved geometrically that whatever shape the water content profiles may have, provided that they are similar, the area between the

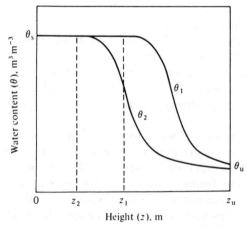

Fig. 6.16. Equilibrium soil water profiles for the purpose of obtaining the unconfined aquifer yield. For symbols, see text.

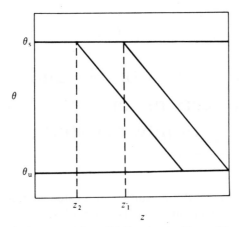

Fig. 6.17. Geometrical construction of soil water profiles to illustrate the calculation of the unconfined aquifer yield.

curves θ_1 and θ_2 is then

$$\int_0^{z_u} (\theta_1 - \theta_2)\, dz = (\theta_s - \theta_u)(z_1 - z_2), \qquad (6.36)$$

where θ_s is the water content at saturation.

The total water storage in the aquifer of depth z_u is $\bar{\theta} z_u$. We have, from Equations (6.35) and (6.36) that

$$(S_1 - S_2) = (\bar{\theta}_1 - \bar{\theta}_2)z_u = (\theta_s - \theta_u)(z_1 - z_2), \qquad (6.37)$$

where $(S_1 - S_2)$ is the change in total water storage of the aquifer when the water table changes its position from z_1 to z_2. The change in total water storage of the aquifer per unit change of water table depth is therefore

$$S = (S_1 - S_2)/(z_1 - z_2) = \theta_s - \theta_u, \qquad (6.38)$$

and is the unconfined aquifer yield, formerly known as the specific yield. Its symbol is commonly S and it is equivalent to the aquifer storage coefficient, defined by Equation (6.33), when the mathematical treatment is simplified, by approximations, to a consideration of solutions of Equation (6.16) or (6.17).

7

The use of isotopes and other tracers in soil water and groundwater studies

Field investigations of the subsurface of a catchment or watershed may be undertaken by methods such as those whose principles have been described in Chapters 5 and 6. Sometimes the pattern of sub-surface flow may be difficult to discover or to interpret by conventional hydrological experiments. Tracing techniques then offer an alternative that can be less costly in effort in the field and can provide results that would be unattainable by other methods.

There are two kinds of tracers. Firstly, there are those that are injected into the aquifer or soil water for the purpose of the experiment. They must be carried with the water which is to be traced and must be capable of being recovered after a reasonable time. The altered concentrations, as measured, should be capable of hydrological interpretation and so the recovery grid of bore-holes should be designed with some existing knowledge of the hydrological system for guidance. The second kind of tracers are those that are not added artificially, but occur naturally as a peculiar feature of the hydrological cycle. There are a number of such environmental tracers, whose concentrations vary and can provide the data for hydrological enquiry. Chloride dissolved in rainwater is an example. It is carried into the soil by soil water infiltration and remains there or in groundwater in concentrations that are greatly increased at locations where most of the water is abstracted from the soil by the process of evapotranspiration.

Some of the useful environmental tracers are radioactive, and of these tritium and carbon-14 are particularly important. They afford an opportunity of age-dating the water that contains them. Other isotopes such as deuterium and oxygen-18 which are stable, undergo fractionation on change of state, by evaporation or condensation, thawing or freezing of the water that contains them. The variations of their abundances with respect to a standard isotopic abundance can afford data for useful hydrological interpretation. In this chapter we give an account of some uses of tracers for soil water studies, but recognize the need to restrict our scope so as not to include large-scale studies of groundwater basins.

7.1 Radioactive tracers

The study of the distribution of tritium in soils and groundwaters has proved to be particularly rewarding in improving our knowledge of hydrological situations. In the early 1950s Libby suggested that large scale investigations could be possible if precipitation, enriched in tritium that diffuses downwards from the tropopause, could be regarded as carrying that tracer into any natural hydrological system. Indeed, tritium produced naturally by cosmic ray bombardments of stratospheric gases (Kaufman and Libby, 1954) and tritium released from H-bomb tests (Libby, 1963) and widely dispersed into the stratosphere and troposphere are both useful in this context. The artificial additions of tritium to the world inventory commenced with the 'Castle' atomic weapons tests of 1954. These and subsequent episodes have been documented and the tritium release estimated by Eriksson (1965).

Tritium is produced naturally by cosmic ray interactions with atmospheric nuclei, principally nitrogen, in the higher stratosphere (Nir et al., 1966). Oxidation enables the tritium atoms to become incorporated into the water vapour of the lower stratosphere, whence they diffuse through the tropopause into the region of weather processes. The amount of natural tritium in the world inventory is exceedingly small, being of the order of 1 kg. Most of it is, in fact, not held in the atmospheric water vapour reservoir but is distributed between the oceanic and terrestrial reservoirs of liquid water.

The small abundance of tritium, even with H-bomb additions, that has to be measured in natural waters demands very sensitive techniques (see Taylor, 1978). Tritium is detected by its emission of β-particles, which have an energy spectrum distributed from a peak in counting rate at about 3×10^3 eV to a maximum energy of 18 keV (Curran, Angus and Cockroft, 1949). Its energy spectrum enables tritium to be readily distinguished from other radioactive isotopes that could be present in water, particularly ^{14}C, ^{40}K and the uranium and thorium series elements. Its concentration is usually expressed in terms of the tritium unit (TU) which is defined as one atom of tritium, ^3H, per 10^{18} atoms of protium, ^1H.

Tritium concentrations in precipitation have been intensively studied since 1955. Representative values are given in Table 7.1. The very great increase in tritium concentration caused by weapons testing, appeared first in northern hemisphere precipitation. It then occurred with lesser intensity, and with a delay that appears to be from one to two years, in the southern hemisphere. These tritium additions decay at a rate determined both by the residence time of gaseous material in the stratosphere (about 3 years) into which most of the bomb tritium was initially deposited, and by the half-life of tritium (12.3 years). Fig. 7.1 shows, in more detail, tritium concentrations of Australian, New Zealand and

Table 7.1. *Tritium concentrations, in TU, in precipi-*
tation to show the effect of man-made additions to
the natural inventory after 1953 (data assembled by
Bosch et al., 1970)

Year	North America (Ottawa)	Europe (mean estimate)
1968	214	150
1967	315	160
1966	590	240
1965	865	580
1964	1565	1300
1963	3032	2500
1962	988	700
1961	219	110
1960	145	145
1959	540	450
1958	515	300
1957	126	125
1956	146	100
1955	45	35
1954	302	300
1953	30	25
1952	20	20
1951	15	15

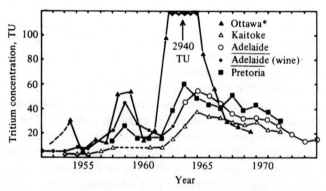

Fig. 7.1. Tritium concentrations in precipitation sampled at Ottawa, Canada; Kaitoke, New Zealand; Adelaide, Australia; and Pretoria, South Africa (from Allison and Hughes, 1977). * Ottawa concentrations are 10 × tritium units shown on the ordinate.

South African precipitations with data from Ottawa to illustrate the contrast with concentrations of tritium in northern hemisphere precipitations.

Rain-fed infiltration and replenishment of soil water, the subsequent percolation of water to the water table and recharge of aquifers cause additions of tritium to such reservoirs in which the tritium content is being reduced by radioactive decay. Therefore, there is a possibility that the age and other hydrological features of soil water and underground water can be interpreted by comparing the tritium concentrations of such water, which has been depleted by radioactive decay, with that of the regional precipitation.

The simplest situation to interpret may be illustrated by Fig. 7.2. If intake of water to a confined aquifer is restricted to the outcrop area only, then the tritium concentration of water sampled from the reservoir at a distance, x, along a groundwater stream line is given by

$$C_R = C_I e^{-t/T}, \tag{7.1}$$

where C_I is the tritium concentration of precipitation that entered into the outcrop t years before the water was sampled and T is the mean life of tritium, 18 years approximately. The time, t, is related to the velocity of groundwater flow, v', by

$$v' = v/\varepsilon = x/t \tag{7.2}$$

where v is the discharge per unit cross-section of the aquifer and ε is its porosity. Since a principal purpose of interpreting tritium concentrations is to obtain an estimate of v without recourse to the expensive field measurements of transmissivity etc., Equation (7.1) could be useful in practice only if C_I is constant, or has a relatively simple time dependency. Furthermore, it is rare to find any shallow groundwater body for which this piston flow description is sufficiently appropriate. Confined aquifers may satisfy the simplifying assumptions but are outside the scope of the present discussion.

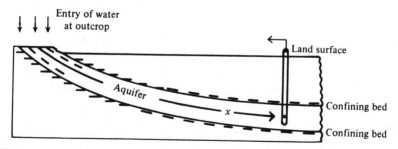

Fig. 7.2. Water flow in a confined aquifer to illustrate the influence of velocity, x/t, upon 'age' of the water.

The piston flow description of the movement of water through the unsaturated zone of the soil towards the water table was shown by Zimmermann, Munnich and Roether (1967) to be tolerably accurate. Their experiment, performed in the field, involved the use of water labelled with either deuterium or tritium so that the location of the layer of labelled water could be discovered by sampling and displaying the spiked layer as a peak in the tracer profile. Dispersion of the peak was accounted for adequately by the influence of molecular diffusion and exchange of water between small and large pores as the labelled layer passed by soil aggregates with a fine pore size distribution (see Section 10.7).

Smith *et al.* (1970) put to good use the peaks of tritium concentration in rainwater that were caused, in the northern hemisphere, by H-bomb additions of tritium during the years of weapons testing in the atmosphere. They measured profiles of tritium in water in a clay soil to a depth of 4 m and in a chalk profile from the surface soil to a depth of 27 m in relatively unweathered chalk, at which depth the water table was encountered. The samples were taken in October 1968. Fig. 7.3 shows their result of the tritium profile in the clay soil.

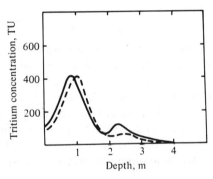

Fig. 7.3. The tritium concentration profile in the soil water of a clay soil in England. ——, observed; - - -, modelled. (After Smith *et al.*, 1970.)

The two peaks of tritium concentrations that occur at about 2.2 m and 0.9 m were stated to be due to the peaks of tritium concentration in precipitation in 1958/59 and 1963/64. The velocity of propagation, v', of the deeper peak is therefore about 0.23 m yr^{-1} and of the shallower peak about 0.18 m yr^{-1}. The water flux would hence be given by $v'\theta$, where θ is the volumetric water content, and from the data of Smith *et al.* the recharge of that clay is suggested to be in the range 115 mm yr^{-1} to 90 mm yr^{-1}, as given by the velocities of the two peaks. By a second method of analysis, a model that best fitted the observed tritium profile, shown as producing the second curve on Fig. 7.3, was set up to account for the total tritium content of the profile, expressed as TU metres. It required an assumption of 90 mm yr^{-1} for the mean annual recharge of water through the soil to the water table, together with a fitted dispersion rate of the tracer peak, to spread it out.

The two methods of analysis of the tritium concentration profile in the soil used by Smith *et al.* are quite distinct. To use the peak of the tritium concentration and to derive from it a velocity of soil water flow in the unsaturated zone could be done for any tracer that could be assumed to move with the same velocity as the water itself. The radioactive decay of the tritium was used only in the second manner of interpretation. Sometimes a peak in tritium concentration is not clearly discernible in the data, particularly in the southern hemisphere. Furthermore, if there are no more weapons tests that inject huge amounts of artificial tritium into the atmosphere, all future interpretations will have to rely solely upon calculations of the rate of decay and rate of turnover of tritium in the soil water.

Allison and Hughes (1974) showed that the second method of Smith *et al.* could be developed for situations where the tritium contents were much smaller than those encountered in recent years in the northern hemisphere. Fig. 7.4 shows the tritium profile of a podzolic soil in southern Australia, sampled in July 1971. The tritium peak is not well defined and the data were analysed on the basis of the total amount of tritium that remains in the profile. The tritium balance can be described as follows. If the passage of water through the soil profile is so slow that all the tritium has decayed before the annual increment to the water table aquifer is finally delivered, then the total tritium content of the profile at any time should equal the annual additions remaining after their decay, calculated for a sufficient number of years. That is

$$\int_0^{z_w} \theta C \ dz = \sum_{n=0}^{3\tau} Q_n C_n \exp -(n\lambda), \qquad (7.3)$$

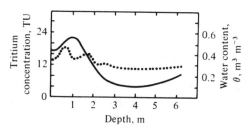

Fig. 7.4. The tritium concentration profile (solid line) in the soil water of a podzolic soil in southern Australia. (After Allison and Hughes, 1974.)

where C is the tritium concentration of the soil water at depth z where the water content is θ m^3 m^{-3}, and C_n is the tritium input function, namely the tritium concentration of the annual increment, Q_n, to the soil store at year n, which is counted backwards from the year of date of sampling: and λ is the decay constant of tritium. The integration is taken to go deep enough to reach a depth in the soil where C is negligible and may be the depth to the water table, z_w. The summation is shown to be taken over three half-lives of tritium (τ), about 37 years.

Allison and Hughes (1974) applied Equation (7.3) in the form

$$\bar{Q} = \left(\int_0^{z_w} \theta C \, dz \right) \Big/ \sum_{n=0}^{3\tau} C_n \exp -(n\lambda) \qquad (7.4)$$

because they considered that the annual increment of tritium $Q_n C_n$ did not vary sufficiently to render invalid the procedure of taking a mean annual increment to the soil water store, $\bar{Q}$, outside the summation. With that assumption and a minor improvement to the tritium balance to include the current year's additions not catered for in the summation, they were able to derive mean annual recharge rates to the extensive water table aquifer under study, for a variety of soil types. For the example given here, the recharge was 50 mm yr^{-1}.

The results of this and later work (Allison, 1975) showed that $\bar{Q}$ varied considerably, from a value of about 50 mm yr^{-1} beneath a deep well-structured clay soil upon which the vegetation was 'improved' pasture, the standard grazed pasture of southern Australia based upon subterranean clover (*Trifolium subterraneum*, L.) and perennial rye-grass (*Lolium perenne*, L.), to 180 mm yr^{-1} beneath a shallow soil possessing a near surface, kunkar horizon overlying a deep porous limestone zone in which the water table occurred at 20 m. At each site the mean rainfall is about 750 mm yr^{-1}.

The interpretation of natural tritium variations in the soil profile is sometimes rendered perplexing by an apparent increase in the tritium concentration of groundwater from a minimum near the water table to larger concentrations at greater depths, such as is shown in Fig. 7.4. The water samples from bore-holes penetrating to the water table is then presumed not necessarily, or wholly, to have infiltrated through the adjacent overlying unsaturated zone, but to have had a more distant origin. The great variability of mean annual recharge from site to site, depending upon soil conditions, as shown by Allison's work, is a widely observed phenomenon. Furthermore, recharge of groundwater is greatly affected by the kind of plant community and land use (Allison and Hughes, 1972), as is discussed further in Chapter 12. The water in the saturated zone beneath the water table can therefore be regarded as deriving from a variety of sources, because of its predominantly horizontal flow. If there are natural mechanisms of mixing or if the sampling procedure can give a mixed water from an interval of depth, then the tritium concentration of such a sample contains information about the distribution of tritium injection into the aquifer from up-gradient recharge sites of varying magnitude.

Suppose we consider the hydrology of an underground reservoir into which each year's mean incremental recharge is completely mixed in a time much less than one year. The concentration of tritium in such a well-mixed aquifer can then be obtained by the following argument. Suppose the total reservoir volume, V_R, has a uniform tritium concentration C_R. The total tritium content is $V_R C_R$. We have

$$\frac{d(V_R C_R)}{dt} = IC_I - DC_R - \lambda V_R C_R, \tag{7.5}$$

where I is the annual recharge, D is the annual discharge of the reservoir, and λ is the decay constant, more often used than the average life of tritium ($\lambda = 1/T$). If the reservoir is at equilibrium

$$\frac{d(V_R C_R)}{dt} = 0,$$

and

$$\bar{Q}C_R = \bar{Q}C_I - \lambda V_R C_R,$$

where we have identified the mean annual recharge, $\bar{Q}$, with the recharge and discharge from the aquifer. Hence,

$$C_R = C_I/(1 + \lambda V_R \bar{Q}^{-1}). \tag{7.6}$$

The ratio $V_R/\bar{Q}$, of the volume to mean annual recharge, is an important parameter, both in attempts to interpret tritium age of reservoir waters and in the general context of water supply evaluation. It has dimensions of time (years). Equation (7.6) can be derived only if C_I is steady. The present fluctuations in tritium concentration of precipitation caused by H-bomb additions require a more elaborate treatment, which leads to the expression for the tritium content of the well-mixed reservoir as

$$C_n = \frac{\bar{Q}}{V_R} \sum_{p=0}^{\infty} C_{(n-p)} \left(1 - \frac{\bar{Q}}{V_R}\right)^p \Gamma^p, \tag{7.7}$$

where C_n is the tritium concentration of the reservoir water at year n, $C_{(n-p)}$ is the tritium concentration of the recharging water at year $n - p$ and Γ is the quantity of tritium remaining, not decayed, per year ($=0.945 \text{ yr}^{-1}$). Here the reckoning of years n and p is not backwards from the present year but forwards from a suitable beginning of time to encompass all the significant decay of tritium, again about three half-lives or 37 years.

Use of isotopes and other tracers

There have been a number of successful applications of Equation (7.7), one, for example, being the assessment of $\bar{Q}/V_R$ for aquifers contributing to base flow of the Rhône and other European rivers by the workers responsible for its elaboration, Hubert *et al.* (1970). The method of application relies upon repeated calculations of the parameter C_n for a variety of trial $\bar{Q}/V_R$ values, and comparison with observed tritium concentrations. Fig. 7.5, taken from the work of Allison and Holmes (1973), shows the tritium concentration to be expected in a well-mixed groundwater reservoir in southern Australia, where the tritium input is related to the distribution of tritium in rainfall shown in Fig. 7.1. In arid regions the ratio $\bar{Q}/V_R$ is very small, and the Polda Basin of South Australia, so examined, had tritium concentrations of about 2 TU or less (see Allison and Holmes, 1973). Real aquifers have characteristics that must fall between the two extremes so far considered, i.e.of piston-type (serial) flow and recharge instantaneously mixed (see Nir, 1964).

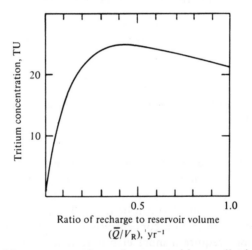

Fig. 7.5. The tritium concentration to be expected in a well-mixed aquifer as a function of $\bar{Q}/V_R$, the ratio of mean annual recharge to reservoir volume, for a tritium input function appropriate for southern Australia, to the year 1970. (After Allison and Holmes, 1973.)

Aquifers underlying terrain of low relief are often themselves flat lying and have a large lateral extent relative to thickness. This situation lends itself to the Dupuit–Forcheimer approximations and to elaboration into a theory of the distribution of tritium concentration with depth. Consider an unconfined aquifer into which surplus rain infiltrates at a steady recharge rate v, distributed uniformly along the water table, such as depicted in Fig. 7.6. At a position l remote from the upstream or down-stream boundaries, water that has been sampled from a shallow depth would have followed a shorter stream line path than water samples from near the impermeable floor of the aquifer. Vogel (1967) used this fact to obtain an approximate expression for the age distribution of water with depth in an unconfined aquifer. At position l (Fig. 7.6), the total discharge through the vertical section is vl, i.e. a steady state is assumed with no change in storage in the aquifer. The total discharge may also be thought of as the local velocity, v', of the water horizontally through a vertical section of the aquifer, multiplied by its depth and porosity, i.e. it is

$$\varepsilon b v' = \varepsilon b \, \mathrm{d}l / \mathrm{d}t. \qquad (7.8)$$

Equating the quantity shown in Equation (7.8) to vl we have

$$\varepsilon b \, \mathrm{d}l / \mathrm{d}t = vl,$$

whence

$$t = (\varepsilon b / v) \ln (l / l_1). \qquad (7.9)$$

In Equation (7.9) the limits of integration identify two lengths, l and l_1, measured from the watershed. The relationship $l_1 < l$ is required, and then t is the time taken for the water front to pass from l_1 to l. Also from the geometry of the problem, it can be shown that

$$l / l_1 = b / b_1,$$

where b_1 is the height, in the section at l, from the impermeable floor to the stream line which originates at the surface of the unconfined aquifer

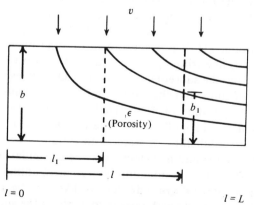

Fig. 7.6. Schematic to illustrate the basis of calculation of tritium concentration as a function of depth $(b - b_1)$ in an unconfined aquifer with surface recharge v. (After Vogel, 1967.)

at l_1. Thus

$$t = (\varepsilon b/v) \ln (b/b_1), \qquad (7.10)$$

and Equation (7.10) may be taken as descriptive of the distribution of age of water, t, as a function of depth $(b - b_1)$, or height above the lower confining bed, b_1. The age appears to be independent of the horizontal dimensions of the aquifer. It is so because the recharge rate, v, per unit horizontal area of the water table is converted into a horizontal flux through the vertical section and the age must depend upon v', the velocity in the horizontal. Because of the assumptions involved, the result must be only approximately correct, but it appears to hold reasonably well at positions remote from the boundaries. It should also be noted that, although it was derived by Vogel in the context of interpreting carbon-14 abundances, Equation (7.10) is merely an expression for the length of time that water has remained underground at a particular sampling location since infiltrating into the ground. It could be put to use to derive the age of water in very large, flat-lying aquifers of arid regions, thus possibly to help in interpreting past climatic excursions from the normal, but there does not appear to be any record of such an attempt in the literature.

The large tritium injections produced by the explosions of H-bombs, when these weapons were tested in the 1950s and 1960s, made tritium hydrology feasible. That artificially large inventory of tritium has decayed and the levels of tritium concentrations in natural waters are now (in 1987) too small for routine use in hydrology, in both the northern and southern hemispheres.

Carbon-14, with a half-life of about 5600 years, is the other radioactive isotope in addition to tritium, which has proved to be a useful hydrological tracer. Its half-life is long enough to be appropriate to investigating the residence times of water in reservoirs of humid, sub-humid and semi-arid environments. It is a constituent of atmospheric CO_2 and its source is located, like tritium, in the stratosphere. Fixation of CO_2 by photosynthesis and the subsequent release of CO_2 to the soil atmosphere by respiration of roots and the oxidation of plant humus provides the pathway for incorporation of ^{14}C derived from the atmosphere at large into the soil water, principally as the bicarbonate, HCO_3^-, ion.

Unlike the direct labelling of water with tritium, ^{14}C is present as a constituent of a solute and its velocity may be dissimilar to the velocity of the water. The equilibrium saturation concentrations of carbonate species in soil water and groundwater are likely to be different from place to place, particularly with respect to depth in the soil profile. This could result in precipitation and dissolution that would render interpretation difficult. Transport of the isotope between solid and dissolved phases dilutes the isotope concentration carried with the ionic convection of HCO_3^- in the water flow, but it should be distinguished from isotope exchange effects which are considered to be insignificant (Pearson and Hanshaw, 1970). The initial quantity of ^{14}C present in HCO_3^- derived from modern plant material may be diluted by subsequent further additions of bicarbonate derived from much older mineral sources as soil water infiltrates deeper and deeper below the surface of the ground where the partial pressure of CO_2 steadily increases. Sometimes precipitations may reverse this trend and reliable interpretations of measured ^{14}C concentrations can be achieved only if the situation of interest is sufficiently well known.

The concentration of the stable ^{13}C isotope can be used to correct for the dilution of ^{14}C by carbon from mineral sources of the anion HCO_3^- which contain no ^{14}C. The $\delta^{13}C$ value is defined to be

$$\delta^{13}C = \frac{R_x - R_{std}}{R_{std}} \cdot 1000, \qquad (7.11)$$

where R_x is the ratio of ^{13}C concentration to ^{12}C concentration in the sample of interest and R_{std} is the similar ratio of the ^{13}C standard. The latter is often provided by the carbon atoms in the calcite of a marine limestone, the Pee Dee Formation of South Carolina (Craig, 1957). The delta-value is therefore the relative deviation of the ^{13}C concentration from an accepted standard, which may be the Pee Dee belemnite or, for convenience, a local standard whose relation to the PDB standard has been determined. The delta-value is expressed per mil (per thousand) rather than per cent since it is often less than one per cent. The natural variations of ^{13}C in atmospheric CO_2 and in CO_2 released to the soil atmosphere are related to fractionation that occurs in the process of photosynthesis. The $\delta^{13}C$ value for bicarbonate derived from the oxidation of plant residues and plant respiration lies in the range -27 to -12 per thousand depending upon the nature of the photosynthetic cycle of the dominant species of the plant community (see Rightmire and Hanshaw, 1973 and Sternberg and DeNiro, 1983).

Pearson (1965) considered the problem of making a correction to the measured ^{14}C values and proposed that the multiplying factor should be $1/P$, where

$$P = \frac{\delta^{13}C_x}{-27}, \qquad (7.12)$$

if the appropriate plant detritus value contributing to soil CO_2 is $\delta^{13}C = -27$ per thousand. The corrected activity of ^{14}C determined by radioactive counting should then be $^{14}A_m/P$, where $^{14}A_m$ is the measured ^{14}C activity. By this means the dilution of ^{14}C present in the soil solution as a constituent of HCO_3^- by further dissolution of inorganic carbonates can be allowed for. The actual ^{14}C activity conferred by the HCO_3^- species initially present in the unsaturated zone and incorporated during recharge can therefore be calculated. The assumption that the isotopic species ^{13}C and ^{14}C behave identically in the chemical reactions appears to be justified.

The presence of ^{14}C in soil water and groundwater can then be treated in exactly the same way as the presence of tritium for interpretation of hydrological regimes. Pearson and Swarzenki (1974) gave an interesting case study of soil water and groundwater in the Ewaso Ngiro River Basin of the arid Northeastern Province of Kenya. Using the best estimates of the correction factor P in Equation (7.12) they were able tentatively to date three occurrences of fresh groundwater, enclosed in a more saline, extensive water body elongated along the river drainage way. Catastrophic floods are presumed to have caused the residual fresh groundwaters. They appear to have occurred at intervals of ~ 3000 years. Many occurrences of fresh groundwater bodies, apparently entirely surrounded by more saline groundwater, are reported from the semi-arid zones. It is possible that recurrence intervals of order of 1000s of years of huge rains are the cause of what has sometimes presented an insoluble puzzle in interpretation.

7.2 Stable isotopes

The stable isotopes of hydrogen and oxygen, namely deuterium (D) and ^{18}O, occur in small natural abundances, that are nevertheless extremely important in hydrological studies. There is one atom of deuterium in about 6500 atoms of hydrogen and one atom of ^{18}O in 500 atoms of oxygen. Water molecules can therefore have the formulae $H_2^{16}O$, $H_2^{18}O$, $HD^{16}O$, $D_2^{16}O$ and $D_2^{18}O$, all of which in their pure forms possess slightly different vapour pressures in equilibrium with the liquid. This property gives rise to fractionation of water when it undergoes evaporation. The isotope fractionation factor, α, is defined to be

$$\alpha = p/p'$$

where p is the equilibrium vapour pressure of the light component, in the present case H_2O, and p' is the equilibrium vapour pressure of the heavy component. Table 7.2, taken from Dansgaard's (1965) review of stable isotopes in precipitation shows some data for HDO (α_D) and $H_2^{18}O$ (α_{18}). It can be seen that the fractionation during evaporation or sublimation increases with decrease of the temperature at which the process occurs.

Table 7.2. *Fractionation factors, α, for evapora-tion of water and sublimation of ice*

Temperature (°C)	α_D	α_{18}
40	1.060	1.0074
20	1.079	1.0091
0	1.106	1.0111
−10	1.123	1.0123

The amount of enrichment (or depletion of the heavy isotope in the issuing vapour) can be measured precisely by the isotope mass spectrometer. The delta-value for oxygen-18 or deuterium abundance is

$$\delta_i = \frac{R_i - R_i^{SMOW}}{R_i^{SMOW}} \cdot 1000, \tag{7.13}$$

where R_i^{SMOW} is the isotopic ratio present in the standard (Standard Mean Ocean Water). It has been found that δ_D-values possess a range from about +100‰ to −300‰, in a great variety of naturally occurring waters.

The oxygen-18 and deuterium variations are strongly correlated. Fig. 7.7 shows the data published by Craig (1961) of the delta-values of these two isotopes in many water samples, most of which were rain-waters or snows. The relation $\delta_D = 8\delta^{18}O + 10$ quantifies the correlation very well, when the samples are rain, snow or surface waters of recent meteoric origin. The slope has been observed to be much less than 8 for older soil waters, waters from aquifers and interstitial waters from dry-lake sediments (see below). Craig's results initiated interest in a study of stable

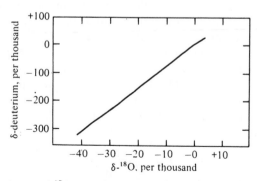

Fig. 7.7. Deuterium and ^{18}O variations in precipitation and surface waters. The large negative values are from the higher latitudes and the values nearer to zero are from lower latitudes or from surface waters. (After Craig, 1961.)

isotopic variations in lakes and rivers for the purpose of hydrological deduction (Gat, 1970).

The site where fractionation actually occurs during evaporation must be at the interface between the liquid and vapour. In the case of a lake, one might expect that a thin layer of heavy water would exist as a skin on the lake's surface. It probably does during very calm periods, but mixing due to waves and overturning at night generally produces a well-mixed water mass. The isotopic composition of this upper mixed layer (epilimnion) is generally uniform, though the actual magnitudes of the delta-values vary from lake to lake in the same region, depending mainly upon the evaporation rate, gross volume of the lake and the net through-flow rate. For example, lakes that possess only a small through-flow relative to the evaporation rate show δ_D values of up to $+20$ per thousand, whereas the local rainfall δ_D-values would be about -30 per thousand. The effects of fractionation are often very substantial.

By contrast with lake processes, when water evaporates from the surface of a wet soil, the liquid water remaining in the pore spaces cannot mix downwards. Zimmermann *et al.* (1967) recognised that it should be possible to observe a build-up of deuterium concentrations in the pore water close to the soil surface. In the steady state, the rate of production of the deuterium-enriched water should just equal the rate of dissipation of that enrichment, such dissipation being achieved both by downwards molecular diffusion in which no actual loss would occur, and by upwards diffusion, and loss across the soil/atmosphere boundary. Subsequently, Allison *et al.* (1983) repeated and elaborated upon the experiment of Zimmerman *et al.* In one of their experiments they used a column filled with sand, which was saturated with water of known δ_D value and maintained fully saturated during 53 days when evaporation proceeded from the upper surface. The results gained from this experiment are shown in Fig. 7.8. It can be seen that the water in the pore spaces of the sand has been enriched in deuterium and at the surface the increase is about 38 per thousand relative to the feed waters entering the bottom of the column whose deuterium composition was -13.6 per thousand. Following Zimmermann *et al*, let us consider the conditions that would eventually produce an equilibrium δ_D-profile in the sand. The upwards convective flux of deuterium in the steady state must just equal the downwards flux of deuterium caused by molecular diffusion. We therefore have the equation

$$\rho E R_i = D' \frac{d(\rho R_i)}{dz}, \qquad (7.14)$$

where E is the evaporation rate ($m\,s^{-1}$), ρ is the spatial density of the water in the sand column ($kg\,m^{-3}$) and D' is an appropriate diffusivity

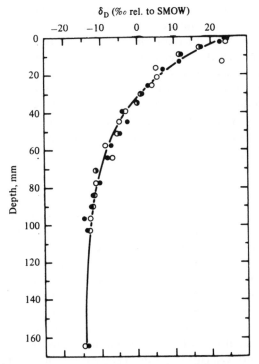

Fig. 7.8. The dependence of delta-deuterium values of the pore water upon depth in a sand column allowed to evaporate into the air. ● δ_D; ○ δ_D derived from the correlation $\delta_D = 4.2 \, \delta_{18} - 3.6$, where δ_{18} is the delta-oxygen-18 enrichment. (Data of Allison *et al.*, 1983.)

$(m^2 s^{-1})$. Assuming that ρ is constant, the solution of Equation (7.14) is

$$R_i - R_i^{res} = (R_i^{sur} - R_i^{res}) \exp(-z/\hat{z}_i) \qquad (7.15)$$

where R_i^{res} and R_i^{sur} are the values of R_i in the incoming feed-water and in the water close to the soil surface. $\hat{z}_i = D'/E$ is the 'mean penetration length' of Zimmerman *et al.*, but referred to as the 'decay length' by Allison *et al.* Through the dependence of $\hat{z}_i$ upon the evaporation rate from the bare soil surface, the fitting of the exponential curve implied by Equation (7.15) to experimental observations of R_i as a function of depth, z, suggests an indirect method of deriving E.

It is more convenient to use delta-values rather than isotopic ratios to express the concentrations of the isotope and Equation (7.15) can be recast as

$$\ln(\delta_D - \delta_D^{res}) = \ln(\delta_D^{sur} - \delta_D^{res}) - \frac{z}{\hat{z}_i}. \qquad (7.16)$$

The diffusion coefficient can be calculated by the relationship $D' = \theta b D^{liq}$, where θ is the volumetric water content, b is the tortuosity factor of the sand and D^{liq} is the diffusion coefficient of HDO species in water, taken to be the self-diffusion coefficient of liquid water. Allison *et al.* adopted a tortuosity factor of 0.66 as used by Penman (1940) and the volumetric water content was 0.36. It did not change with depth in the column, since the water table was maintained at the surface. By suitable fitting of the theoretical curve supplied by Equation (7.15) to the experimental profile, Allison *et al.* obtained a best estimate of $\hat{z}_i$. Then, from the relationship

$$E = \frac{D'}{\hat{z}_i} \qquad (7.17)$$

the evaporation rate E can be derived. They calculated the evaporation rate to be 1.12 ± 0.05 mm day^{-1}. The value measured by the rate of input of water to the bottom of the column was 1.14 mm day^{-1}.

The process of evaporation from the surface of a soil in the field departs from the controlled conditions set up in idealized laboratory experiments, which nevertheless serve as excellent conceptual aids. The three stages of drying, described later (Section 10.4), guided Allison and his co-workers in their experiments. When the soil becomes very dry, the vapour/liquid interface where vaporization occurs retreats downwards from the soil surface. This is the third stage of drying, characterized also by very low rates of evaporation.

Fig. 7.9, also taken from the work of Allison and colleagues, shows a peak in the δ_D values of the interstitial water contained in a relatively dry column of attapulgite granules. The location of the peak at about 40 mm depth below the surface shows the place where most of the vaporization occurred. The thickness of that zone was stated by Gardner and Hanks (1966) and Fritton *et al.* (1967) to be generally about 10 mm. It moves downwards as drying continues.

The procedure to calculate the evaporation rate is still valid for such a profile as shown in Fig. 7.9, if the lower part of the curve is used below the peak. Allison *et al.* calculated the evaporation rate to be 1.01 ± 0.01 mm day^{-1}, which compared well with the measured inflow rate at the water reservoir of 1.10 mm day^{-1}. Such a rate is, in fact, quite large when compared with expected field rates. The explanation is that the attapulgite has a very porous type of granule, the mean water content in the column was 0.70 (volumetric ratio) and the hydraulic conductivity was still large though the column was unsaturated.

The first application of these procedures to the estimation of evaporation in the field was soon reported by Allison and Barnes (1983). Their measurements of the deuterium delta-profiles of the interstitial water in

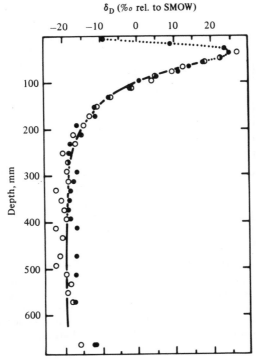

Fig. 7.9. The dependence of delta-deuterium values of the pore water upon depth in a column packed with attapulgite (clay) granules. Note the peak near the surface that indicates the zone of maximum evaporation. Experimental points designated as for Fig. 7.8 but $\delta_D = 2.64\ \delta_{18} - 12.5$. (Data of Allison *et al.*, 1983.)

the sediments of Lake Frome, South Australia (a dry lake) are shown in Fig. 7.10. Using Equations (7.17) and (7.16) they calculated the evaporation rate from the bare lake surface to be 63 ± 8 mm yr^{-1}. Lake Frome is in an arid part of Australia, where the mean annual rainfall is only about 100 mm yr^{-1}. In addition to the water loss from the brine solution in the sediments, whatever rain water falls upon the dry surface also evaporates and that amount should be added to the estimate of 63 mm yr^{-1} supplied from the groundwater source to complete the evaporative flux of this lake's water budget.

Deuterium and oxygen-18 delta-values have been used mainly as tracers in humid climatic zones to identify particular groundwater bodies and to distinguish separate sources of underground water. The quantitative, interpretative analysis of δ_D and δ_{18} profiles in soils is a new application and is likely to enable evaporation rates from soils in the arid zones to be derived with precision for the first time. An excellent

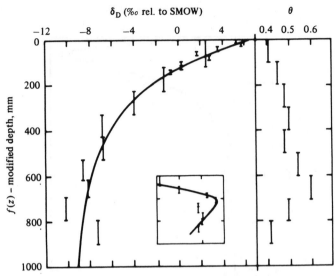

Fig. 7.10. The depth profiles of deuterium delta-values and water content, θ, of lake sediments produced by evaporation from a (water-free) lake surface in the arid zone. The inset shows the deuterium delta-profile close to the surface. The depth is indicated by $f(z)$, which is a modified depth to take account of varying θ with depth. (Data of Allison and Barnes, 1983.)

review entitled *Stable isotope hydrology* was produced by the International Atomic Energy Agency in 1981 (IAEA, 1981).

7.3 Other techniques for tracing water flow

The movement of water through porous media, both through soils in the unsaturated zone and through aquifers generally, is so slow that turbulence is absent and stream line tracing has often been attempted. The field scale is so much larger than the size of laboratory experiments that we may conveniently discuss the latter applicaton and then proceed to a consideration of field studies.

In the laboratory, Gurr *et al.* (1952) used chloride, present in the soil as NaCl, for distinguishing liquid water flow from a parallel and contemporary flux of water vapour in an experiment to study thermo-osmosis. Chloride is a good tracer because its negative charge ensures that it is unaffected by the adsorptive properties of most clay minerals. The exchange of cation species, calcium for sodium etc., on the clay complexes is thought to have a negligible effect upon the velocity of the accompanying anion, which is swept along with the speed of the water stream. It should be noted that Gurr *et al.* found it unnecessary to add chloride to the loam soil of their water redistribution studies. The natural chloride content of the soil was sufficient to provide a tracer when considerable deposition of salt by evaporation within the soil pores caused a piling up of salt at the warmer end of the experimental column.

In the field too, the natural salt content of soils may sometimes be used for tracing the water flow. Chloride is usually the preferred component for analysis of the suite of soil salts. Suppose we have a chloride concentration profile such as that shown in Fig. 7.11. The total chloride content of the soil above some level, often the water table, below which chloride content does not vary vertically is

$$\text{total chloride} = \int_{z=0}^{z} C\theta \, dz, \text{kg m}^{-2}, \qquad (7.18)$$

where θ is the water content, kg m^{-3}, and C is the chloride concentration of the soil water kg kg^{-1}. If there had been no deposition of salt by upward capillary rise, the chloride content would be just

$$C_0 \int_0^z \theta \, dz, \qquad (7.19)$$

where C_0 is the initial chloride concentration of the soil water, assumed to be uniform.

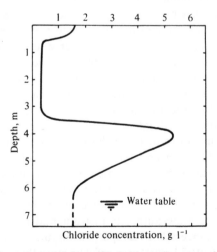

Fig. 7.11. Chloride concentration profile in the soil water of a deep sandy soil, to show the peak near the ground surface caused by evaporation of soil water and the peak at about 4 m caused by root absorption of water and the exclusion of chloride. (After Holmes, 1960.)

Therefore the integrated flux of water flow to deposit the salt by evaporation would be

$$Q = \frac{1}{C_0} \int_0^z C\theta \, dz - \int_0^z \theta \, dz. \qquad (7.20)$$

The idealized chloride concentration profile shown in Fig. 7.11 is meant to illustrate two features of salt distribution in soils in a sub-humid environment. These are that there may be two zones of accumulation. Near the surface the salt accumulation is caused by evaporation of water out of the soil. Near the water table if it happens to occur within reach of plant roots another zone of accumulation of salts may occur that is caused by the absorption of water by the plant roots and the selective rejection of the soil solutes.

Such a simple view ignores other mechanisms for distribution of salt in the soil profile, but in a short time when rain or irrigation percolation has not occurred the integration of the chloride (or other salt tracer) profile can be useful as a guide to the magnitude of vertical soil water flux. Furthermore, the accumulation of chloride in the soil profile, when suitably interpreted by comparison with the provenance of salt fall in the rain, can be an indicator of the integrated evapotranspiration. Anderson (1945) calculated the water harvest from several catchments of southern Australia using the technique of the chloride balance. Peck and Hurle (1973) used a similar tecnique to estimate the amount of underground drainage that enhanced the salt content of streams in semi-arid regions of Western Australia, as already described in Section 6.2.

The use of artificially added tracers in field experiments to study the direction and velocity of flow of soil water and groundwater can be rewarding if the experiment is well controlled. The direction downwards of flow of rain water in unsaturated zones of soils is vertical (see Childs, 1969, p. 334) and a strong refraction of the stream lines occurs at the water table. Therefore the addition of a tracer for study of conditions in the unsaturated zone is redundant for the purpose of tracing direction, though it may be useful to discover the speed of water flow. Below the water table both direction and rate of discharge of water through the aquifer can be indicated by well-chosen injection and sampling sites.

If it is intended to use radioactive materials incorporated into the tracing substance, very strict surveillance of the experiment should be insisted upon by the State authority instrumental in regulating the use of radioactive substances. The isotopes, chromium-51 (half-life 28 days) and cobalt-58 (half-life 72 days) are useful because they can be chemically incorporated into large anions like EDTA complexes that are not absorbed. Bromine-82 (half-life $1\frac{1}{2}$ days) and iodine-131 (half-life 8 days) have been used as anions. Stable tracers that can be detected in water in small concentrations by liquid chromatographic techniques may replace radio-active tracers for some purposes. Bowman (1984) evaluated six tracers and found that three anionic tracers (introduced as organic acids) move with the velocity of the water and could become very useful if the technique is elaborated. They were o-TFMBA (orthotrifluoromethyl benzoic acid), PFBA (pentafluorobenzoic acid) and 2,6-DFBA (2,6-difluorobenzoic acid). Although it is not possible to detect such tracers at the very low concentrations possible for radioactive species, larger amounts of stable tracers can be used without incurring health hazards.

The problems to be overcome in using tracers for soil water and groundwater studies are two-fold. Firstly, actual flow velocities are usually slow, being measured in order of $m\,yr^{-1}$ and, secondly, the direction of flow may be difficult to pick up by any two-dimensional bore-hole grid. A useful technique was suggested to the IAEA panel on the use of isotopes in hydrology (IAEA, 1968), whereby copper rods inserted into a shallow aquifer reacted with the halide-ions, bromide or iodide, by sorbing them. Subsequent radioactive scanning of the rods could reveal, by the distribution of the radioactivity of the sorbed halide ion, the principal direction of flow of the groundwater, by appropriate interpretation of the grid of the rod locations.

Testing and calibration of exotic isotopic or stable tracers can be done in the field, using a second, standard tracer. It is also well-established practice to test new techniques in a soil water or groundwater body that has been well explored by conventional means.

8

Soil structure

8.1 Definition and description

The solid phase of soil consists of particles of various shapes and sizes packed together in various ways, as discussed in Chapter 1. The packing may be close or open, and the particles may behave either as individuals or as clusters in domains and aggregates. The amount of pore space, its continuity, and the size of the pores vary in a complementary way. These are all aspects of the structure of soil which is defined as the arrangement of the solid particles and of the pore space located between them (Marshall, 1962a).

The agents of soil formation discussed in Chapter 1 all influence the structure of soil. The type of parent material affects the primary particles of the solid phase; swelling and shrinking help to rearrange them; soil animals and plant roots do so too; and chemical and biological processes mobilize and deposit materials that hold them together as aggregates. The resulting structure and especially the size, shape and arrangement of the aggregates that can be separated along cracks and natural surfaces of weakness are noted as basic characteristics by soil morphologists. They are consequences of the development of the profile over a long period. But in many ways structure is quite ephemeral since it changes rapidly under common management practices. Soil can, for example, be made less dense by tillage or more dense by compaction, and the fine aggregates of a prepared seed-bed may collapse when flooded with water (Fig. 8.1).

Structure directly affects many of the properties of soil. Water retention and conductance are dependent on pore space and pore sizes, as discussed in Chapters 2 and 4. It influences tillage operations because the properties of individual particles are more or less masked in stable aggregates which can thus give a favourable physical condition to soil that would otherwise be intractable. It affects the environment for roots through its effects on water and oxygen supply and soil strength. Growth of plants can be severely retarded or wholly prevented by structure that is grossly

(a)   (b)

Fig. 8.1. Effect of wetting and drying on structure. (a) Slaking and crusting of an unstable soil. (b) Persistence of original structure in a duplicate sample treated with a soil conditioner to stabilize the aggregates. The containers are 10 cm in diameter (D. S. McIntyre, CSIRO, Adelaide).

unfavourable to water or air movement or resistant to seedling emergence or root growth. A poorly aggregated soil confines the range of water content within which plants grow satisfactorily. They are unduly limited on the wet side by poor aeration and on the dry side by restricted root exploration. Hence defects in structure that are not ordinarily serious may nevertheless reduce yields under adverse seasonal conditions of excess or deficient rainfall (Low, 1973; Letey, 1985).

8.2 Aggregation of particles

Structure as we have defined it includes the size, shape and arrangement of the aggregates formed when primary particles are clustered together into larger separable units. Standardized terms are used in describing natural aggregates (called 'peds') in the field for purposes of soil classification. Aggregates are of highly irregular shape and size but they are nevertheless distinguished according to whether they have axes that are roughly equal (blocky structure), a short vertical axis and a horizontal arrangement of aggregates (platy structure) or a long vertical axis (prismatic structure). Within each of these three classes aggregates are further distinguished according to the distinctness and morphology of the faces and according to dimensions. Blocky aggregates that are small and rounded (<10 mm diameter), as is often the case in surface soil of grassland, are described as having granular or (when particularly porous) crumb structure. Large prismatic aggregates with rounded tops (as found in the subsoil of solodized–solonetz soils) are described as having columnar structure. Sands of single-grain structure and clays of massive structure are described as apedal.

Natural shapes and sizes may be greatly modified under agricultural management. Our interest in aggregates is therefore not only in the geometrical arrangement of the particles but also in the stability of that arrangement. A model proposed by Emerson (1959) provides a basis for considering how the particles are held together so that they behave as a unit (Fig. 8.2). Clay is here arranged in domains or packets of oriented crystals, as described in Chapter 1, and these domains are attached to one another and to the larger particles of sand and silt by bonds as indicated in the diagram. The link may be electrostatic as between the positive edge charges on one domain and the negatively charged face of another at D or it may be through the medium of organic matter as at A, B and C.

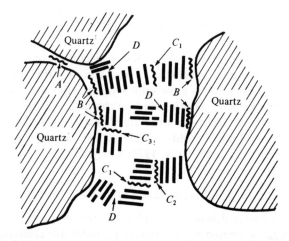

Fig. 8.2. Possible arrangements of quartz, clay domains, and organic matter in an aggregate (Emerson, 1959). Type of bond: A, quartz–organic matter–quartz; B, quartz–organic matter–domain; C, domain–organic matter–domain (C_1, face–face; C_2, edge–face; C_3, edge–edge); D, domain edge–domain face.

8.3 Biological agents in aggregation

The roots of grasses are largely responsible for the stable soil crumbs associated with old pastures. Their fibrous roots press soil particles together as they expand during growth (Fig. 8.3). They ramify the soil thoroughly separating some parts and compressing others, drying it, and incorporating organic matter within it. Fungal hyphae entangle the particles too and help hold them together. Further, some of the products from

Fig. 8.3. Compression of soil by a root as shown by the impressions made by root cells (in parallel bands) on the wall of a root channel. Microbial filaments and oriented clay plates are also shown. (Bryan Smith, CSIRO, Adelaide, using a replica technique with transmission electron microscope.)

the decomposition of roots and other organic matter by micro-organisms act as binding agents. These include polysaccharides whose long, linear, flexible molecules link particles by bridging gaps between them (Greenland, 1965; Theng, 1982). Polysaccharides are ordinarily decomposed readily by micro-organisms but they can nevertheless constitute a considerable part of the soil organic matter. Features of the root–soil interface as shown by electron microscopy are described by Jenny and Grossenbacher (1963) and Foster, Rovira and Cock (1983) and an account of the role of micro-organisms in soil aggregation is given by Lynch and Bragg (1985).

The way in which organic matter helps in forming the basic units of micro-aggregation from which larger aggregates can be built up by the mechanisms described above is not well understood. Fig. 8.2 gives one hypothetical approach to the arrangement and bonding of particles. Edwards and Bremner (1967) concluded from experiments using sonic vibration to disperse and then reaggregate soil that the basic units are micro-aggregates smaller than 250 μm in diameter consisting of complexes of clay and organic matter linked by polyvalent metals (e.g. Ca, Mg, Fe). Humic substances produced by microbial decomposition of organic matter have been shown by Chaney and Swift (1986) to be capable of giving the necessary stability to such units when adsorbed by soil. The basic units are presumed to have a degree of permanence because of tight bonding and inaccessibility of the organic matter to microbial attack. Certainly much of the organic matter in soils has persisted for a long time according to radiocarbon dating (Scharpenseel, 1973). The mean residence time for ^{14}C in the surface soil (A_p horizon) of chernozemic (black) and podzolic (grey-wooded) soils in Canada was found by Campbell *et al.* (1967) to be 870 ± 50 and 250 ± 60 years, respectively. On fractionating the organic matter of the chernozemic soils chemically, they found a mean residence time of 1400 years in the oldest and 25 years in the youngest of the fractions separated. Part of the organic matter has a fairly rapid turnover in surface soils and so provides nitrogen and other nutrients for plants. But a large part of it is stable and protected within pores too small (<1 μm diameter) for micro-organisms to have access to it. Evidence of this is given by Rovira and Greacen (1957), who found that when aggregates were broken up the activity of micro-organisms increased greatly because organic matter became exposed to their attack.

8.4 Physical agents in aggregation

The cracks and surfaces of weakness that separate natural aggregates are probably in a large part due to movements of soil in shrinking and swelling. The joints where shrinkage cracks develop are particularly subject to distortion and, as demonstrated by Weymouth and Williamson (1953), any movement in moist clay causes preferred orientation of clay particles along shear joints. The resulting orientation at crack surfaces no doubt helps these to persist as surfaces of weakness. Further, water flowing between separated surfaces can deposit clay and other materials carried in suspension to form skins of oriented clays. Buol and Hole (1959) described such skins in a number of subsoil horizons in Wisconsin, USA, as being about 0.08 mm thick and having a higher proportion of clay, iron oxide and organic matter than the rest of the soil. An example is given in Table 8.1 from the subsoil of a grey-brown podzolic profile. Surfaces modified by stress or by deposition of clay are called cutans by Brewer (1964), and orientation of clay in them can be detected with the polarizing microscope. In undisturbed soils, cutans provide zones where naturally occurring aggregates can be separated from one another.

Croney and Coleman (1954) demonstrated the effect that the degree of drying has on the structure of a slurried clay soil. In Fig. 8.4, from their data, clay dried from a slurry (curve A) did not recover its original water content when re-wet to its original suction (B). The higher the

Table 8.1. *Properties of clay skins* (*Buol and Hole, 1959*)

Property	Clay skin	Whole soil material
Clay content (<2 μm), %	87	24
Free iron (Fe$_2$O$_3$), %	3.95	1.90
Organic matter (carbon content × 1.72), %	3.09	0.71

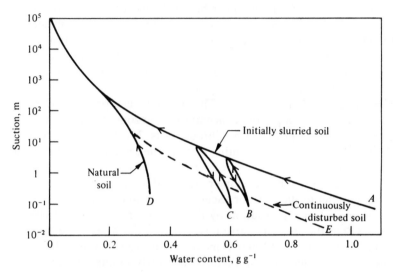

Fig. 8.4. Dependence of moisture characteristic on structure as affected by prior degree of drying (Croney and Coleman, 1954).

suction to which it was subjected in drying, the greater was this effect (compare *B* and *C*). It can be seen also that when dried still more thoroughly it recovered the structure of the natural unslurried soil (*D*). Thorough drying has the effect of consolidating aggregates of clayey material to a higher density than they would reach by any external pressure likely to be experienced by soils under natural conditions and the resulting cracks between them may remain effective for water transmission for some months after re-wetting, as noted in a review by Marshall (1962*a*).

The pattern of cracks at the surface of a uniform drying clay appears in Fig. 8.1*a* as a set of polygons of roughly similar size and shape. This type of pattern on various size scales is to be expected according to Kopp (1966) and under ideal conditions it would take the form of hexagonal shapes for which the cracks meet at an angle of 120° to one another, as described for homogeneous soil by Hénin *et al.* (1960, p. 166). Using the Griffith crack theory (Section 9.7) as a basis for studying the cracking of cement-stabilized soil, George (1970) has shown that a uniform brittle material will tend to crack into hexagonal units on shrinking. Subsequent cracks will occur at right angles to existing cracks but seldom crossing them because the flaw from which the new crack is propagated allows release of strain energy only on the side where it is located.

For a regular polygonal pattern to form, the soil needs to contract uniformly towards equally spaced centres in a manner similar to the cooling of lava to form columnar basalt. However, uneven distribution of flaws, plant roots, and water content often prevents this. Johnston and Hill (1945) and Fox (1964) observed that major cracks run midway between the rows of row-crops. Although fallowed ground or a uniform plant cover allows polygonal cracking to occur, as noted by Johnston and Hill, according to Fox the major cracks tend to go around rather than through individual plants. The major cracks presumably start to develop away from high root concentrations in zones where the soil lacks the reinforcement provided by fibrous roots. Near the plants, cracking occurs on a smaller scale resulting often in the separation of the soil into granular aggregates enmeshed within the root system. The local pattern may also be affected by the slope of the land which can cause additional cracks parallel to the contour (Scott, Webster and Nortcliff, 1986).

Freezing affects aggregation by the expansion of water on changing to ice within the soil. In coarse-textured soil, the water freezes in place but in fine-textured soils there is movement of water towards the freezing sites so that ice lenses form there. With slowly falling temperature, the liquid water is withdrawn from areas where it has not readily frozen because of the suction at which it is held by clay surfaces. Compression and heaving of the soil occur as the ice lenses grow. The process is described by Yong and Warkentin (1975) and by Miller (1980). Compression of the somewhat dried soil near the lenses helps aggregates to form in fine-textured soils. But Kay, Grant and Groenvelt (1985) found that, although the soil surface may be displaced several centimetres, the pores that are created apparently collapse after melting and draining have occurred so that there is no reduction in bulk density of soil under zero tillage. Repeated freezing and thawing in cracks can cause crumbling of cloddy soil according to Vershinin *et al.* (1959). Cracks enlarged by expansion on freezing can take in more water when the ice has thawed and then open further on freezing again. Richardson (1976) showed that three freeze/thaw–wet/dry cycles on a sandy loam that he had puddled severely gave it better structural stability than that of the untreated soil. But the effects of freezing and thawing can be unfavourable too (Hénin *et al.*, 1960, p. 100) especially if rain or poor drainage causes puddling on thawing. Czeratski (1971) distinguished three conditions of frost structure: no frost structure (in sandy soils), laminar frost structure which may slake on thawing, and polyhedral which occurs in swelling soils and can be stable on thawing. He outlined soil treatment to protect laminar frost structure and to make use of polyhedral frost structure, as for example in relation to minimum tillage.

In agriculture, when soil is disturbed periodically under tillage, it is fractured into clods and fragments as well as having its natural aggregates separated. It is pulverized to a range of aggregates sizes that varies with the purpose of the tillage operation. A fine seed-bed of small aggregates of about 0.5 to 5 mm diameter, similar to those of Fig. 8.1b, favours germination and plant growth (Dobrzanski, Witkovska and Walezak, 1975). Schneider and Gupta (1985) found that a geometric mean diameter between 1.0 and 6.8 mm was suitable for corn. On the other hand ploughed land left in fallow withstands erosive damage from wind or water better if left in a coarse cloddy condition.

The size distribution of tilled soil can be measured by sieving but, since aggregates are often fragile, gentle handling is necessary. A rotary sieve developed by Chepil (1962) is satisfactory for this purpose. The size distribution of dry aggregates of a tilled soil subject to wind erosion is illustrated in Table 8.2 where material collected from wind drifts is compared with the soil that remained in position. Some of the finest material, carried off from the area altogether as dust, was not collected. The soil is a silt loam and is one of four black soils of Kansas, USA, examined by Chepil (1953).

Table 8.2. *Size distribution of dry aggregates of a tilled soil subject to wind erosion* (*Chepil, 1953*)

Material	Size fractions, diameter in mm			
	>6	6–0.8	0.8–0.4	<0.4
Residual soil, %	25	14	20	41
Drifted material, %	0	1	12	87

8.5 Pore space

The fraction of the bulk volume of soil that is occupied by water and air ranges in the selected porosity data of Table 1.1 from about 0.3 to 0.6. It is high in a wet clay whose particles are separated by water, and low in a soil composed of particles of a range of sizes that pack closely together. The pore space exists because of inevitable gaps in the packing of particles, as discussed in Chapter 1, and because of disturbances including those due to roots, soil animals, swelling, cracking on shrinking, and tillage that alter the spacing of aggregates or particles. It can be regarded as a largely interconnected system of pores of various sizes and as such it allows soil to act as a medium for the movement of water and air. Pores are also referred to as 'voids' but this use of a word meaning 'empty spaces' seems incongruous when we are so much concerned with the water they contain.

Pores in soils range in size over several orders of magnitude, as is illustrated by examples in Table 8.3. The distribution of pore space according to size of pore is usually measured from the curve relating water content to suction (moisture characteristic). To do this an effective size of pore can be calculated from the suction using Equation (2.10):

$$p = -\rho g s = -(2\gamma \cos \alpha)/r. \tag{8.1}$$

Here p is the pressure of the water, γ is the surface tension of water ($72.75 \, \text{mJ m}^{-2}$ at $20 \, ^\circ\text{C}$, as in Table 2.2), r is the tube or pore radius, ρ

Table 8.3. *Examples of pores of different sizes ranging over several orders of magnitude. Size is expressed as an equivalent radius unless otherwise stated*

Size, m	Example
10^{-2}	Spaces as large as this are commonly formed between the clods of newly ploughed soils. Cracks in dry clay soils can reach widths of this order of size.
10^{-3} (1 mm)	Pores of about this size and smaller are formed between aggregates of finely tilled soil as for a seed-bed (see Fig. 8.1b).
10^{-4}	Pores between spherical particles 0.65 mm in diameter in closest packing have this size (Dallavalle, 1948, p. 131). Roots will not extend into rigid pores smaller than this (Wiersum, 1957).
10^{-5}	Pores larger than about 15 μm (corresponding to 1 m suction) are drained in most soils that can be said to be at field capacity.
10^{-6} (1 μm)	Pores down to this size are accessible to bacteria.
10^{-7}	Water in pores of about this size or larger is available to plants in non-saline soil. (Corresponds to 1.5 MPa, 150 m or 15 bars suction.)
10^{-8}	When micropores are treated as slits between parallel plates, about half the pore space in dried aggregates of clay soil can commonly be attributed to plate separations of 10 nm or less (Sills, Aylmore and Quirk, 1974).
10^{-9} (1 nm)	This is about the thickness of 3 layers of water molecules on a clay surface.

is the density of water (0.9982 Mg m^{-3} at 20 °C as in Table 1.7), g is the acceleration due to gravity (9.80 m s^{-2}), s is the suction and α is the contact angle which is here assumed to be zero. As an example we may calculate the radius of pores corresponding to a suction, s, in metres, using the values given above:

$$998.2 \times 9.80 \times s = 2 \times 72.75 \times 10^{-3} \times r^{-1} \quad \text{or} \quad r = 1.5 \times 10^{-5} s^{-1} \text{ (metres)}.$$

Thus a suction of 1 m will withdraw water from pores with an effective radius larger than 15 μm. On this basis, a moisture characteristic curve can be used to show the amount of pore space (as given by the water content on a volume basis) that has pores smaller than a given effective size as in Fig. 8.5. In applying this method of determining the size distribution of pores, it is necessary to recall the following conditions on its use as discussed in Chapter 2. (1) The soil should be rigid and not change its volume with water content. (2) The pores are not tubes of circular section so that an effective rather than an actual size is measured. (3) A drying rather than a wetting soil is used in the measurements because α is then more likely to be zero. However, a drying curve is influenced more by the size of openings than by the size of the pore.

This method is mainly used over the range covered by suction and pressure-membrane procedures (approximately 10^{-3} to 10^{-7} m radius) but if necessary the range can be extended to smaller pore sizes by using Equation (2.6) or (3.21) in the form

$$2\gamma/r = -\rho RTM^{-1} \ln{(e/e_0)}. \tag{8.2}$$

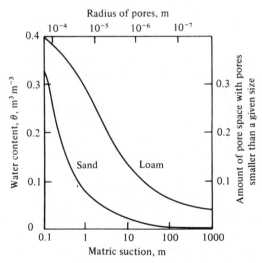

Fig. 8.5. Size distribution of pores as given by the moisture characteristics of two soils.

Soil structure

R is the gas constant, T is the temperature in kelvin, M is the mass per mole of water and e/e_0 is the relative vapour pressure of the soil water. Using the values given in examples calculated for Equations (2.6) and (8.1), it can be seen that, at 20 °C,

$$1/r = -9.28 \times 10^8 \times \ln (e/e_0) \text{ metre}^{-1}$$

and hence, when $e/e_0 = 0.85$ for example, $r = 6.6 \times 10^{-9} = 6.6$ nm. Equation (8.2) is not applicable if the vapour pressure, e, is seriously affected by the solute content of the water in the soil. The conditions on the use of Equation (8.1) apply here also.

Liquids other than water can be used for determining the size distribution of pores in dry soil. In particular, mercury is used for this purpose in porous rock samples for pore sizes ranging from 100 μm to 10 nm and is being increasingly used for other materials. The method depends on measuring the amount entering progressively smaller pores of the dry material as the pressure, p, of the mercury is increased. Equation (8.1) is applied as before, with γ taking a value of 480 mJ m^{-2} for the surface tension of mercury and α a value of 140° for its contact angle, but both these quantities are subject to some uncertainty. Error can arise especially from departure of the contact angle from its assumed value. Because mercury does not wet soil, positive pressure has to be exerted to inject it as can be seen from Equation (8.1) in which cos 140° is negative. Mercury may not be as satisfactory as water for the pore-filling liquid for soil studies relating to water. However, Nagpal, Boersma and De Backer (1972) found the mercury injection method more rapid than the moisture characteristic method and obtained satisfactory agreement between the two for pore sizes down to a radius of 100 nm in sandy soils. Size distribution curves for soils of medium texture had a similar shape by the two methods but were displaced from one another indicating that bulk density may have been affected by mercury pressure. A correction for this was applied.

An advantage of mercury is that a soil of fine texture is not subject to volume change such as occurs as shrinkage during water withdrawal. Other methods that avoid the use of water are desorption of nitrogen or a non-polar liquid and microscopic examination of thin sections. However, the problem of volume change is not wholly overcome in these methods because samples have to be pre-dried, thus affecting the structure of moist soils of fine texture. Special techniques such as freeze-drying used to minimize this are reviewed by Lawrence (1977).

In clay soils, the finest pores are probably more like slits between parallel plates than the tubes of circular cross-section assumed in applying Equations (8.1) and (8.2). If this is taken to be the case, Sills *et al.* (1974) have shown that about half the pore space in most clay soils can be accounted for in slit-like pores with a thickness between plates of 10 nm or less. They measured the size distribution of pores on this model using modified forms of Equation (8.1) for mercury injection and of Equation (8.2) for adsorption of nitrogen in the pores of dry soil. The modification made to Equation (8.1) for mercury injection is to substitute d, representing the plate separation, for r, in accordance with the geometry of the liquid–gas interface between parallel plates.

Nitrogen sorption is undertaken on dried soil cooled to a low temperature of 78 K at which a liquid–gas interface can be formed for nitrogen. In applying the equation to the desorption of nitrogen, Sills *et al.* allowed for the thickness, t, of the layer of nitrogen adsorbed on the surface of the plates, as shown in Fig. 8.6, which illustrates the parallel plate model as used by Innes (1957). With this model, $d - 2t$ is substituted for r in Equation (8.2). The thickness may be determined with the aid of Equation (1.15) or from measurements made on coarse particles of comparable material in which plate separation is unimportant. The use of an adsorbed film thickness to correct the radius of curvature in the equation is a simple device to allow for the effect of surface adsorption by the solid on the liquid–gas interface. Philip (1977) has instead treated the interface as a surface of uniform matric potential resulting from combined adsorption and capillary force fields, and he concluded that the correction by means of t was inadequate.

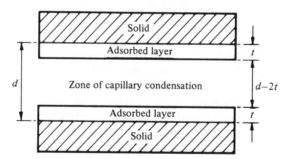

Fig. 8.6. Parallel plate model for estimating plate separation, d, from nitrogen adsorption (Innes, 1957).

The microscopic methods used for examining soil particles are also used for pores. The optical microscope has a resolving power of about 250 nm, the scanning electron microscope 10 nm and the transmission electron microscope 1 nm, according to Stoops (1973), who summarizes the principles and uses of optical and electron microscopy as applied to soils. These instruments provide details about the size, shape and arrangement of particles and pores. Terms covering all types of pores have been

Soil structure

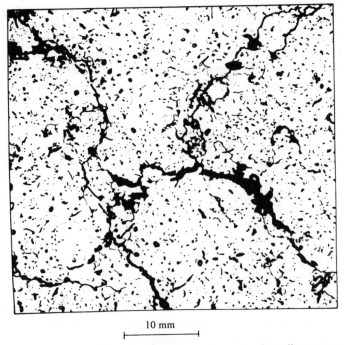

10 mm

Fig. 8.7. Pores (in black) in a cross-section of a soil of medium texture at a depth of 10 cm. Much of the pore space is in pores too small to be represented (Lafeber, 1965).

defined by Brewer (1964, Chapter 9) in order to help in the identification of these features. Because of the great complexity of soil structure, quantitative measurements of the spatial distribution of pore space by these means are laborious. Consequently size distribution is best measured by the physical methods we have outlined. However, microscopy provides details about shape and arrangement that are not given by those methods. The relatively large pores shown in Fig. 8.7 from observations made with an optical microscope include slit-like planar pores (cracks) arranged in a polygonal pattern and channels roughly circular in cross-section made by roots. These differ obviously in shape and arrangement from the packing pores found between aggregates in disturbed soil (Fig. 8.1*b*) or between particles in a sandy soil.

8.6 Structure and permeability

In Section 4.2 we showed that on combining Darcy's law with Equation 4.7 for the flow of a fluid through a capillary tube, the permeability of a model porous material made up of parallel tubes of uniform size could be expressed as a function of its porosity, ε, and pore radius, r, as

$$k = \varepsilon r^2/8. \qquad (8.3)$$

Here k is the intrinsic permeability which as shown in Equation (4.3) can be converted to the hydraulic conductivity, K, by the expression

$$K = \rho g k / \eta \qquad (8.4)$$

where ρ is the density of water, η is its dynamic viscosity and g is the acceleration due to gravity.

Equation (8.3) is the starting point for many proposals for relating permeability to porosity and pore size in porous media. For lack of means for measuring pore size, the first attempts were based on particle properties. Slichter (1899) examined the geometry of the pore space between regularly packed spherical particles of uniform size and derived an expression for permeability in terms of porosity and particle size. Kozeny (1927) used the mean hydraulic radius (cross-section divided by wetted perimeter) as given by ε/S where ε is the porosity and S is the surface area of the particles per unit volume of porous material. His equation is

$$k = a^{-1}\varepsilon^3 S^{-2}. \qquad (8.5)$$

The factor a was considered to be a constant with a value of about 5 that combined pore shape and tortuosity factors. Tortuosity due to the sinous path of flow was represented by the ratio of the effective length, l_e, to the apparent length, l, of the path and was given the value $\sqrt{2}$ that would apply for flow at an angle of 45° to the apparent direction (Carman, 1937; 1956). The tortuosity factor for the flow of fluids is $(l_e/l)^2$ as explained by Childs (1969, p. 199). Equation (8.5) was widely used but it fails for materials with a wide range of pore sizes because it has the effect of averaging radii rather than the square of radii as required by Equation (8.3).

With the development of methods described in the previous section for measuring the size distribution of pores from the moisture characteristic, it became possible to measure the effects of size of pores on permeability (Baver, 1938), and equations relating permeability to pore size distribution and porosity were proposed.

Purcell (1949) represented the pore space by parallel tubes of equal length whose circular cross-sections covered the range of pore sizes found in the porous material. The equation he established from this model can be put in the form

$$k = b\varepsilon \int_0^\varepsilon r^2 \, d\theta, \qquad (8.6)$$

where ε is the porosity and r is the radius of pores in a portion $d\theta$ of the pore space as determined from the moisture characteristic of the material (see Fig. 8.5). This type of equation enables k to be estimated for material with diverse sizes of pore and for any degree of saturation. In dealing with unsaturated material, the unfilled pore space is excluded from the estimate and so the water content, θ, replaces ε in Equation (8.6). Since k is difficult to measure in unsaturated media, it would be useful to be able to estimate it in this way. However Purcell found that the factor b varied widely in tests on this equation and changes to remedy this followed as discussed below.

Childs and Collis-George (1950) independently developed an equation for calculating k from the size distribution of a material. In this they allowed for interconnections between pores of differing size by taking the radius of the smaller of two pores in a sequence to be the effective size of that sequence. If the two surfaces exposed by cutting through a porous material are thought to be randomly rejoined, the permeability resulting from the random juxtaposition of the pores of different sizes is given by them as

$$k = M \sum_{\rho=0}^{\rho=R} \sum_{\sigma=0}^{\sigma=R} \sigma^2 f(\rho)\, \delta r f(\sigma)\, \delta r, \qquad (8.7)$$

where σ and ρ are the radii of two pores forming a sequence, σ being the smaller, $f(\sigma)\,\delta r$ and $f(\rho)\,\delta r$ are the fractions of the cross-sectional area occupied by pores of radius r to $r+\delta r$ and M is a matching factor found by experiment. Childs and Collis-George successfully applied their equation to three non-swelling materials at various degrees of saturation with water, using one value for M. For unsaturated materials, the largest radius, R, in the summation is taken to be that of the largest pore that is full of water.

Following the Childs and Collis-George model of the interaction between the pores of two surfaces of a cut and randomly rejoined porous material, Marshall (1957, 1958) obtained an average cross-sectional area of necks for pores in sequence. Unit area of the two surfaces exposed by a cross-section through the material was treated as having n portions each of the same area, $1/n$, porosity ε, and containing pores of one radius $r_1, r_2 \ldots$ or r_n, with $r_i > r_{i+1}$. When one surface, A, was rejoined to the other, B, each portion of A was matched with n portions of area $1/n^2$ in B each containing one only of the pore sizes $r_1, r_2 \ldots$ and r_n. In rejoining, all pore space was assumed to communicate across the junction but the smaller pore size in each sequence was taken to be the size of that sequence. The cross-sectional area of all such sequences was further reduced by a factor ε to allow for the poorer fit between pores as porosity decreases. The average area of cross-section πr_t^2 of necks joining all the

n^2 portions of surface B to A was thus shown to be

$$\pi r_t^2 = \pi \varepsilon n^{-2} \{ r_1^2 + 3r_2^2 + \cdots + (2n-1)r_n^2 \}. \tag{8.8}$$

On substituting r_t^2 for r^2 in Equation (8.3), Marshall obtained the permeability equation

$$k = \tfrac{1}{8} \varepsilon^2 n^{-2} \{ r_1^2 + 3r_2^2 + \cdots + (2n-1)r_n^2 \}. \tag{8.9}$$

In this equation $r_1, r_2, r_3, \ldots$ and r_n represent the mean radius of the pores (in decreasing order of size) in each of n equal fractions, ε/n, of the pore space as determined from the moisture characteristic of the material. For unsaturated material, the water content, θ, is substituted for porosity, ε, and n is the number of fractions of the water-filled pore space. The hydraulic conductivity, K, can be substituted for k through Equation (8.4) and the suction, s, can be substituted for r through Equation (8.1). With these substitutions, Equation (8.9) may be written as

$$K = \{ \gamma^2 \theta^2 / (2\rho g \eta n^2) \} \{ s_1^{-2} + 3s_2^{-2} + \cdots + (2n-1)s_n^{-2} \}. \tag{8.10}$$

where γ is the surface tension, ρ is the density and η is the viscosity of water. In this equation we have the condition $s_i < s_{i+1}$ and s has the dimension of length as is usual for the moisture characteristic. It may be noted that matric suction is called capillary pressure in petroleum technology and chemical engineering where much work has been done on permeability.

Many modifications have been proposed for these equations. Tortuosity was introduced into those related to Equation (8.6). It was no longer treated as a constant but as a porosity-dependent factor or it was determined from the electrical conductivity (Burdine, 1953; Wyllie and Spangler, 1952). The main changes affecting Equation (8.9) were also related to porosity (Millington and Quirk, 1959, 1961; Mualem, 1976).

Comparisons between measured and estimated permeability using Equation (8.9) were made on sands and porous rocks over a range of k that covered several orders of magnitude (Marshall 1958, 1962b). This was done without help from any matching factors. The correspondence was good enough to show that porosity and pore size as represented in Equation (8.3) do provide a basis for permeability estimates in such materials. However in practice absolute estimates are too unreliable to be useful and a matching factor is needed. If a measured and an estimated value are matched at one water content, values can be estimated more reliably at other water contents when adjusted by the ratio between the matched pair. Matching is usually done at saturation where measurement is easiest. Hence instead of representing K in absolute terms as we have so far done in the equations, it is usual to set them up for the relative hydraulic conductivity, K_r. This is the ratio of the hydraulic conductivity, K, of the unsaturated material to that, K_{sat}, at saturation as given by the model.

The equations that we have discussed so far use tabular data that have been obtained directly from the moisture characteristic as the basis for calculating hydraulic conductivity. Instead it is possible to set up an arbitrary equation to represent the moisture characteristic with parameters obtained by curve fitting as discussed in Section 2.1. For this purpose Brooks and Corey (1966) adopted the equation

$$S_e = (P_b/s)^\lambda \quad (s \geqslant P_b) \tag{8.11}$$

in which s is the suction, P_b is the bubbling pressure (the suction at which air enters the saturated soil), S_e is the effective saturation given by $(S - S_r)/(1 - S_r)$ where S is the degree of saturation (as defined in our Equation (1.10) and S_r is the residual saturation (the degree of saturation at which the water is 'immobile'). The parameters λ and P_b are determined by plotting log S_e against log s, and S_r is chosen arbitrarily so as to give the best fit of the data to a straight line. On combining Equation (8.11) with a permeability equation of Burdine (1953), Brooks and Corey derived their expressions for relative hydraulic conductivity,

$$K_r = K/K_{sat} = S_e^{(2+3\lambda)/\lambda} \quad \text{or} \quad K_r = (P_b/s)^{2+3\lambda} \quad (s \geqslant P_b). \tag{8.12}$$

Another equation, used by van Genuchten (1980) to represent the moisture characteristic, is

$$\theta = \theta_r + (\theta_{sat} - \theta_r)/[1 + (\alpha s)^n]^m. \tag{8.13}$$

Here θ_{sat} is the saturated water content, s is suction, θ_r, α and n are parameters to be determined, and $m = 1 - 1/n$. The residual water content, θ_r, can be given some arbitrary low value but is better obtained by curve fitting as are α and n. On combining this with the hydraulic conductivity equation of Mualem (1976), van Genuchten obtained the following equation for relative hydraulic conductivity,

$$K_r = [1 - (\alpha s)^{n-1}\{1 + (\alpha s)^n\}^{-m}]^2/[1 + (\alpha s)^n]^{m/2}. \tag{8.14}$$

Equations of the type represented by (8.13) and (8.14) have the advantage over those that use tabular data that they can be applied directly in the mathematical analysis of hydraulic problems.

These and other formulae for estimating the hydraulic conductivity of unsaturated soils have been reviewed by Mualem (1986).

The use of these equations for soils poses some difficulties. Errors inherent in the models and in the moisture characteristic can be serious. In particular the medium is assumed to be isotropic with the pore space randomly distributed so that there are no continuous channels. These conditions do not hold in soils. These obstacles are not necessarily overcome by matching at saturation because cracks and channels not allowed for in an equation may contribute greatly to the actual conductivity at saturation but not at unsaturation. However in suitable soils and with matching done at some degree of unsaturation when necessary, fairly good estimates of K_r can be obtained for unsaturated soil according to tests by Field, Parker and Powell (1984), Schindler, Bohne and Sauerbrey (1985), Alexander and Skaggs (1986) and Rab, Willatt and Olsson (1987) and by others cited by them.

The question of how to deal with flow through soils with cracks and channels has received some attention. Routes of water flow can be identified by using tracers like methylene blue or by depositing gypsum in channels as has been done by Bouma, Belmans and Dekker (1982). A problem with a dye is that it can stain a surface although the pore has not been filled with water. An interesting study of the contribution of the larger pores to flow in a swelling soil has been made by McIntyre and Sleeman (1982). They found that the part of the profile having the greatest hydraulic conductivity also had the greatest amount of pore space as channels and interconnected vugs as shown by microscopy. For the whole profile, with a range of hydraulic conductivity from 1 to 8 mm/day near saturation, they found the conductivity to be exponentially related to the amount of water contained in the larger pores, mainly smaller than 0.5 mm in diameter, where it did not contribute to swelling.

Bouma (1986) has shown how morphology as studied in a soil survey can help in selecting the location and size of a representative test area to encompass a crack pattern. Where this is not practicable, the two parts of the conducting system, the macropores and the micropores, can be examined separately. For example Wang and Narasimhan (1985), in a study of a fractured rock, obtained moisture characteristic data for cores of unfractured material to which an equation of van Genuchten (1980) could be applied for estimating unsaturated hydraulic conductivity. From this they worked towards a prediction of flow through the fractured rock as a whole at various levels of unsaturation. Treatment of the material in this way necessarily involves conjecture about shape, continuity, and degree of saturation of the cracks and channels. Other ways of handling similar problems of flow through soils with cracks and channels are discussed in a review by Bevan and Germann (1982).

8.7 Stability of structure of wet soil

The stability of aggregates and pores decreases on wetting dry soil particularly if this is done rapidly. There are a number of reasons for this. (1) The strength of soil decreases with water content, as will be discussed in the next chapter, because of reduced cohesion and the softening of cements. (2) If macroscopic swelling occurs during adsorption, this will cause uneven strains throughout an aggregate so that its structure will be distorted and weakened. This effect of uneven wetting will be greatest when dry clay soil wets rapidly (Panabokke and Quirk, 1956). (3) Rapid wetting of dry soil can also cause damage by air trapped by water that fills the outside pores of aggregates before advancing inwards. If air, compressed by this advance, reaches a pressure greater than the tensile strength of the soil, it escapes explosively, breaking off fragments in doing so, as can be observed when dry aggregates are immersed suddenly in water. This causes slaking of many soils (Yoder, 1936; Hénin and Santamaria, 1975), particularly those of low or medium clay content.

The pressure, p_a, of the trapped air in a cylindrical pore of radius, r, when the water on the other side of the meniscus is at pressure p, is given by

$$p_a = 2\gamma/r + p. \qquad (8.15)$$

This follows from Equation (2.8), if the radius of curvature is assumed to be equal in magnitude to the radius of the pore. Bolt and Koenigs (1972) suggest that it is the relatively large pores that are likely to explode. These fill first but air is probably forced back into them as smaller pores fill. For aggregates immersed in water, p can be taken as zero and, according to actual measurements discussed by Bolt and Koenigs, p_a may reach pressures of the order of 100 mbar or 10 kPa in suitable cases. But if they are wetted under suction (as on a suction plate), p is negative and p_a may not then be of consequence. In any case the largest outside pores may not fill when water is under suction and the air can then escape during wetting. Under field conditions, surface soil can wet rapidly with p near zero but, in subsoils, wetting is usually slower and under a suction. Hence explosive damage is most serious at the surface.

Because of these effects of water, wet aggregates are generally fragile. Bare, tilled soil exposed to rain is especially subject to structural damage from rapid wetting and the impact of raindrops on the softened soil. If it slakes, the loss of large pores and aggregates causes reduced hydraulic conductivity and, on drying, crusting of the soil surface (Fig. 8.1a). The consequences in reduced water infiltration, increased run-off and erosion, and reduced emergence of seedlings can be great. Special attention is therefore paid to the stability of structure when the soil is wet.

Methods for finding how stable aggregates are in water are mostly variations of a wet sieving method introduced by A. F. Tiulin in 1928 and modified by Yoder and others. References to these methods are listed by Kemper and Koch (1966). In a method described by Kemper and Rosenau (1986) aggregates of a selected size (1 to 2 mm diameter) are put onto a sieve with openings of 0.26 mm which is moved up and down mechanically in water. The dry weight of soil remaining on the sieve, expressed as a percentage of the original dry weight, is reported as aggregate stability. A deduction is made from both weights for sand grains too large to go through the sieve. Wet sieving is widely used for comparing effects of soil treatments and cultural practices on aggregate stability. To obtain reproducible results the method has to be rigorously standardized particularly as to sample preparation, wetting procedure, and the handling and treating of wet aggregates. Slow wetting by capillarity from water under a suction is commonly used to allow air to escape freely and so avoid explosive damage to weak aggregates. The percentage of stable aggregates is usually much lower when the samples are immersed rapidly. This is shown in a comparison of wetting methods undertaken by Kemper and Koch (1966) and summarized in Table 8.4. Kemper and Koch favoured vacuum wetting for good reproducibility and found this to be highly correlated with capillary wetting. They evacuated their samples over water, thus allowing them to adsorb water vapour before being immersed. Soils that collapsed on immersion at atmospheric pressure were excluded by them from this comparison.

Table 8.4. *Effect of method of wetting on the percentage of aggregates stable in water (after Kemper and Koch, 1966)*

Soils	Wetting rapidly by immersion	Wetting slowly from water under suction	Wetting while evacuated
Average of 16 soils from western USA	41	74	70
Average of 4 stable soils from south eastern USA	83	98	99

Other ways of measuring and reporting stability will be mentioned briefly. A sample of the whole soil rather than of aggregates of a selected size can be used. If a nest of sieves is used, a mean size can be determined for water-stable aggregates as in the method set out by van Bavel (1950). Middleton (1930) shook soil in distilled water in a sedimentation cylinder end-over-end 20 times and then determined from the settling rate, the amount of material smaller than 50 μm. The ratio of this to the amount determined by thorough dispersion in a particle size analysis was called the dispersion ratio. McCalla (1944) measured the number of falling water droplets required to cause aggregates to disintegrate. Emerson

(1954) measured the hydraulic conductivity when successively more dilute solutions of sodium chloride were passed through a bed of aggregates. These swell under the influence of the sodium ion and the test of stability is to find how far the concentration can be reduced before flow is seriously impeded by their collapse.

Methods can be selected to suit any particular problem and they can be adjusted in their severity to suit the degree of stability in the particular soils being examined. For most purposes it is sufficient to report results for a single size, as in Table 8.4. More complex methods are scarcely justified because of the arbitrary nature of them all. In a novel departure from these quantitative methods, Emerson (1967, 1983) introduced a classification scheme for water stability of aggregates based on observation of the slaking, swelling, or dispersion of natural and remoulded aggregates when immersed in water. The tests are designed as a simple means for finding the likely behaviour of soil under tillage, in erosion, in earth dams, and in subsoil drainage works, without having to make detailed analysis to find the amount and kind of clay, exchangeable cations and organic matter, all of which affect stability. Greenland, Rimmer and Payne (1975) examined 180 soils by this method and found that most could be readily classified by these tests and that the eight broad classes appeared to be related to behaviour under stress in the field.

8.8 Effects of soil constituents on stability

Relations between the stability of wetted aggregates and some constituents of soil are illustrated in Fig. 8.8 by data of Kemper and Koch (1966) for 500 soils from the western part of the USA and Canada. There is a strong influence of clay content (Fig. 8.8a), since stability arises internally from bonds between clay plates, packets of clay plates (domains), and other particles (Fig. 8.2). However, the presence of clay particles does not in itself ensure stability. Sodium in the exchange complex of the clay can make it quite unstable (Fig. 8.8f). Also if the clay crystals are arranged in an open structure, as described by Smart (1975), they may collapse under shear, causing land slides in a so-called sensitive clay. At the other extreme there are unusually stable clays which possibly owe this stability to crystal intergrowths. These will not disperse completely with the chemical and mechanical treatments customarily given in dispersing soil for particle size analysis (Norrish and Tiller, 1976).

The role of organic matter in stability ranges, as we have seen, from entanglement by fungal hyphae and roots to binding by the decomposition products and secretions of roots, micro-organisms and soil animals. Stability increases greatly with content of organic matter in the soils represented by the curve of Fig. 8.8b, especially among those with less

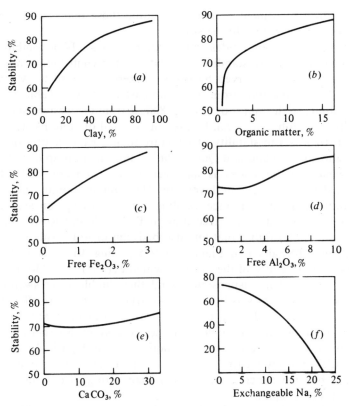

Fig. 8.8. Relation of aggregate stability to various constituents of soils from the arid, semi-arid and sub-humid regions of western USA and Canada. Soils with exchangeable sodium percentage >4 are excluded from (*d*) (Kemper and Koch, 1966).

than 2 per cent. A strong relation between water stable aggregation and content of organic matter has been shown in most of the investigations that have been made. Tisdall and Oades (1982) and Chaney and Swift (1984) have listed some of these and have reported results in this field. Some of the organic matter in soil may be protected from microbial attack by its close association with clay as outlined hypothetically in Sections 8.2 and 8.3. The stability given to small aggregations of clay and organic matter is then long lasting. But organic matter in the larger more porous aggregates, built up of smaller aggregates, silt and sand particles, is not protected. Stability is there conferred by the various organic materials described in Section 8.3 including fungal hyphae, fibrous roots and polysaccharides all of which are subject to microbial activities. These larger aggregates are more fragile and their stability varies with management practices including those that cause gains or losses of organic matter to the soil.

Iron and aluminium oxides can act alone or possibly in combination with organic matter to stabilize aggregates. Their contribution to stable structure has been observed by many workers (reviewed by Marshall, 1962a). The stable soils from the south eastern USA in Table 8.4 owe their stability to this but those in Fig. 8.8c, d from the western states are less affected. Iron oxides can be present as small discrete particles and so contribute less to stability of structure than organic matter does, as Deshpande, Greenland and Quirk (1968) found in a group of red soils. Effectiveness of the oxides as stabilizers of structure thus depends on their distribution. As interlayers in Na-montmorillonite, iron and aluminium hydroxides reduce swelling and as coatings on aggregates they confer some stability, according to El Rayah and Rowell (1973). Blackmore (1973) found that an adsorbed film of iron oxide stabilized aggregates. El-Swaify and Emerson (1975), using dried discs of illite, kaolinite and mixtures of the two into which $Fe(OH)_3$ and $Al(OH)_3$ had been precipitated in suspension, found that double layer swelling in dilute NaCl solutions was inhibited. $Al(OH)_3$ was the more effective of the two in reducing slaking, probably because it made a larger area of contact with the clay particles. In the process of weathering, iron, aluminium, silicon and calcium are mobilized and deposited as oxides, hydroxides, silicates or carbonates in sites where they sometimes accumulate and become concretionary rather than acting as cements in soil aggregates (Fig. 8.8e).

According to the principles discussed in Chapter 1, soils will tend to swell and disperse more readily the lower the valence of the exchangeable cation and, within each valence group, the smaller the ion. These principles for single ion systems apply qualitatively also to the mixed ion systems of natural soils. When Na^+ constitutes a sufficient part of the exchange complex, aggregates may be unstable even if the larger part of the complex consists of the divalent cations Ca^{2+} and Mg^{2+}. As shown in Fig. 8.9, an exchangeable sodium percentage (ESP) of 5 was required to initiate dispersion and swelling in soils examined by Emerson and Bakker (1973) in which Ca^{2+} was the other exchangeable cation, but an ESP of 3 was sufficient when it was Mg^{2+}. Soils with a high ESP are termed sodic soils (U.S. Salinity Laboratory Staff, 1954) and special steps to be discussed below have to be taken to improve their stability. As originally defined, sodic soils are those in which $ESP \geqslant 15$, but many soils are affected at an ESP of about 6 (Loveday, 1976).

Because of the effect of electrolyte concentration on the extent of the double layer (Table 1.5) the tendency to swell is reduced the higher the concentration of solutes in the water. The combined effects on structure

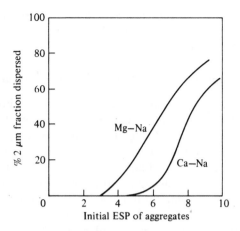

Fig. 8.9. The effect of exchangeable sodium (ESP) in the presence of exchangeable calcium or magnesium on dispersion of aggregates (Emerson and Bakker, 1973).

of the proportions of Na^+ and Ca^{2+} in the exchange complex and of salt concentration of the solution are illustrated in Fig. 8.10, from work of Quirk and Schofield (1955). The stability of a bed of aggregates is represented here by the effects of their swelling and collapse on hydraulic conductivity, as in the method of Emerson (1954) discussed earlier. Any combination of ESP and electrolyte concentration in the region below the curve allows stable structure to be maintained while in the region above the curve it causes loss of stability. This curve probably applies

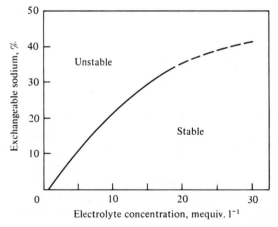

Fig. 8.10. Concentration of electrolyte required to keep a soil stable at different exchangeable sodium percentages (Quirk and Schofield, 1955).

to most sodic soils but it does not have universal application because organic matter or other agents can help maintain stability in spite of adverse effects of sodium ions.

8.9 Practices and treatments affecting structure

Many agricultural practices affect soil structure. Machines and farm animals compact the soil and tillage loosens it in breaking it up into aggregates of a desired size. Tillage has also adverse effects because it exposes aggregates to rain drops, it shears the worked soil and compacts and shears the soil below the tillage depth to form a plough sole. It also encourages a more rapid decomposition of organic matter, and the resulting decrease in the amount in the soil can be expected to affect the stability of aggregates in water. Decrease in both the stability and the organic matter in soils under annual tillage has been widely observed (Harris, Chesters and Allen, 1966; Low, 1972) and an example is given in Table 8.5.

Table 8.5. *Effect of period under annual cereal crops on the water-stable aggregation and organic matter in the tilled horizon (Clarke and Marshall, 1947)*

Period under cultivation Years	Aggregation		Total nitrogen %
	0.05 mm* %	0.2 mm† %	
0	26	62	0.222
1	20	52	0.170
3	18	52	0.160
20	13	41	0.135

* Particles <0.05 mm in aggregates >0.05 mm by a sedimentation method.
† Particles <0.2 mm in aggregates >0.2 mm by a wet-sieving method.

If long-cultivated land is put under pasture, the stability of its structure increases, as shown for example by Mazurak and Ramig (1962) for a chernozem soil in Nebraska, USA. Improvement is usually slow and in this Nebraska experiment it continued for 20 years under a perennial pasture. Harris *et al.* (1966) discuss the large numbers of experiments conducted to find rotations of crops and pastures that will maintain structure stability. Many other treatments are used also. Covering bare soils with mulches of leaves, crop residues and dung protects it from direct impact of rain drops and provides material for the operations of earthworms and micro-organisms. The rate at which the aggregates are wetted will also be reduced by the interception of the rain by the cover. Similarly damage from wetting can be reduced in irrigated culture by avoiding flooding of soil with unstable structure, as demonstrated by Kemper, Olsen and Hodgdon (1975).

Management practices such as these together with avoidance of tillage can help growth as demonstrated by Tisdall, Olsen and Willoughby (1984) for irrigated peach trees on a soil of unstable structure. Systems of 'minimum', 'zero', or 'no-till' tillage, in which weeds are destroyed by herbicides instead of by tillage, reduce effects of disturbance but on the other hand bulk density may increase progressively in the untilled soil due to compacting. This may adversely affect root growth (Section 11.2), hydraulic conductivity and at times aeration also (Mielke, Doran and Richards, 1986) so that conventional tillage may have an advantage for crops that need to have good early growth and strong root development (Soane and Pidgeon, 1975). However the loss of pore space is mainly in the larger pore sizes and this is compensated by the continuity of pore space for fluid movement when earthworm tunnels and root channels remain undisturbed (Baeumer and Bakermans, 1973; Boone and Kuipers, 1970). Minimum tillage is especially successful as a soil conservation measure (see Section 9.9) and it consumes less energy with faster land preparation than normal tillage (Cannell, 1985).

Various additives are available for improving structure or its stability. Compost, farmyard manure, and green manure help maintain organic matter content in a tilled soil but they do not provide all of the benefits of the fibrous root system of grasses. MacRae and Mehuys (1985) concluded that green manuring (the ploughing-in of a cover crop grown for the purpose) did not necessarily improve structure.

In sodic soils it is common practice to reduce the exchangeable sodium percentage (ESP) by replacing Na^+ by Ca^{2+}. Gypsum ($CaSO_4 \cdot 2H_2O$) is most widely used for this purpose. An example of the effect of a relatively small application is given in Table 8.6 for a sodic soil in the Voronezh area, USSR, with an exchange capacity ranging from 13 to 23 mequiv. per 100 g soil in the depths sampled. Other sources of Ca^{2+} that are sometimes used are lime and calcium chloride. Sulphuric acid has been applied in California and Armenia to reduce the ESP by mobilizing calcium from lime (if it is present in the soil) and aluminium from the clay. Elemental sulphur serves the same purpose after it is oxidized by micro-organisms in the soil. The use of chemical amendments to correct high ESP is most successful in irrigated land because high value crops warrant the cost and because of the greater effectiveness with leaching water available. However, gypsum is used in dry-land farming in a number of countries including the USSR and Australia. Also Pak and Tsyurupa

Table 8.6. *Effect of gypsum on the exchangeable cations of a solonetz soil 5 years after an application of 6 tonnes/ha:* (a) control; (b) treated (Pak and Tsyurupa, 1976)

Depth cm	Exchangeable cations, % of exchange capacity					
	Ca^{2+}		Mg^{2+}		Na^+	
	a	b	a	b	a	b
0–15	36	48	50	42	14	9
15–30	35	48	48	42	17	10
30–50	29	52	52	34	19	12

(1976) describe how sodic soils with lime and gypsum at a suitable depth are ploughed deeply in the USSR to make use of the calcium they contain naturally. This is said to be effective under irrigation and also under dry-land farming if extra water can be provided during reclamation to remove soluble salts and products of exchange. In reclaiming sodic soil that has also a high salt content, as in land being reclaimed from the sea and in some arid lands under irrigation, it is usually advisable to add gypsum or other amendment before the salt is leached out. If leached first with non-saline water, the clay may disperse, making the soil less permeable and more difficult to treat subsequently. In other situations also, the electrolyte concentration can affect the stability of the structure. This is the case when gypsum is applied as shown in the experiments of Shanmuganathan and Oades (1983) who found that as the electrolyte concentration decreased following an application the clay gradually dispersed. They advocated small annual applications to take advantage of the electrolyte effect on stability. Methods of reclamation of sodic soils are dealt with by U.S. Salinity Laboratory Staff (1954), Szabolcs (1971), and Kovda and van den Berg (1973).

Various synthetic organic polymers with long flexible molecules are effective in stabilizing structure by bridging across adjoining particles in a manner similar to the naturally occurring polysaccharides. In 1951 the Monsanto Chemical Company in the USA launched a hydrolysed poly-acrylonitrile and vinyl acetate maleic acid as soil-conditioning polymers called 'Krilium'. They have remarkable ability to stabilize soil structure when only a small amount, usually in the range 0.01 to 0.1 per cent by weight, is added to prepared soil. Preparation consists of bringing the surface soil to the desired distribution of aggregate sizes as in Fig. 8.1b. The impact was great, as indicated by an issue of *Soil Science* (**73**, No. 6) devoted to the topic but the cost of the materials confined their use to intensive crops. Experimental work to prevent crusting, to aid water entry, and to reduce erosion proved their capabilities and stimulated continuing research in this field. Later, polyvinyl alcohol (PVA) and polyacrylamides (PAM) have come into use and have shown promise as soil-conditioners. Wallace and Wallace (1986) state that, with some of the newer polymers, lower rates of application and less exacting needs for soil preparation are now possible. Application in irrigation water allows greater efficiency of distribution according to Wallace *et al.* (1986) who have set out soil tests for determining rates of application by that means.

Bitumen, Portland cement, and hydrated lime, which are widely used as soil stabilizers in pavement construction, have also been tried out as soil conditioners in agriculture. The principles and use of these materials in engineering fields are well reviewed by Ingles and Metcalf (1972). Bituminous emulsions enable bitumen to be lodged around contact points between particles where it is deposited as a stable link as the soil dries. For agricultural purposes as a soil conditioner, De Boodt (1972) describes how they can be applied to moist soil brought to the desired aggregate size. The emulsion can also be modified to make the soil more hydrophobic so as to prevent crusts forming at the surface under the action of rain. Portland cement reacts at the surface of aggregates to form a rigid skeleton. Amounts of cement varying from about 3 per cent by weight in sands to 10 per cent or more in clays are used to give the stability required for compacted soil in pavement construction. Ahuja and Swartzendruber (1972) found that 1 per cent was effective for agricultural purposes but, even at this low rate of application, costs preclude its use. Hydrated lime, $Ca(OH)_2$, is used by road engineers to make soil less plastic and easier to handle and because it reacts with silica from clay to confer stability on compacted soil (Ingles and Metcalf, 1972, p. 137). Stocker (1975) found it reacted with the edges of clay crystals without otherwise changing their crystal structure. Aggregate stability was greatly increased by applications of about 4 per cent hydrated lime. Quicklime (CaO) has an additional effect according to Hartge and Bohn (1984) caused by the hydration to $Ca(OH)_2$ occurring within the mix.

The treatments discussed in the last two paragraphs have not yet reached the stage where they are economically useful in other than highly intensive agriculture. However, continuing activity in this field offers promise of future applications for soil management on a larger scale. This work has the objectives of stabilizing structure to prevent poor aeration, restricted water entry, and high resistance to seedling emergence and root growth. Along with this goes the need to be able to recognize soils that will respond usefully to treatment, in order to avoid unnecessary use of any conditioners that may be developed.

9

Deformation of soil

9.1 Consistency

The state of soil – whether it is solid, plastic or liquid – is referred to as its consistency. Soil in the plastic state has the property of flowing, after a yield or threshold stress has been exceeded. It differs from a Newtonian fluid such as water which requires no yield stress for flow to occur, and it differs from soil in the solid state in that, under stress, it yields without fracturing. These differences provide the basis for simple tests to find the water content at the upper (wetter) and lower (drier) limits of plasticity of a soil that contains sufficient fine material to show plastic behaviour. The tests as devised by Atterberg (1911) and modified by Casagrande are described fully by Sowers (1965), American Society for Testing and Materials (1984) and British Standards Institution (1975). For these tests, soil samples from which particles larger than 2 mm in diameter have been removed are remoulded into a paste after wetting. In standard engineering methods only material smaller than 0.4 mm is used.

The lower limit of plasticity is determined by finding the water content at which soil, rolled into a thread 3 mm in diameter, begins to crumble. This was called the rolling-out limit by Atterberg but is now more usually called the plastic limit. The upper limit of plasticity of a soil (called its liquid limit) is determined either by finding the water content at which a groove formed in the soil will just close under the influence of standardized impacts, or by the method preferred by the British Standards Institution using a cone penetrometer. The cone is set with its point touching the soil. It is then released and the liquid limit is that water content at which the dropped cone penetrates to a depth of 30 mm. The two methods give similar results but according to Wires (1984) the penetrometer method is simpler, faster and more precise. The difference between the liquid and the plastic limits is called the plasticity index. Croney and Coleman (1954) have found that the suction of the soil water in the continuously disturbed soil at the plastic limit is about 10 m and at the liquid limit about 3 cm, and Greacen (1960b) found values of about 7 m and 10 cm, respectively, in five clay soils. In general, liquid and plastic limits increase with clay and organic matter content. They are also affected by type of clay mineral, as is reflected in the correlations obtained by Farrer and Coleman (1967) which were higher with specific surface area and cation exchange capacity than with clay content (Table 9.1). Their work was done on soils of low organic content using water absorption for the measurements of surface area.

Table 9.1. *Correlation coefficients between liquid limit, plastic limit and other properties (Farrer and Coleman, 1967)*

	Plastic limit	Cation exchange capacity	Clay content	Specific surface area
Liquid limit	0.83	0.90	0.82	0.91
Plastic limit		0.81	0.60	0.79
Cation exchange capacity			0.66	0.90
Clay content				0.64

Although these tests originated from agricultural research in Sweden, they have found their greatest use in the classification of soil for engineering purposes. Used in conjunction with particle size analysis, they provide a means for separating soils into classes to help in assessing material to be used in constructing roads and airfields. Results are expressed in g per 100 g dry soil. Tables and charts showing engineering classifications of soils with liquid limits ranging from 10 to 100 and plasticity indexes from 0 to 60 are given in books on soil mechanics (e.g. Lambe and Whitman, 1969, p. 35). Materials with low liquid and plastic limits and particles of widely distributed sizes can be compacted to a high density and will behave well. Among the least satisfactory materials for engineering purposes are those with a plasticity index that is low relative to the liquid limit, such as a silty clay or an organic soil with a liquid limit that may be about 70 when the plasticity index is only 20.

In addition to their use in soil classification, liquid and plastic limits serve as a guide to the water content at which a soil can be handled for some particular purpose. In this they act as markers in an approximate consistency scale just as field capacity and permanent wilting point do for the availability of water (Fig. 9.1). For example, tillage implements can be expected to smear a soil readily if it is wetter than its plastic limit, but below the limit it will behave as friable material. However, results

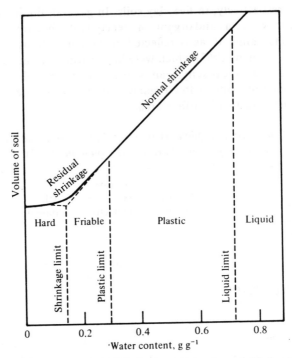

Fig. 9.1. Consistency states and shrinkage stages of remoulded soil illustrated by values appropriate to a soil of high clay content.

have to be treated cautiously because severe manipulation of the sample in determining the plastic limit causes a degree of dispersion and destruction of structure that is greater than that occurring in most tillage operations.

Two other consistency limits, the shrinkage limit and the sticky point, are also in occasional use. The shrinkage limit is the hypothetical water content when a pat of remoulded soil ceases to shrink assuming it to remain saturated until then. As shown in Fig. 9.1 this limit separates the friable consistency of moist soil from the hard consistency of dry soil. Its measurement is described by the American Society for Testing and Materials (1984) and by Archer (1975). The sticky point is the minimum water content at which soil will adhere to a metal spatula. It lies near the liquid limit in most soils and, like the liquid and plastic limits, it increases with clay and organic matter content. Its measurement is described by Sowers (1965).

Examples of the range of values to be expected for shrinkage, plastic and liquid limits of soil in agricultural use are given in Table 9.2. Sandy soils are non-plastic and are not represented in the table. Archer (1975)

Table 9.2. *Consistency limits for a range of soil textures (examples from data of Archer, 1975)*

Texture	Clay content, %	Consistency limits, % water by mass		
		Shrinkage limit	Plastic limit	Liquid limit
Sandy loam	12	14	16	21
Sandy clay loam	23	18	25	40
Clay	51	13	36	83

noted for the group of soils from which these examples are taken that the liquid limit was of a similar magnitude to the field capacity. His data also show that the water content at 150 m suction (taken as representing permanent wilting point indirectly) was greater than the shrinkage limit in soils of high clay content and less in those of low clay content.

9.2 Strength

The strength of soil affects its behaviour in load bearing, compaction, tillage, and root penetration. In a given soil it usually tends to increase with increasing bulk density and decreasing water content. Strength is imparted to soil by cohesive forces between particles and by the frictional resistance met by particles that are forced to slide over one another to ride out of interlocked positions. Reversible elastic strain is relatively limited in soils and our concern is mainly with the larger permanent deformations that occur when the stress exceeds the strength of the material and the soil fails by fracture or plastic flow. At that stage soil does not recover its original size and shape when the stress is removed.

The shear strength, τ, of a soil is expressed empirically in terms of cohesion, c, and intergranular friction by the Mohr-Coulomb equation

$$\tau = c + \sigma \tan \phi, \tag{9.1}$$

where τ is the shear stress required for failure to occur, σ is the stress normal to the shear plane and $\tan \phi$ is the coefficient of internal friction. In Equation (9.1), τ, c, and σ have dimensions of force per unit area. The angle ϕ is called the friction angle and has the significance shown in Fig. 9.2. According to this equation τ will equal c when there is no normal stress acting on the shear plane. Also in sand free from cohesion, τ will be proportional to the normal stress because the friction caused by sand grains sliding, rolling or riding over one another is proportionally

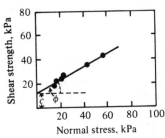

Fig. 9.2. Evaluation of cohesion ($c = 13$ kPa) and friction angle ($\phi = 27°$) from measurements of shear strength at different normal stresses on a silty clay loam at a water content of 0.27 g g^{-1} (Payne and Fountaine, 1952).

increased by it. If the shear strength of samples of a soil is measured at failure under different normal stresses, c and tan ϕ can be obtained for the soil as the intercepts and slope of the line when shear strength is plotted against normal stress, as in Fig. 9.2. In different materials c can range from zero in clean cohesionless sand to about 30 kPa in clays, while ϕ can be zero in saturated clay soils and as high as 45° in densely packed sand with angular grains. Both are to be regarded as parameters of an empirical equation rather than as physical properties of the soil.

Sallberg (1965) describes methods for measuring shear strength. It can be measured directly as in Fig. 9.3a on a sample in a horizontally split box, one half of which moves at a constant rate while a normal load is applied to the surface of the soil. The data in Fig. 9.2 were obtained by Payne and Fountaine (1952) in the field by shearing a cylinder of soil in torsion when different loads were applied at its surface. In soft clay soil in which cohesion is the main component, shear strength can also be determined in the field by using a vane (Fig. 9.3b) pressed into the soil. The torque is measured when the rod to which four vertical blades are attached is rotated to shear the soil.

Shear strength is most reliably measured in the triaxial compression method (Fig. 9.3c) in which a cylindrical soil sample contained within a thin rubber membrane is subjected to an axial load while confined laterally by water or air at a pressure σ_3. The load is increased until the soil fails at an axial stress σ_1. In Fig. 9.4, σ_1 and σ_3 are plotted on the abscissa and a circle (called a Mohr circle) is drawn through these points as shown. Further tests are then run with other values for σ_3 and circles are constructed as before. In books on soil mechanics it is shown that the line drawn tangentially to these circles is a failure envelope for the soil. This line represents Equation (9.1) and from it the parameters c and ϕ for the soil can be determined. Examples of values of c and ϕ obtained in this way are given in Table 9.3 for a soil containing 12 per

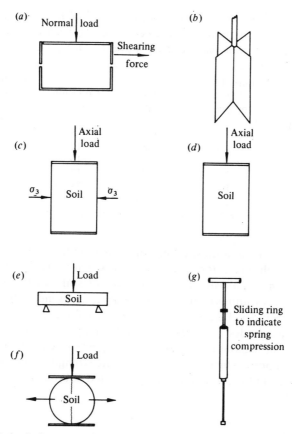

Fig. 9.3. Methods for measuring strength: (*a*) direct shear; (*b*) vane; (*c*) triaxial compression; (*d*) unconfined compression; (*e*) rupture; (*f*) indirect measurement of tensile strength; (*g*) penetrometer (Proctor penetrometer is illustrated).

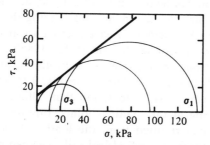

Fig. 9.4. Method of obtaining the failure envelope from measurements by triaxial compression. The soil is a loam with water content of 0.12 g g^{-1}, suction of 30 kPa, bulk density of 1.6 g cm^{-3}. Values derived from the line: $c = 12$ kPa; $\phi = 39°$. (Farrell and Greacen, 1966.)

Table 9.3. *Cohesion and angle of friction calculated from triaxial compression tests on a soil at different initial conditions of bulk density and suction (Farrell and Greacen, 1966)*

Suction m	Bulk density Mg m^{-3} or g cm^{-3}	Void ratio	Cohesion kPa	Friction angle (°)
3	1.5	0.77	8	37
	1.6	0.66	12	39
	1.7	0.55	16	41
7	1.5	0.77	12	39
	1.6	0.66	17	40
	1.7	0.55	23	41

cent clay and having a plastic limit of 17 per cent. A marked increase of cohesion with increasing suction and bulk density is shown.

The strength of cohesive lumps or cores of soil can be tested in a number of ways without confining the sample, using methods discussed by Gill and Vanden Berg (1967). In the unconfined compression test, (Fig. 9.3d), a cylindrical sample is loaded axially until it fails, as in the triaxial test, but without having any confining pressure applied to it. The modulus of rupture which is related to tensile strength can be measured by loading soil in the form of a beam supported at both ends (Fig. 9.3e). It is given by

$$3Fl/2bd^2$$

where F is the force required to break a beam of width, b, and depth, d, when l is the distance between the supports. The method has been used for measuring the strength of soil crusts (Richards, 1953). The tensile strength of soil can be measured directly by attaching end plates to a cylindrical sample with a cement and loading the soil axially in tension, as was done by Farrell, Greacen and Larson (1967) or by a pneumatic fracture method of Snyder and Miller (1985b). But it is usually measured indirectly from the load required to rupture a sphere or cylinder of soil compressed between two plates as in Fig. 9.3f. The process was examined by Frydman (1964) who found that, although compression caused some flattening at the loading points, rupture was due to the tensile stress, T, which resulted from the load and caused cracks to develop down the middle of a cylindrical soil specimen placed horizontally between the plates. The tensile strength of natural aggregates can be measured in the same way when treated as spheres (Rogowski, 1964; Dexter and Kroesbergen, 1985). The tensile strength of a cylinder of length, l, and radius r, which is ruptured by a loading force, F, is given by

$$T = F/\pi rl \qquad (9.2a)$$

and for a sphere of radius, r, by

$$T = 0.45F/\pi r^2. \qquad (9.2b)$$

The problem of allotting a value to r for an aggregate of a characteristically irregular shape has been examined by Dexter and Kroesbergen who have given several methods for doing this. Results of rupture tests on individual aggregates, expressed as tensile strength or in other ways, have been used in studies of cementing agents (Rogowski, Moldenhauer and Kirkham, 1968), friability (Utomo and Dexter, 1981b), the energy of break up (Skidmore and Powers, 1982), and the efficiency of tillage implements (Hadas and Wolf, 1984b).

An indirect measurement of soil strength can be made in the field by forcing a probe into the soil by means of numbered blows from a hammer, or more commonly by pressing it in at a steady rate against a spring or a proving ring that enables the required pressure to be measured. Penetrometers used for this purpose are fitted with a tip larger than the shaft and of a diameter suited to the resistance offered by the soil (Fig. 9.3g). Tips are of various shapes including flat discs and cones. Cone penetrometers are described fully by Bradford (1986). A cone penetrometer was used for example by Cassel (1982) in a study of tillage effects on mechanical impedance. Since the moisture status strongly affects results, the soil was first brought to field capacity to reduce the effect of variation of water content over the field. Further examples of penetrometer use are given in Section 11.2 where the effect of soil strength on root growth is examined. A theory for the resistance offered to fine probes used for that purpose has been proposed by Farrell and Greacen (1966). Resistance is due to the combined effects of shear strength, compression ahead of the probe, and metal–soil friction.

9.3 Effect of water on strength

Soil strength usually decreases with increasing water content as illustrated in Fig. 9.5. One reason for this is that bonds that hold the particles together in structural units are weakened as more water is adsorbed. Strength is imparted by the bonds linking clay crystals into clay packets and the packets into aggregates. These include van der Waals forces, attraction between oppositely charged surfaces, organic matter in various forms and inorganic cements. The bond strength is reduced by water through the softening of cements and the increased separation of particles

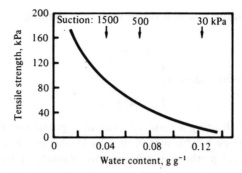

Fig. 9.5. Effect of water content on the tensile strength of a soil containing 12 per cent clay compressed when moist to a bulk density of 1.7 g cm^{-3} (Farrell *et al.*, 1967).

as water is adsorbed. Two other mechanisms also affect the relation between water content and strength. Soil may be strengthened by a negative pressure in its pore water. Cracking acts in an opposite way to the general trend by weakening soil as it dries. We shall now look further at the effect of water pressure on strength and will leave the effect of cracks until Section 9.7.

When a saturated compressible soil is loaded its shear strength depends on whether the water is able to drain from it as it consolidates. Under conditions where drainage is prevented, pressure will develop in the water and the applied load will not be wholly supported by the particles. On the other hand, if drainage allows the pressure to dissipate, more of the load will be carried by the particles which will then be under greater intergranular stress at contact points giving it greater shear strength. Hence when the pore-water pressure, p, is positive, the intergranular or effective stress normal to the shear plane in Equation (9.1) and Fig. 9.3*a* is less than the stress, σ, due to the applied normal load and is given by

$$\sigma' = \sigma - p. \qquad (9.3)$$

The concept of effective stress (Terzaghi, 1943) applies also to soil where p is negative causing the grains to be pressed together under the influence of capillary retention. If the soil is saturated, as it may be when its pores are small enough to prevent air entering at a suction of $-p$, Equation (9.3) is applied directly. In this case the effective stress increases progressively with suction. However Equation (9.3) does not apply to unsaturated soil because of the reduced fraction of the shear plane over which the water acts when air enters the system. This is illustrated in Fig. 9.6 where the effective stress due solely to negative pressure in the water is shown to be equal to the suction over the range in which the

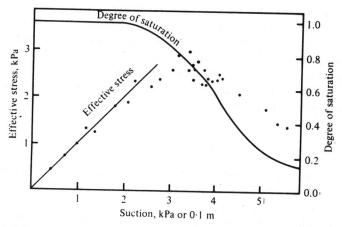

Fig. 9.6. Effect of suction and degree of saturation on the effective stress of a beach sand drying from saturation. The experimental points for effective stress follow the line $\sigma' = |p|$ until the degree of saturation decreases. (From data of Towner and Childs, 1972.)

sand remains saturated. Beyond that range when air enters the sand Equation (9.3) fails.

To cover unsaturated conditions, Bishop and Blight (1963) modified Equation (9.3) to

$$\sigma' = \sigma - \chi p \tag{9.4}$$

where χ, determined experimentally from triaxial compression tests, ranged from one at saturation to zero which was reached at different degrees of saturation in different materials. Theoretical treatment of unsaturated conditions stems from the work of Haines (1925) and Fisher (1926) on the tensile force between adjoining grains that results from the ring of water between them. Subsequent work has dealt with effects of water content (Aitchison, 1961), hysteresis (Towner and Childs, 1972) and cracking (Snyder and Miller, 1985a).

The available evidence shows that the effect of suction on intergranular stress and hence on the shear and tensile strength of soil can be considerable. It is however limited in dry soils by the onset of cracking. Also some of the water in soils is adsorbed on clay surfaces where it is not held by forces of capillarity (Section 2.1). Snyder and Miller have considered these aspects and have concluded from their results that the maximum tensile stress to be expected from this source in unsaturated soil is about half of the suction. Intergranular stress of this kind is probably an important source of the strength of uncemented soil material such as surface crusts and hard setting soils according to Mullins and

234   *Deformation of soil*

Panayiotopoulos (1984). Using mixtures of sand and kaolin formed into cylinders by draining from saturation to a range of suctions in split moulds, they found that intergranular stress induced by suction was largely responsible for strength at suctions greater than 10 kPa and that strength increased with suction to a maximum at some suction greater than 100 kPa. The maximum tensile strength was 30 kPa for a mixture having 8 per cent of kaolin.

9.4 Compression

When an unsaturated bed of aggregates is compressed it is found (Day and Holmgren, 1952) to deform plastically if wetter than the plastic limit, reducing the pore space between the aggregates progressively. If drier than the plastic limit, the aggregates flatten where they contact one another so that the load is supported by a larger area of contact and further compression is resisted. In both cases the larger pores are reduced in size and in the plastic material they may be isolated (Davis, Dexter and Tanner, 1973) thus severely reducing the hydraulic conductivity of a bed of aggregates. The effect of compression on the size distribution of pores is shown in Fig. 9.7.

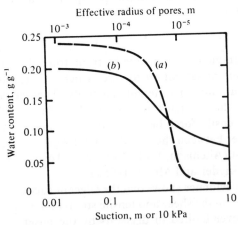

Fig. 9.7. Size distribution of pores in a silty sand at (*a*) low and (*b*) high bulk density (After Croney and Coleman, 1954.)

The compression of soil is described empirically in books on soil mechanics by the equation

$$e = e_0 - I \log (\sigma/\sigma_0), \tag{9.5}$$

where e is the void ratio (the volume of the pores divided by the volume of the solid particles), σ is the pressure applied by a piston uniaxially to the soil in a container and I is the compression index for the soil as given by the slope of the line on a semi-logarithmic plot as in Fig. 9.8. In this equation e_0 and σ_0 represent an initial void ratio and pressure, respectively. Similar expressions are available for compression in terms of porosity, bulk density and specific volume as well as for void ratio (Harris, 1971). In later work, Braunack and Dexter (1976) have empirically expressed the volume, V, of a bed of aggregates subjected to a uniaxial pressure, σ, as

$$V/V_0 = 0.4 + 0.6 \exp\{0.017(\sigma/y) - 0.38(\sigma/y)^{1/2}\} \qquad (9.6)$$

where V_0 is the initial volume and y is the tensile yield strength of an individual aggregate. In their work the tensile strength was measured by crushing aggregates between parallel plates, where failure occurs under tension. In the loam soil they used, y varied from 3.6 kPa at zero matric potential to 84 kPa at a potential of -50 MPa (or 5000 m suction) over a range of water content from 0.34 to 0.03 g g^{-1}. They found y to be little influenced by size of aggregate. Possible applications of this equation to load bearing in agriculture are given by the authors.

Under a given pressure, bulk density may increase substantially if the soil is also subjected to shear. This was well demonstrated by Bodman and Rubin (1948) and is illustrated in Fig. 9.8a by data obtained with their type of apparatus. In their method, normal and shear stresses are applied to a bed of aggregates contained in a ring shear machine.

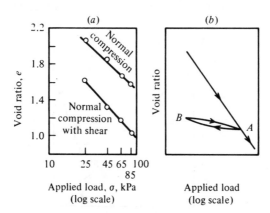

Fig. 9.8. Compression of soil under an applied normal load (Greacen, 1960a). (a) Unsaturated bed of aggregates of a clay soil compressed with and without shear force. (b) Diagrammatic representation of the consolidation of a saturated clay with an unloading and reloading hysteresis loop.

When failure of aggregates has occured during compression, the process is largely irreversible. However, after soil has been compressed and the load removed, there is usually some recovery of volume or 'rebound' as illustrated by the line *AB* in Fig. 9.8*b*. Day and Holmgren (1952) considered that in their soils this was probably due to swelling occurring as water became redistributed between sheared and unsheared zones. Water that has been forced out of contact zones during compression is presumably readsorbed by the tightly packed and oriented particles in these zones. Rebound due to particle elasticity is of relatively small magnitude. Since unloading of a compressed soil causes little increase in volume, the soil can now be made to carry a load up to that previously carried, without much decrease in its volume (line *BA* in Fig. 9.8*b*).

9.5 Compaction by animals and machines

An increase in the bulk density of soil resulting from loads applied for short periods is referred to as compaction. This contrasts with compression under a static load which, when accompanied by the slow expulsion of water from a saturated soil, is referred to as consolidation. Under intensive agriculture and grazing, surface horizons are subjected to the compacting effects of machinery and animals which can exert pressures of the order of 100 kPa. Surface horizons usually have a relatively small bulk density because of disturbances such as those due to soil animals and plant roots and those caused by tillage. Hence pressures of about 100 kPa can be expected to cause compaction and this may adversely affect water and air movement, seedling emergence and root penetration. An example of compaction by machinery is given in Fig. 9.9. Full details

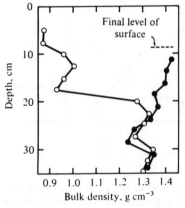

Fig. 9.9 Bulk density profile before (O) and after (●) the passage of a tractor wheel (Soane, 1970).

of this subject are to be found in a book published by the American Society of Agricultural Engineers (1971). The effects on soils have also been covered in papers by Soane *et al.* (1981) for wheeled agricultural vehicles and by Willatt and Pullar (1984) for animal treading.

When a soil is subjected to a given effort of compaction, the bulk density reached depends on its particle size distribution and its water content. For road construction, an optimum water content is sought at which soil can best be compacted to obtain the high density and strength required for this purpose. This is determined by packing a sample of the soil in a cylinder under a set number of standardized blows from a hammer in the method of Proctor (1933). This is done for a series of water contents and, from a curve like those shown in Fig. 9.10, the optimum water content corresponding to the maximum bulk density (expressed on the dry basis) is obtained. At water contents above this optimum, the pore space fills as compaction proceeds and the process is then resisted by the water. Below the optimum it is resisted by the greater cohesion and internal friction between particles and aggregates. Hence a well marked optimum can usually be determined for controlling the water content of each soil during construction of a roadway. The determined maximum density is then used as a guide to the compaction to be expected from rolling. The minimum density shown also by these curves is possibly due to the low cohesion between particles of the driest samples during compaction.

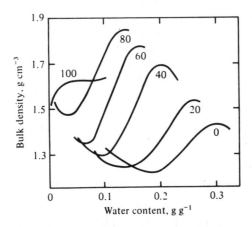

Fig. 9.10. Effect of water content on the compaction of mixtures of a fine sand and a silty clay. Percentages of the fine sand in the mixtures are shown and the resulting composition corresponds approximately to the following textures: 0–silty clay, 20–clay loam, 40–sandy clay loam, 60–sandy loam, 80–loamy sand, 100–fine sand (Bodman and Constantin, 1965).

The effect of particle size distribution on maximum bulk density is evident in Fig. 9.10. The materials used by Bodman and Constantin (1965) in this work were made up by mixing a sand and a silty clay in different ratios as shown. They were compacted with a kneading compactor. The greatest maximum density was obtained with the mixture (containing 80 per cent of the sand) that corresponded to a loamy sand in texture. They showed from a theoretical discussion of the packing of binary mixtures that, near this ratio, the pores in a matrix of contacting sand particles would be occupied by the finer particles if ideally packed. In mixtures with less fines, the pores of the matrix would not be fully occupied by particles. In those with more fines, the fine material constitutes the matrix with sand distributed through it. On the basis of the two types of matrix they were able to estimate from packing theory a maximum density for each of the mixtures. The coarser the sand used in the mix the closer the experimental maxima were to the calculated maxima.

In the packing method of Bodman and Constantin, high density was sought and this is also true of the Proctor method. It is found, on modifying the Proctor method by expending less energy on compaction, that the decreased maximum bulk density that results corresponds to an increased optimum water content for compaction. The Proctor method with and without modification has been applied to cases of agricultural interest. Weaver and Jamison (1951) showed that the optimum water content for compaction can be unfortunately near the lower plastic limit where conditions are also optimal for ploughing.

It follows from the curves in Fig. 9.10 that farm animals and machines do not necessarily cause much increase in bulk density when the soil is wet. Also, instead of compacting the soil, the hoof or wheel sinks deeply, displacing the soil upwards alongside. Aggregated soil can be reduced to a puddled condition by traffic in these circumstances. Gradwell (1966) showed that incautious grazing in winter reduced the ability of soil to store water in the readily available range during the following summer because of puddling. This and other aspects of soil management relating to compaction are reviewed by Larson and Allmaras (1971).

9.6 Deformation by swelling and shrinking

Horizontal movements of soil are made more obvious than vertical movements by the crack pattern displayed at the surface. However, vertical movements are substantial in a clay soil in a climate with dry summers and wet winters. They decrease with depth as in Fig. 9.11 and are usually minor below 2 m where there is little change in water content between winter and summer. Movements like those of Fig. 9.11 seriously

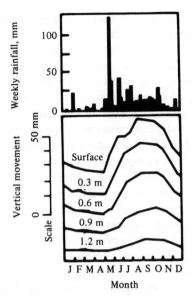

Fig. 9.11. Vertical movements at various depths in a soil of high clay content over a period of 12 months at Adelaide, Australia (Aitchison and Holmes, 1953).

affect engineering structures. Buildings or pavements are subjected to strain caused by unequal soil movements when the soil near the exposed edge of the structure changes in water content at a greater rate than elsewhere under it. Foundation failures resulting from this are of great economic importance in many parts of the world. Stability is gained by placing footings of structures such as houses deeply enough to avoid dangerous seasonal movements or by protecting the outer edge from seasonal changes in water content. Since special design can add significantly to the cost of a house, it is an advantage in vulnerable areas to find by soil examination whether the soil is of a type that requires this to be done (Aitchison, 1973).

A saturated clay can be made to lose water and consolidate either by expelling water under an external pressure on the soil or by subjecting the water to an increased suction. If air does not enter when the soil shrinks on drying, the bulk density will be increased equally by a given effective stress, whether this arises from an external pressure or from a suction of the same magntiude. This equivalence of compression and drying has been discussed by Terzaghi (1943) and demonstrated for clay soils by Aitchison and Donald (1956). It is evident also from the equivalence of water content of soil under a suction on a porous plate and under a gas pressure of the same magnitude in a pressure membrane apparatus. In the latter case, so long as the soil remains saturated, the gas exerts external pressure on the soil.

For soils in general, there is a full range of behaviour from that of the wet clay drying over the 'normal' stage of shrinking with no air entering the system to the rigid sandy soil in which all the water lost on drying is replaced by an equal volume of air. For those soils or aggregates that shrink to some degree, shrinkage behaviour is affected by previous history in ways that resemble the compression behavior illustrated in Fig. 9.8*b*. This is demonstrated in Fig. 8.4 by data of Croney and Coleman (1954). Curve *A* with its successive branches to *B* and *C* shows that during the drying of a slurried clay soil the water content and hence the void ratio at a given suction depend on the greatest suction to which the soil has been subjected previously. Similarly curve *A* does not converge with curve *D* until the slurried clay has been subjected to an effective stress in drying equal to that experienced at some time in the history of the natural soil, either in drying or compression. A further aspect of this work was the demonstration in curve *E* that, when the slurry was subjected to shear while shrinking by continuously disturbing the soil, the water content was lower at a given suction. This may be compared with the lowered void ratio in Fig. 9.8*a* in soil that is subjected to shear while being compressed. It is of interest that Croney and Coleman found that the liquid limit (76 per cent) and the plastic limit (27 per cent) of the soil fell on curve *E*. In both these procedures the soil is continuously manipulated.

Since curves *E* and *D* are approximately linear over the plastic range, the relation between water content, θ, and suction, s, was represented for clay soils shrinking between liquid and plastic limits by Greacen (1960*b*) in an equation that is similar to that for compression under a load. It may be rewritten for saturated clay in the form of Equation (9.5) as

$$\theta = \theta_0 - A \log (s/s_0) \qquad (9.7)$$

where θ_0 and s_0 represent an initial water content and suction.

9.7 Break up of soil

The fracture of intact soil or large clods of soil into smaller pieces requires energy to be expended in breaking cohesive bonds. The process differs from failure by plastic deformation in that new surfaces are created by extending flaws, small cracks and pores in the surfaces of weakness discussed in Chapter 8. This extension requires the expenditure of an amount of energy equal to the product of the new area of surface and the surface tension of the solid. According to the theory of brittle fracture developed by A. A. Griffith in 1921, energy involved in straining a brittle material under a tensile stress is converted to this surface energy when a crack is extended. The critical tensile stress required to extend a crack depends on its length and is given (Jaeger, 1964) by the Griffith theory as

$$\sigma = (E\gamma L^{-1})^{1/2} \tag{9.8}$$

where σ is the tensile stress perpendicular to a crack of length, $2L$, E is Young's modulus (ratio of tensile stress to strain) and γ is the surface energy per unit area of the material. George (1970) has discussed this theory especially in relation to the propagation of shrinkage cracks in soil stabilized by Portland cement.

The degree to which soil of friable or hard consistency will break up for a given expenditure of energy on it can be measured by dropping it from a measured height onto a hard surface. The energy available for enlarging cracks and fracturing the soil is

$$\Phi = mghn, \tag{9.9}$$

where n is the number of times a soil sample of mass, m, is dropped from a height, h, and g is the acceleration due to gravity. Marshall and Quirk (1950) found that for a given input of energy as represented by the cumulative distance of dropping, nh, the size to which the soil is reduced is not greatly affected by h. Data for one of their soils plotted in Fig. 9.12 show, for example, that soil initially in large lumps was reduced to a median size of about 25 mm, irrespective of whether the soil was dropped twice from 1.2 m or 4 times from 0.6 m. This occurred with drop heights that were so great relative to strength that cracking occurred freely at each drop. In the circumstances highly efficient use of the input energy can be expected in creating new surfaces so that cracks extend progressively in successive drops. With drops that are small relative to the strength, the efficiency of the process would no doubt be reduced in accordance with the Griffith criterion of a critical tensile strength for crack extension. The drop height (usually about 1 m) and the number of drops can be varied according to the friability, but unnecessarily great heights should be avoided. Over a range of heights up to 12 m, Hadas and Wolf (1984a) found some deviation from the expected degree of fracture which they attributed to air resistance and possibly also to the loading rate of compressive stress at high rate of fall.

The drop-shatter method provides a measure of the energy required to break soil composed of clods and aggregates of miscellaneous sizes and irregular shapes into smaller fragments. The reduction in size for a given input of energy corresponds to friability which has been defined as the tendency for a mass of unconfined soil in bulk to crumble and break down under applied stress into smaller fragments, aggregates and

242   *Deformation of soil*

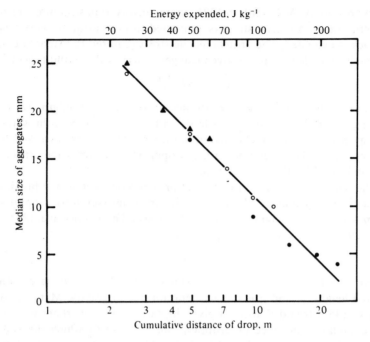

Fig. 9.12. Effect of cumulative distance of drop (height × number of drops from
that height) on diameter of aggregates of a dry clay soil as represented by the
median size. Dropped from height of 0.6 m (▲), 1.2 m (○) or 2.4 m (●).

individual particles (Bodman, 1949). Tensile stress was the cause of
fracture in the drop-shatter method over the range of matric suction
examined by Farrell, Greacen and Larson (1967) of 250 m to 5 m. The
soil needs to be fairly dry for this to be so (Bateman, Naik and Yoerger,
1965). The highest water content for this mechanism to hold, particularly
if the soil is of massive structure, is probably in the neighbourhood of
the plastic limit. The method has been used for quantifying ped size and
stability for soil survey, for measuring clod strength in tilled soil (Utomo
and Dexter, 1981a) and as a means for examining energy efficiency of
tillage machines (Section 9.8).

9.8 Tillage

Tillage is any physical manipulation of soil that changes its structure,
strength or position in order to improve conditions for crop production.
Particular objectives in tilling include preparation of a seed-bed, destruction of weeds, improvement of soil–water and soil–air relations, reduction

of impedance to roots and burying of crop residues. Tillage tools, including ploughs, cultivators and harrows, are designed to shatter, cut, loosen, invert or mix the soil and to smooth or shape its surface depending on the objective. They may also have unwanted effects, including especially the compaction of soil under the ploughshare and the smearing or puddling of soil when tilled in the plastic state.

The reaction of soil to the mouldboard plough is described by Nichols, Reed and Reeves (1958) and Gill and Vanden Berg (1967). Soil cut by the forward edge of the ploughshare is sheared and turned as it slides over the metal surface. Energy is consumed in pulverizing and turning, and by sliding friction. The ploughshare (or other tool) tends to build up a soil body at its forward edge, especially when blunted by wear. Also soil tends to adhere in places to the metal and soil has to slide over soil, increasing the sliding friction. Both these effects add to the draught required to pull the plough. Moreover, the soil body at the forward edge adds to the compacting effect of the ploughshare on the underlying soil which can lead to the development of compaction pans or plough soles.

The behaviour of soil in tillage depends strongly on its moisture status since strength, compactibility and adhesion vary with water content. Adhesion of soil to a metal tool varies in much the same way as cohesion. It increases with suction so long as the degree of saturation remains high and then decreases to zero as the degree of saturation decreases. Fountaine (1954) showed that adhesion of soil to a metal plate was about equal to the suction for the low range of suctions illustrated in Fig. 9.13. In sand, adhesion tended to fall below the suction when air entered the system and reduced the area over which the water could act. But soil

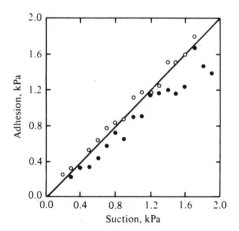

Fig. 9.13. Effect of suction on adhesion of soil to a metal plate (Fountaine, 1954). O, clay loam; ●, sand.

with sufficient clay content can remain nearly saturated at higher suctions than were used by Fountaine and so the linear relation held well for clay loam.

Adhesion affects tillage in two ways. By clinging to wheels and other parts of machines, a sticky soil affects performance. The second effect is on friction between soil and a tillage tool. When one body slides over another, the force required to maintain the motions is proportional to the normal force between the two bodies. The proportionality constant is called the coefficient of sliding friction. Fountaine and Payne (1954) found that the force required to slide a metal plate horizontally over saturated soil was increased equally by a mechanical load placed on the slider or an equivalent suction induced in the soil water. They concluded that suction acted as a normal force between slider and soil, affecting friction accordingly. Gill and Vanden Berg (1967), in a critical review of this and other work, concluded that adhesion can be treated as part of the normal stress when soil is forced to slide over the surface of a tool. The coefficient of sliding friction is then given by

$$\mu = S/(\sigma_a + \sigma),\qquad(9.10)$$

where S is the sliding stress necessary to maintain motion, σ_a is the adhesive stress and σ is the normal stress due to a mechanical load.

The behaviour of soil in relation to tillage operations is summarized in Table 9.4 for four moisture conditions. This table embraces the effects that water has been shown to have on consistence, strength, compaction and adhesion. In general, the moisture condition corresponding to the friable consistency requires the least expenditure of energy in tillage as is well shown for a number of soils by Bakhtin *et al.* (1969). Their data also show effects of structure on tillage behaviour that cannot be detected by consistency measurements made on remoulded soil.

The traditional mellowing of tilled soil has been examined by Utomo and Dexter (1981*a*). Under certain circumstances, exposure increased the amplitude of the fluctuations in the water content and so decreased

Table 9.4. *Effect of water content of a cohesive soil on its behaviour in tillage (after Archer, 1975)*

Moisture status	Dry	Moist	Wet	Very wet
Consistency	Hard	Friable	Plastic	Liquid
Resistance to tillage tool	High	Low	High	Low
Ability to carry traffic	High	High	Low	Very low
Adhesion	Very low	Low	High	Low
Resistance to compaction	High	Moderate	Low	High
Resistance to smearing	Very high	High	Low	Very low

the strength of the clod due to cracking. The proportion of smaller aggregates of desirable size produced by a second implement pass was increased as a consequence. In such cases, a second pass should be delayed several days to take advantage of the mellowing.

The performance of tillage tools in breaking up soil has been examined using the drop-shatter method of Marshall and Quirk (1950). Gill and McCreery (1960) and Gill and Vanden Berg (1967) compared the amount of energy expended on a tillage implement with the amount required to shatter soil to the same mean size by dropping it on a hard surface. A curve for the soil like that of Fig. 9.12 was used to assess the latter from the cumulative distance of drop, $\sum h$, required to do this. The method thus provided a basis for assessing the efficiency of different ploughs and of different ways of using them. An example showing the effect of speed of ploughing on efficient use of energy with two types of plough is given in Table 9.5. The efficiency of mouldboard and chisel ploughs, discs, land planes and heavy rollers for working a soil into seed-bed condition was examined in this way also by Hadas and Wolf (1983).

Table 9.5. *Effect of type of plough and its forward speed on efficiency in use of energy* (*Gill and Vanden Berg, 1967*)

Implement	Forward speed m s^{-1}	Mean size of soil cm	Input energy, kJ m^{-3}		Efficiency Φ_d/Φ_p
			Ploughing Φ_p	Dropping Φ_d	
Mouldboard	1.0	4.7	46	37	0.80
plough	1.9	5.4	70	35	0.50
	2.9	5.4	94	35	0.38
Rotating vertical	0.2	2.4	253	48	0.19
plough	1.2	3.1	126	44	0.35

9.9 Soil erosion

Erosion of soil by wind and water is both a natural and a man-made phenomenon. On the one hand, geological erosion over long periods of time has provided the parent material for soils on which much of our agriculture depends. On the other, use of forest and grassland by man for agriculture has changed the equilibrium between the slow rates of soil formation and erosion that occur in nature. Where climate and topography make bared soil vulnerable, tillage and intensive grazing by domesticated animals accelerate the rate of erosion. Historical examples of this are given by Bennett (1939). Scientific enquiry was started about 100 years ago by E. Wollny in Germany. Also at about that time the work of Dokuchaev in Russia on the origins of soil made it clear that soil is renewable only slowly and is consequently a rather fragile resource. Nevertheless the subject received only occasional attention for many years because accelerated erosion was not widely recognized as a universal problem until about 50 years ago. By that time, erosion of land developed for commercial farming in North America, Africa, Australia and elsewhere had been made apparent to all by its most obvious manifestations in gullying and dust storms. Vigorous research and education programmes were launched in the United States. These have been effective there and have had a great impact throughout the world.

A common approach to both water and wind erosion has been to protect soil by maintaining a plant cover and, when this is not possible, by keeping plant residues on the surface. Tillage machinery was developed in wheat growing areas of North America to anchor residues and cut off roots of weeds without burying the trash. McCalla and Army (1961) in a review of stubble mulching noted that this had been effective. Since then minimum and zero tillage methods have been introduced following the development of suitable herbicides (see Section 8.9). By means of such practices, about one quarter of the cropped land in the USA has been brought under some form of conservation tillage in which plant residues cover 30 per cent or more of the surface (Allmaras and Dowdy, 1985). The increasing acceptance of these practices in many countries has been a notable development in the conservation of soil. The protection afforded may be illustrated by the reduction in the amount of soil eroded from vulnerable cropland in the Darling Downs, Australia, under natural rainfall in trials conducted by Freebairn and Wockner (1986) over several years with and without stubble mulching. Loss of soil ranged from 29 to 62 t ha^{-1} $year^{-1}$ without retention of stubble but stubble mulching reduced this to less than 5 and zero tillage to 2. We note here that 1 tonne per hectare equals 0.1 kg m^{-2}.

Water erosion

The main causes of erosion of soil by water are the detachment of soil by the impact of falling raindrops and the transport and further detachment of soil by flowing water (Ellison, 1947). Splash resulting from impact of the raindrop can cause some net movement of soil down the slope but its transport is mainly into and down the rills that form where the flow concentrates. This concentration itself causes increased erosion as the rills enlarge down slope and converge in severe cases to form incised gullies. The importance of raindrop impact has been demonstrated by many experiments. Young and Wiersma (1973) found that when the kinetic energy of simulated rain falling on bare soil was reduced by 89 per cent without reducing the intensity of the rain, soil loss was reduced by 90 per cent or more in a loam, silt loam and sandy loam. They also showed by means of pre-formed rills in their plots that 80 per cent of the soil removed from the area between rills reached a rill before leaving the small plot. Back from the rills, raindrops falling in very shallow water create disturbances that allow the transport of soil particles even though the water moves only slowly. Moss and Green (1983) found that the rate of transport of soil by this mechanism increases with increasing depth of water to a maximum at a depth that is about 2 or 3 times the diameter of the raindrops. Raindrop splash from adjacent bare soil can also contribute material that is carried off in this way. The resulting rain-flow or interrill erosion is less obvious than rill and gully erosion but it can cause serious loss of soil from the fertile surface.

Raindrops vary in size up to about 5 mm in diameter and the terminal velocity of these largest drops is about $9 \, \text{m s}^{-1}$. The kinetic energy of rain increases with intensity and for an intensity of $50 \, \text{mm h}^{-1}$ it is of the order of 20 J per mm of rain falling on 1 m^2 of soil. Further details are given by Hudson (1981) and Smith and Wischmeier (1962) from their own work and that of Laws (1941), Mihara (1951) and others. When a large raindrop strikes wet soil, particles and aggregates are detached and splashed over the surrounding surface. Water infiltrating into the soil carries with it detached material, clogging pores and creating a condition of low conductivity near the surface (Duley and Kelly, 1939; McIntyre, 1958a), so enhancing run-off. McIntyre found that the saturated hydraulic conductivity of a silt loam was reduced from $10^{-5} \, \text{m s}^{-1}$ to about $10^{-8} \, \text{m s}^{-1}$ in a zone 1 or 2 mm thick at the surface after 30 mm of simulated rain had been applied in drops whose energy was equivalent to that of 3 mm raindrops. A dense layer about 0.1 mm thick was formed as a skin seal at the surface apparently by the compacting action of raindrops. Below that was a washed-in layer where porosity was decreased by material trapped in the pores by straining or by other mechanisms described by Vinten and Nye (1985). As well as causing increased run-off and erosion, a surface zone formed in this way can act as a crust that impedes the emergence of seedlings (Section 11.2).

Many attempts have been made to assess the erodibility of soil from the instability of its aggregates in water as measured by dispersion (Middleton, 1930), by wet sieving, or by the effect of the impact of single drops of water (McCalla, 1944; Bruce-Okine and Lal, 1975). However, although the water drop method has some promise, none of these methods can specify erodibility of soil satisfactorily according to Bryan (1968) in a review of stability methods. Other properties such as texture, bulk density, organic matter and hydraulic conductivity are involved as well as water stability. Multiple regression analysis involving many soil properties has been used with some success by Wischmeier and Mannering (1969). But recourse has usually been made to measuring erodibility directly from the rate of loss of soil from standard bare plots of a particular slope and length, as discussed below.

An equation has been developed by Wischmeier and Smith (1978) that predicts the rate at which soil will be lost from a particular soil under given circumstances. It is used to identify the cropping systems and soil conservation practices that may be required to reduce the loss of soil below a rate that can be tolerated. This is usually taken to be 7 to 11 tonnes of soil per hectare (or 0.7 to 1.1 kg m^{-2}) per year (Wischmeier, 1970). Their 'universal soil loss' equation is

$$A = RKLSCP, \qquad (9.11)$$

where A is the mass of soil lost from unit area per year (or other specified period), R represents the erosive character of the rainfall ('erosivity'), K is the erodibility of the soil, and L, S, C and P are, respectively, the ratios of soil loss under the given conditions of slope length, steepness of slope, cropping management, and special erosion control practices to soil loss under the specified conditions of the standard plot. In this equation, R is treated as a dimensionless index number and all other factors except K are dimensionless ratios. These terms are described more fully as follows.

The erosivity, R, is usually obtained as an average value for a locality calculated from rainfall records as $\sum EI$ where E is the kinetic energy of rain from an individual storm and I is the intensity (depth per hour) for the most intensive period of the storm. EI is summed for all storms for a year or other desired period. The erodibility, K, is the amount of soil lost per unit EI from a plot of specified dimensions of bare fallow with a slope of 9 per cent. It is obtained as an average value since the amount of soil lost per unit EI varies in different storms (Wischmeier, 1976). The factor for slope length, L, is the ratio of soil loss from the given length of slope to that from a slope whose length is that of the standard plot (22.2 m). Similarly, the factor for the slope, S, is the ratio of soil loss from the given slope to that from the standard (9 per cent). Empirically established relations between soil loss and the length and steepness of slope are used in evaluating these two ratios for topographical effects. Allowance can also be made for a convex or concave curvature which respectively increases or decreases soil loss (Wischmeier, 1970; D'Souza and Morgan, 1976).

The factors C and P for cropping management and special control practices cover the ways in which erosion can be modified by farm operations. Together they provide a basis for guidance on land which, from the factors *RKLS*, is seen to be vulnerable if unprotected. Erosion may be most effectively reduced by preventing undue exposure of soil to raindrops so as to reduce detachment and maintain a satisfactory infiltration rate. This can be done by maintaining a plant cover, leaving crop residues on the land or bringing in mulching materials such as straw onto the surface. Damage from run-off can be reduced by reducing the flow velocity. This is accomplished by tilling along the contour instead of up and down slope, by strip cropping along the contour using alternate strips of pasture or suitable crops, or by constructing banks to divert water to grassed waterways. The factor, *C*, for cropping management is the ratio of soil loss from the cropped land to soil loss from the standard plot. It is always <1 because the standard plot from which *K* is obtained is continuously under fallow and is worked up and down slope to give conditions most favouring soil loss. Since *K* is defined in this way for a continuously fallowed soil, effects of recent cropping history are included (somewhat anomalously) in *C* rather than *K*. Illustrative examples of *C* are given in Table 9.6 for various cultural systems of maize in Indiana, USA. It can be seen that rate of soil loss increases with successive years of cropping and decreases with increasing cover on the soil. Further, Wischmeier describes other tillage systems for maize that reduced *C* to about 0.1. He also illustrates the effect of changing surface conditions and plant cover at different stages throughout the year. The factor *P* for special erosion control practices is derived by experimentation in the same way as *C*. For example, *P* is about 0.5 for land with a slope of 7 to 10 per cent when conventional tillage is carried out on the contour (Wischmeier, 1970).

Table 9.6. *The ratio of soil loss from pasture land ploughed up for maize to the loss under continuous fallow* (*data from Wischmeier, 1970*)

	Year after pasture		
	First	Second	Third
Conventional tillage, residue removed	0.24	0.46	0.55
Conventional tillage, residue left	0.19	0.34	0.43
Residue left, plough-minimum tillage	0.12	0.24	0.34

Despite the empirical nature of the factors, Wischmeier (1976) quotes an average error in predicting mean annual soil loss from 189 field plots that is within 13 per cent of the measured loss. However, the data basic to the equation are not always available and judgement is then called for in supplying the deficiencies. Hudson (1981), for example, shows how the equation can be employed as a guide in countries which have not the resources to obtain all the data necessary for its use. Situations outside the present scope of the equation include erosion of range land, road cuttings and banks, slumping, gully erosion, and tunnel erosion. Tunnel or piping erosion (Downes, 1946; Crouch, 1976) occurs where water causes tunnels to develop in an easily dispersed B or C horizon leading eventually to the collapse of the surface soil. For a review of conservation measures that can be taken in most of these cases, reference can be made to Hudson (1981), who also gives a good account of field plots and rainfall simulators used in research on soil conservation.

One consequence of soil erosion is the transport and deposition of eroded material into streams, lakes and reservoirs resulting in the pollution of water, the flooding of land due to siltation of water-courses, and the loss of storage capacity of reservoirs. The effect of siltation on water storage and the effect of catchment conditions on the rate at which siltation occurs can be illustrated by the data of McQuade, Hartley and Young (1981) for a reservoir located in a winter wheat area of low rainfall near Orroroo, Australia. From 1911 to 1944 the average rate of siltation was 23 000 m^3 per year, from 1944 to 1971 it was 5000 and from 1971 to 1980 it was about 2000. The decrease with time is ascribed by McQuade *et al.* mainly to the change from a fallow–wheat rotation to fallow–wheat–pasture by 1944. Conservation practices are being adopted in the catchment but they have come too late for the reservoir which has lost two thirds of its storage capacity.

Wind erosion

The development of our understanding of wind erosion was in a large part due to Bagnold, working in England and northern Africa on desert sands, and to Chepil, working in Saskatchewan, Canada, and Kansas, USA, on eroding soils. Bagnold (1941) showed erosion to be basically due to grains being moved up from the surface of a sand bed and then accelerated horizontally by wind before dropping back to the bed. The impact of these grains hitting the bed at an angle of 10° or less forces other grains upward. Rising more or less vertically, these go through the same bounding movement and the number participating increases as the process intensifies downwind. This movement was called 'saltation' by Bagnold, the term already employed for the bounding movement of sand under running water.

Grains in saltation are responsible for two other kinds of movement. One of these is the creep of larger grains pushed along the surface by impacts and the other is the suspension of small grains disturbed by impact and carried up by eddies in the turbulent flow of air. This material is removed by the wind as dust. While the height of saltation is restricted to about a metre and is usually less than 30 cm, dust is carried up to heights of several kilometres in dust storms. The sizes of grains moved by creep, saltation and suspension overlap, but as a rough guide they can be considered to be, respectively, in the ranges 2 to 0.5, 0.5 to 0.1 and <0.1 mm in diameter. Dust sampled close to the source by Delaney and Zenchelsky (1976) and hundreds of kilometres from it by Walker and Costin (1971) had a relatively high content of organic matter. This is in part attributable to fragments of plant material which are readily carried off because of their low density.

The threshold velocity, v_t, of air at height, z, required to start the chain reaction of saltation by moving exposed grains of diameter, d, and density, ρ, is given by Bagnold (1941, p. 101) as

$$v_t = 5.75a \left(\frac{\rho - \rho_a}{\rho_a} gd \right)^{1/2} \log (zk^{-1}) \qquad (9.12)$$

where ρ_a is the density of air (1.2 kg m^{-3}), k is a length representing the roughness of the surface, g is the acceleration due to gravity and a is a constant for an erodible material. The value of a is increased considerably if the roughness is contributed to by projections and barriers of non-erodible material, as discussed by Chepil and Woodruff (1963, p. 244). Movement of a grain results, at or above the threshold velocity, from a higher pressure on the upwind than on the downwind side and a relatively low pressure on the upper part of the particle due to the Bernoulli effect from the speeding up of the air flowing over the top of the grain. Bisal and Nielsen (1962), observed microscopically that grains 0.1 to 0.5 mm in diameter can be lifted directly from a bed without having first to roll along the surface.

Wind erosion can be controlled by keeping soil covered with plants or plant residues that reduce the velocity of the wind at the soil surface, by maintaining it in a cloddy condition if tilled so as to expose aggregates too large to be moved by the wind, or by tilling it in strips across the direction of the prevailing erosive wind so as to restrict the distance over which saltation can build up. The protection afforded by aggregates larger than 1 mm in diameter and by cover is illustrated in Fig. 9.14 by the increased wind velocity required to initiate erosion in a wind tunnel when aggregates and cover were increased. Here the plant cover was stubble 15 cm high. Plant residues can be retained by using stubble mulching machinery that leaves trash anchored but not buried, by minimum tillage, or by retiring the land from cropping. It may be noted that wind velocity for Fig. 9.14 was measured at a height of 30 cm, and since the velocity increases with height it is accordingly greater than the velocity needed to move the grains.

Deformation of soil

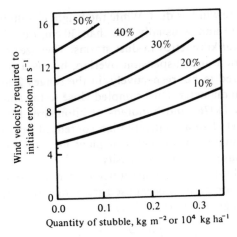

Fig. 9.14. Velocity of wind measured at a height of 0.3 m required to initiate erosion of the surface beneath a standing stubble. The curves are distinguished as the percentage of the soil in aggregates >0.1 mm diameter (Bisal and Ferguson, 1970).

The strength as well as the size of the aggregates has to be considered, because to remain large the aggregates have to be both strong enough to withstand abrasion when dry and sufficiently stable to avoid collapse if wetted by rain. Aggregates of sandy soils are not as strong nor as water stable as those of soils with higher clay content and are consequently more easily eroded. Rain can confer some resistance to wind erosion by means of the crust that remains after the soil has dried, but abrasion by grains in saltation can break this down. The strength conferred on grain and aggregate clusters by water films makes soils immune to wind erosion so long as they remain moist and this confines the problem mainly to low rainfall areas. Chepil and Woodruff (1963) cover the contributions of Chepil in this field of research. Chepil used repeated dry sieving with a rotary sieve to measure abrasion and modulus of rupture to measure strength. For clusters too weak to be handled in a modulus of rupture test, Smalley (1970) concluded that erodibility decreases with increasing tensile strength, which for weakly cohering grains was inversely proportional to the cube of the diameter of the grains forming the cluster.

The amount of soil lost by wind erosion under specified conditions and conversely the conditions required to restrict erosion to a limited amount are given by a wind erosion equation developed by Chepil and described by Chepil and Woodruff (1963) and Woodruff and Siddoway (1965). The mass of soil lost per unit area per year is expressed as

$$E = f(I, C, K, L, V) \qquad (9.13)$$

where I is the erodibility of the soil, C is a climatic index, K represents the roughness of the surface, L is the length of the field in the direction of the prevailing erosive wind and V is an index of the vegetation cover. I is determined as mass per unit area per year from an inverse relation of erodibility to the percentage of the soil in particles or aggregates >0.84 mm as measured on a rotary sieve. K is based on an average height of ridges and clods that trap abrading particles. In L, allowance is made for the protection extending a distance $10h$ from a windbreak of height, h. V covers the quantity, kind and row orientation of plants and residues. E is assessed with the aid of charts and tables and a computer program is also available (Skidmore, Fisher and Woodruff, 1970).

10

Management of soil water

Within any given period the change, ΔW, in the amount of water stored in a zone between the surface and any particular depth in the soil can be represented as

$$\Delta W = P + I - (A + D + E). \tag{10.1}$$

Here the amounts added as precipitation, P, and irrigation, I, are balanced against the amounts lost in run-off from the surface, A, in underground drainage, D, and in evaporation from soil and plants (evapotranspiration), E, during the given period. Values can be negative for A when water runs on to the surface and for D when water comes to the root zone from below. Quantities are usually expressed in mm.

The amount of water in a soil horizon is given as a depth by

$$W = \theta z$$

where z is the thickness of the horizon and θ, its volumetric water content, is the depth of water per unit depth of soil. A soil could ideally retain about 200 mm of water for plant use to a depth of 1 m. When such a soil is thoroughly wetted at the time of seeding, it retains enough water to enable a crop to grow to maturity provided none is lost to the atmosphere except through the crop. This has been demonstrated for maize in USA by Harrold *et al.* (1959) and it serves to illustrate the scope for conserving water in soil. Similarly, irrigation in Egypt before the nineteenth century relied upon the annual flood of the Nile without any supplemental watering. The crops, typically wheat and barley, were sown after the flood receded in about October and were harvested about mid-April. The evapotranspiration over this period could hardly have been less than 300 mm. The deep alluvial soils developed from the deposits of clay and silt, renewed each year, appear to have been able to contain that amount of water.

After water has been applied to a dry soil and the supply has ceased, it continues to move downward at a rate that decreases with time. The transmission zone of soil that is saturated or nearly so during infiltration becomes a drainage zone and the hydraulic conductivity then decreases with the decreasing water content. The hydraulic gradient responsible for downward flow may change with time also, as discussed in Section 5.2, but the effect on drainage rate is likely to be relatively small when the water can drain freely. The consequent decrease in the rate of water loss from the drainage zone is illustrated in Fig. 5.6.

In this dynamic situation it is not easy to define the amount of water that is stored within the root zone long enough to be of use to plants. The conventional method is to determine the water content of each horizon of the soil, free from plants and covered to prevent evaporation, two or three days after the soil has been fully wetted (Veihmeyer and Hendrickson, 1949). This is called the field capacity and it is taken to represent the upper limit of storage of water for plant use. After the drainage period, the water content no longer decreases rapidly in sandy and well-structured soils because of a marked decrease in hydraulic conductivity in that period as air enters the larger pores. But in most soils, field capacity is ill-defined because subsequent drainage losses are not negligible. In applying the concept to field crops, the error in assuming negligible drainage is reduced when the transpiration rate is high because the loss of water to the atmosphere reduces both the hydraulic conductivity and the hydraulic gradient responsible for downward flow from the root zone. In cases where the drainage rate is of consequence after two or three days, the draining period can be extended as has been done in work cited later in this Section.

The method is not a critical one because of uncertainty about subsequent drainage losses and also because wetting procedure can affect the result (Cassel and Nielsen, 1986; Bresler, McNeal and Carter, 1982, p. 84). It is used as an approximate guide in determining the amount of rain or irrigation water needed to recharge the soil for crops and also the amount of any excess over this that will go towards the leaching of salt in irrigated land or towards the yield of water from a catchment.

The water content of a sample with undisturbed structure that has been wetted and then brought to a selected matric potential in the laboratory is often used as a measure of the upper limit of storage in a soil horizon. The selection has been made by finding the matric potential that corresponds best to water contents at field capacity or, originally, to water contents that were obtained by a centrifugal method called the moisture equivalent which was at that time used routinely as an index of storage at field capacity. On this second basis a potential of $-\frac{1}{3}$ bar or -33 kPa was first selected. But the water content at this potential is too low in sandy soils and a better general match is usually obtained at −10 kPa (Marshall and Stirk, 1949; Cassel and Nielsen, 1986). Results from these methods cannot be referred to as field capacity because the horizons are treated independently. Field capacity is measured for a profile whose horizons may differ from each other thus affecting the drainage of the profile as a whole and hence the water content and matric potential of the individual horizons. Hence no single value can be expected to represent the matric potential at field capacity in all circumstances.

Not all the water in soil is available to plants. Briggs and Shantz (1912) proposed that there was a soil water content below which plants could not extract enough water from a soil to avoid wilting. Later the lower limit of available water was termed the permanent wilting point and was measured by finding the water content at which an indicator plant such as sunflower wilts and does not recover turgor when placed in a humid atmosphere overnight (Veihmeyer and Hendrickson, 1949). The corresponding matric potential was found to be about −1.5 MPa (or −15 bars) and accordingly the lower limit is commonly taken to be the water content at that potential as determined in a pressure membrane apparatus (Richards and Weaver, 1943). There is of course no sharp cut-off in availability at this particular value. Drought tolerant plants can use soil water below this limit over long periods. But, as pointed out by Veihmeyer and Hendrickson (1927, 1955), the matric potential decreases rapidly in most soils at this stage of drying for little further yield of water to agricultural crops. It was at one time claimed that water over the whole range from 0 to −1.5 MPa was equally available, but experiments discussed in Section 12.2 show this to be untenable. Both growth and transpiration are found to decrease as soil dries towards the permanent wilting point.

The difference between the upper limit (field capacity) and the lower limit (permanent wilting point) is the range of available water content. For field applications the total amount of water as a depth contained at each limit can be calculated from $W = \theta z$ for each horizon. The difference between the two sums for all horizons to a chosen depth is the amount of available water at field capacity within that depth. Values so obtained are a useful field guide but care needs to be taken in applying them because as we have seen both field capacity and availability are ill-defined and the so-called limits may not be applicable for particular soils or plants.

The average water content at the upper and lower limits is given for soils of different textures in Table 10.1 from extensive data of Ratliff, Ritchie and Cassel (1983). In their study, the upper limit (called drained upper limit by them) was measured after draining for a period of two days or more after thorough wetting. The lower limit was measured when plants suffered water stress as shown by dormancy or premature death. Both limits were measured under field conditions. For the upper limit, drainage was allowed to continue until it became practically negligible. Thus their methods differ in some ways from the conventional methods for the limits. The data show that both the upper and lower limits increase with fineness of texture. The difference between them (the range of available water content) is least in the sands and does not vary greatly in the finer textures. However these are average values; properties other than texture can affect individual cases (see Section 2.1). A comparison of the field results with those obtained by water extraction in Table 10.1 shows that the latter are a useful approximate guide to the water content at the two limits. However, as noted above, a higher potential would give better agreement for soils of sandy texture at the upper limit. There the relative difference is large.

Table 10.1. *Water as a volume fraction stored in soils of different texture (from data of Ratliff, Ritchie and Cassel, 1983)*

Texture	Upper limit		Lower limit	
	By field drainage	At ψ of -33 kPa	By plant indicator	At ψ of -1.5 MPa
Sand	0.12	0.09	0.04	0.03
Sandy loam	0.24	0.21	0.10	0.10
Loam	0.25	0.25	0.11	0.14
Silty clay loam	0.31	0.33	0.18	0.19
Silty clay	0.35	0.37	0.22	0.25

10.2 Control of infiltration

Precipitation is an item in Equation 10.1 that is not amenable to control except through the uncertain process of cloud seeding. However, after it has reached the ground, rain or snow can be managed in various ways for different purposes. During a rainy period with $I = 0$ and $E \approx 0$, Equation (10.1) can be written as

$$P = \Delta W + D + A.$$

Surface run-off occurs whenever $P > \Delta W + D$. Water ponded in surface depressions and retained by intercepting leaves is neglected here.

If full use is to be made of rainwater to recharge the soil or to replenish a groundwater system, it is necessary for the potential infiltration rate (infiltration capacity) of the soil to be great enough to allow infiltration of as much rain as possible. Infiltration is helped by large pores and high porosity of the surface soil so that sands and well aggregated soils can maintain high infiltration rates provided the rest of the profile allows transmission of the water. Since structure of soil at the surface can be greatly altered by farming practices, these are the common means of controlling infiltration and run-off.

Tillage opens up the surface soil and lets water in freely at first but, as discussed in Chapter 9, the structure of bare soil is vulnerable to damage from falling rain. The consequent breakdown of aggregates can lead to sealing of the immediate surface, with serious reduction in the potential infiltration rate. Treatment with organic matter, gypsum, and other materials as discussed in Chapter 8 can make the structure more stable, and crop residues, plastic covers and various mulches offer protection of bare soil from raindrop impact. In a review of soil and water management in dryland agriculture of the semi-arid tropics, Kanwar (1976) notes the value of crop residues in protecting bare soil but states that residues are in demand for other purposes in India.

Interception of raindrops by plants and their litter is the most widespread way of protecting soil and maintaining satisfactory infiltration. Forested upland slopes of permeable soil in humid areas have little overland flow according to Nutter (1973) because water enters the protected soil and moves down slope in the subsurface as interflow. One consequence of this is that its velocity may then be of the order of 1 m day^{-1} instead of 1000 m day^{-1} by overland flow. This reduces the peak flow from a storm which is then less likely to cause damage by erosion or flood further down.

The amount of rainwater infiltrating into soil is increased by ponding on the surface. Natural surface irregularities provide temporary storage of water during storms and this may be aided artificially by constructing contour banks or terraces (Zingg and Hauser, 1959) or by cutting furrows. In a process called vertical mulching, deep furrows or channels are filled with plant residues to assist infiltration. On a larger scale, water is spread or ponded over intake beds or injected into wells to recharge aquifers.

In regions with limited supplies of water for stock or domestic purposes, run-off is enhanced by deliberately reducing infiltration on a catchment area (Myers, 1967; National Academy of Sciences, 1974). Slopes are made to shed water by baring a subsoil of low permeability and forming it into parallel cambered strips in the so-called roaded catchments of Western Australia, or by spraying with asphalt or sodium salts to seal the surface or by covering with plastic sheets. Infiltration can also be reduced by adding substances that increase the angle of contact between water and soil. When the angle is zero, soil wets readily, but a large contact angle resulting from the addition of hydrophobic substances such as waxes or silicones aids the shedding of water from a slope. For example, paraffin wax has been successfully used in Arizona (Fink, 1984). If the liquid–solid angle, α, is greater than 90°, the pressure, p, of water has to reach a positive value before it can enter a cylindrical capillary tube of radius, r, as is shown by Equation (2.7):

$$p = -2\gamma(\cos \alpha)/r,$$

in which γ is the surface tension of the water. Since α is about 110° for paraffin wax, a soil surface treated with wax can be made to shed water even though the pores are not sealed. Moreover, because the pores are not tubes with parallel sides, the internal geometry gives rise to an apparent contact angle that can be greater than the true contact angle. Farrell and Philip have examined this (Philip, 1971*b*) and Bond and Hammond (1970) have demonstrated microscopically that a bed of sand whose individual sand grains had a true contact angle <70° could have an apparent contact angle as large as 120° in its pores.

Some organic coatings occurring naturally on soil particles are able to prevent water entering the pores of dry soil. Soils that behave like this are called water repellent, using a term that has currency in the textile technology of waterproofing. Wet and dry patches of soil with fingering of water down the profiles result when rain falls on such soils. The causal agents are believed to be chiefly fungi. Sandy soils in grassland or wild lands are most vulnerable but tilled soil can sometimes be affected, resulting in patchy growth and reduced yield of cereal crops caused by failure of seeds to germinate in the dry spots (Bond, 1972). Details about water repellent soils are given in the proceedings of a symposium on the subject (Debano and Letey, 1969). Contact angles greater than zero probably affect the advance of water quite commonly during the wetting of dry soil although the effects are easily observed only in severe cases of water repellence.

Control of contact angles offers some interesting possibilities for water management. Experimental work on run-off is the most promising application so far but there are also possibilities for evaporation control as discussed in Section 10.4.

10.3 Run-off

Although there are methods of management of the soil surface that can increase the infiltration rate (or decrease it), most catchments out of

which water flows in rivers and streams are not managed in that sense.
The yield of water in response to rainfall (or snow-melt) is a natural
property of the catchment that has been outside the influence of man – or
at least so it has been thought until recently.

The storm hydrograph, an example of which is shown in Fig. 10.1, is
used to display the features of stream discharge as a function of time.
The dry weather flow of a perennial stream (hydrograph D) shows a
slow, quasi-exponential decrease as the storage of water underground in
the basin becomes depleted and potential gradients towards the stream
become flatter. When rain falls, it replenishes the groundwater, which,
with surface run-off, initiates a consequent increase in the stream discharge, which is defined technically to be the quantity flowing per unit
time ($m^3 s^{-1}$). The measurement of discharge is often accomplished by
stream-velocity gauging, to give the mean velocity, $\bar{U}$. The product $\bar{U}A$,
where A is the relevant cross-sectional area of the stream equals the
discharge, Q.

The water that swells the stream during and after rain comes from
three sources. In increasing order of amounts, these are the rain that
falls directly upon the stream surface, the amount that flows over the

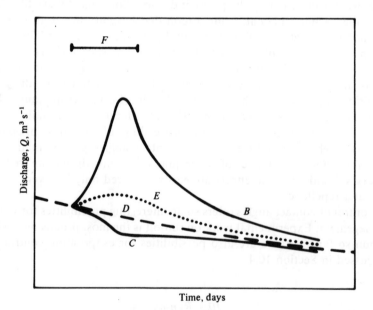

Fig. 10.1. The stream hydrograph, B, following a rainstorm of duration F, to
illustrate the variety of possible base-flow contributions to the discharge, C, D
and E. (See text for further details.)

surface of the ground (surface run-off) and underground flow. It should be noted here that the term run-off is used frequently to denote the yield of water from a catchment (in, for example, $m^3 \, yr^{-1}$) and does not then indicate any particular, hydrological process. Note also that the terms catchment, basin and watershed are used synonymously in the hydrologic literature.

Intense rain-storms may cause water to flow over the surface of the ground as a thin sheet-flow, wherever the infiltration capacity of the soil is less than the precipitation rate. Quantitatively the process may be described thus,

$$R = P - I, \tag{10.2}$$

where P is the instantaneous precipitation rate and I is the infiltration rate. But as one may observe, not all of the water that becomes ponded temporarily upon the ground surface continues to remain there or flows far enough to be collected into stream flow. The actual run-off, R, measured in the stream, given by the difference between curve B of Fig. 10.1 and any one of the other base-flow curves, is dependent upon the spatial variations of the parameters of Equation (10.2). Furthermore the infiltration rate depends strongly upon the time that has elapsed from the beginning of the storm (see Section 5.1, where $di/dt = I$ here).

It is difficult to incorporate sufficiently detailed specifications into a water budget like Equation (10.2), to enable R to be predicted accurately. Not only is the infiltration equation likely to be uncertain (see Fig. 5.5 and the discussion in Section 5.1), but also the proportion of a catchment that actually contributes to the run-off at the storm scale-of-time is small and varies in an unknown way. Betson (1964) was one of the first hydrologists to question in print the rather simple approach that had been fashionable to that date, but many hydrologists had been sceptical of rainfall–run-off models that ignored the partial area of contributing parts of a watershed. The prediction of stream discharge by use of the records of the distribution of precipitation upon a watershed now employ functions and parameters that are statistical in nature and are derived from curve fitting procedures. There is little new that can be revealed about the hydrologic processes of the catchment by such study.

Finally in this discussion of run-off, the separation of the storm hydrograph into its components helps our concepts of the role of groundwater storage in the whole basin. The dry-weather flow (curve D of Fig. 10.1) is maintained by the diffuse seepage of water into the stream along the whole of its length. Such base-flow is maintained for as long as there is groundwater storage in the basin held at a hydraulic head higher than the invert level of the stream bed. Obviously, the whole of the basin up to the groundwater divide contributes to that flow. The discharge of such a stream increases continuously and progressively from place to place in the down-stream direction. Such a stream is referred to as a gaining, or effluent stream. Conversely, a losing or influent stream is one from which water seeps away from the stream bed to a lower water table. The ephemeral streams of arid zones are of that kind.

The recession of the hydrograph of a perennial stream continues until a new rain-storm supervenes. If that storm should occur over the whole of the basin, then it would produce a steepening of the potential gradients everywhere in the vicinity of the stream and a response such as curve *E* could be observed, by some suitable technique, in the base-flow. A response such as curve *C* is caused by a rise in the stream level that is larger than the local rise in the water table and is caused by a more intense storm in an area of the basin remote from the observing station. For a further enquiry into hydrologic science the reader could consult books by Ward (1975), Rodda ed. (1976) or Eagleson (1970).

10.4 *Control of evaporation*

The main evaporative loss of water from soil is by transpiration from plants. If weeds are destroyed by tillage, herbicides or hand labour, unnecessary losses are reduced and more water is available for the crop. In the bare fallows of dryland agriculture, all plants are removed so that water may be stored for use by a subsequent crop. The efficiency of storage in fallows is not high. In regions with mainly summer precipitation, only about one-fifth of the rain falling on a fallow is stored until the time of sowing wheat in the Great Plains of the USA (Viets, 1972), the Canadian Prairies (Staple, 1960) and in Queensland, Australia (Waring, Fox and Teakle, 1958). Water already in the soil at the end of the wet winter of a Mediterranean-type climate can be stored through the dry summer by fallowing before the plants dry the soil. Here again the storage efficiency is low, as shown by data of French (1963) for wheat crops in southern Australia. But although the amount stored in these examples may be of the order of only 50 to 100 mm, the gains are often substantial because each mm improves the yield of wheat by an average of about $10 \, \mathrm{kg \, ha^{-1}}$, according to the North American data reviewed by Staple and Viets.

Evaporation directly from soil is responsible for losses not only from completely bare soil but also from soil incompletely covered by leaves of a crop growing on it. During the whole growing period of an annual crop, as much as half of the water evaporated to the atmosphere may come directly from the soil (Harrold *et al.*, 1959).

Bare wet soil loses water by evaporation at about the same rate as a free water surface. As shown in Chapter 2, the vapour pressure of water in moist soil differs little from that of free water over a wide range of water content. Accordingly the rate of evaporation from initially saturated soil is at first fairly constant under given atmospheric conditions (Fig. 10.2) and it depends on these conditions rather than on the degree of saturation of the soil. This constant stage is followed (in soil that is not freely supplied with water from a water table) by a falling rate stage when soil conditions take control by limiting the rate of supply of water to the surface as the hydraulic conductivity decreases in the drying soil. The lower the potential evaporation rate, as measured from a free water surface, the longer does the constant rate persist. The constant and falling rate stages of drying are well defined in laboratory experiments (Fisher, 1923; Penman, 1941) and their existence has been explained theoretically by Philip (1957d). When the surface soil becomes sufficiently dry, a third stage of drying is reached in which water can no longer move to the surface as a liquid. This is a stage of greatly reduced evaporation due to the retreat of the zone of evaporation below the surface. Water then moves as water vapour at a slower rate through the layer that has dried. This third stage is complicated by higher thermal gradients than exist in the earlier stages when much of the incoming energy is expended in evaporation at the wet surface. All three stages of drying have been found by Idso *et al.* (1974) to occur under field conditions.

During the drying of fairly compact soil, evaporation appears to occur mainly in a thin zone at the top of soil that is moist enough to conduct water in the liquid phase. Marshall and Gurr (1954) found that most of

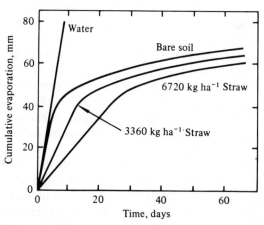

Fig. 10.2. Effect of cover of straw on evaporation from soil (Bond and Willis, 1969).

the water lost below a depth of 1 cm in moist soils with a wide range of initial water content had moved upwards in the liquid phase before evaporating from the top 1 cm. Evidence of this was obtained from the amount of salt deposited there. Gardner and Hanks (1966) and Fritton, Kirkham and Shaw (1967) found at a more advanced stage of drying that this evaporation zone remained about 1 cm thick as it retreated downwards a few centimetres below the surface.

From this discussion it follows that controls devised to reduce the rate of evaporation directly from soil must depend primarily on (1) reducing the potential rate of evaporation at the surface by modifying atmospheric conditions there or (2) reducing the amount of water retained near the surface when water is added to soil. If the evaporation rate can be reduced in either of these ways, plants will be able to make use of the water before it is lost and water added to bare fallows will have the opportunity to move to greater depth, where it will be longer preserved.

Mulches of straw or other materials covering the surfaces as in stubble mulching or minimum tillage reduce the potential evaporation at the surface of the soil by reducing the amount of radiant energy absorbed and minimizing air turbulence. The rate of evaporation during the constant rate stage is therefore reduced (Fig. 10.2) and much water can be saved over periods of the order of one or two weeks (Bond and Willis, 1969). Over longer periods the benefits decrease and in fallows the method is therefore most successful when rains are frequent enough to cause deeper penetration of the water stored from previous falls (Gardner, 1959). Wheat straw retained by stubble mulch tillage (which undercuts plants without burying them) increased storage by 10 to 70 mm in the long-term field experiments of Greb, Smika and Black (1970) on fallows in the Great Plains of USA. The effectiveness of a mulch increases with its bulk as shown in Fig. 10.2 but Tanaka (1985) did not find any significant difference between stubble mulching and herbicide treatment for fallow although the latter left more residue. There is also an effect of packing as shown by Unger and Parker (1976) who found that cereal straw was more effective than the same quantity of sorghum or cotton residues because it packed down better. Covers of plastic sheets, paper, asphalt and gravel and reflective surfaces are highly effective in controlling evaporation but they are applied only in intensive culture, engineering works or experimentation.

Tillage controls evaporation mainly through the control of weeds. Tillage of soil after rain to create a loosened soil mulch that will not conduct liquid water is not effective because a lot of water is lost before the soil can be tilled. Early misconceptions about this were responsible for much unnecessary tillage (Keen, 1931, p. 271). Nevertheless structure does affect rate of evaporation in various ways and research continues on possibilities of applying this. Burov (1954) showed that the flux of water through aggregates during the drying of soil was least when they were sized between 0.25 and 3 mm. This was confirmed by Holmes, Greacen and Gurr (1960), who found that evaporation from soil subjected to radiation and wind was greater for both untilled and cloddy soil than for soil of fine tilth containing aggregates mainly about 2.5 mm in diameter. Liquid transfer was held responsible for losses from the untilled soil and turbulent transfer of vapour for losses from the cloddy layer. To make use of structure in controlling evaporation, Hadas (1975) suggested that surface soil should be brought to the desired condition with water stable aggregates of the right size before it is wetted. Tillage can also save water when cracks that open to the surface are closed over by harrowing (Gavrilov, 1961). Adams *et al.* (1969) showed that evaporation from a simulated shrinkage crack increased with wind speed and was reduced greatly by both straw and gravel mulches. The effective diffusion coefficient of water vapour in soil with cracked or cloddy structure can be 100 times that of molecular diffusion because of micro-oscillations in the air pressure caused by wind (Farrell, Greacen and Gurr, 1966).

Any treatment that ensures deeper penetration of water in the profile helps protect it from evaporation. Thus, much of the extra water saved by adding straw mulches to fallows was at a greater depth than 60 cm in the experiments of Greb *et al.* (1970). Similarly a given amount of water penetrates further into a coarse than into a fine textured soil and is less readily lost in semi-arid areas (Alizai and Hulbert, 1970). By forming a mulch of large aggregates that had been treated to make them water-repellent, Fairbourn and Gardner (1975) caused less water to be retained at the surface and so decreased the rate of evaporation, especially in the first day after water was applied. Hillel and Berliner (1974) have shown that for this purpose the hydrophobic aggregates need to be large enough (2 to 5 mm) to allow water to infiltrate freely. Water can also be made to penetrate more deeply in selected sites by diverting it there from neighbouring areas. This is done both on a watershed basis as run-off irrigation and on a small scale for individual fruit trees as in experimental work of M. Evenari in the Negev Desert in Israel (National Academy of Sciences, 1974). On a small scale, shedding of water can be aided and evaporation reduced by placing plastic covers over ridges between crop rows, but about 90 per cent of the area needs to be covered, according to Willis, Haas and Robins (1963).

The controls we have discussed concern in the main the constant rate stage of evaporation when the soil is wet and when water movement within it is usually treated as an isothermal process. However, thermal gradients associated with diurnal changes in soil temperatures affect both

Management of soil water

the amount and direction of movement of water in soil. Their significance during the drying of soils is discussed in Section 4.6.

10.5 Irrigation

About 200 million hectares of land are irrigated throughout the world. Most of this is in arid and semi-arid regions where the crops depend primarily on irrigation but some is also in sub-humid areas where irrigation supplements the natural rainfall. About half of the world's irrigated land is in China and India. An amount of water of the order of 1000 km^3 is required annually for irrigated land and great engineering projects have been undertaken for storing and distributing it. However, results have not always compensated for the costs, partly because of problems associated with the soil. Comprehensive reviews of irrigation and its problems are given in books edited by Hagan, Haise and Edminster (1967) and Kovda et al. (1973).

Soils of widely different physical properties are used for irrigation. Sands can be successfully used for oranges and clays for rice. Soils of high hydraulic conductivity can be irrigated satisfactorily with sprinklers and those of low by flooding. The physical problems that are most characteristic of irrigated soils in semi-arid areas arise from the presence of too great an amount of soluble salts in the soil (saline soil), of sodium ions in the exchange complex (sodic or alkali soil) or from both (saline-sodic soil). Salinity is dealt with separately in Section 10.6, where it is shown that prevention and cure lie in controlling water movement and water quality. Sodicity has been dealt with in Section 8.9 where it was shown that exchangeable sodium percentage (ESP) greater than 15 was associated with serious instability of structure and low hydraulic conductivity. Sodic effects are recognizable with ESP as low as 6. Sodic soils are commonly treated with gypsum to exchange Ca^{2+} for Na$^+$ because gypsum deposits are often conveniently located near irrigated areas in semi-arid regions. Other treatments include calcium chloride, sulphur and sulphuric acid, as discussed in Section 8.9.

Irrigation water has the advantage over natural precipitation of being usually under control as to timing and amount of application. However, the irrigation efficiency, as represented by the fraction of the water obtained from the reservoir or other source that is used on the farm in evapotranspiration, is often less than 0.5. Data by Erie (1968) for 22 irrigation projects examined by the US Bureau of Reclamation show an average water conveyance efficiency from water source to farm of 0.62 and a water application efficiency of 0.58 for the fraction of the water delivered to the farm that is used in evapotranspiration. The irrigation efficiency overall (obtained as the product of these two fractions) is thus only 0.36 for this group of projects. Conveyance losses are due to seepage, evaporation and transpiration by vegetation on the way to the farm. At the farm, losses occur when more water is applied to some of the area than is needed. Unnecessary percolation both at the farm and on the way to it is not only wasteful but it can contribute to a high water table and to secondary salinization of soil.

Conveyance efficiency is improved by lining of supply channels to reduce seepage and by controlling vegetation where they are not lined. For high water application efficiency at the farm no more water should be applied than is required at any part of the farm. However, because water cannot be distributed over an irrigation area exactly as desired, some parts inevitably receive too much if an insufficiency is to be avoided in others. Since insufficient applications of irrigation water can severely depress yields, the general tendency is to apply more than is necessary, especially if the water is cheap. The benefits of a liberal supply are immediately seen whereas the warning signs of serious damage from water tables and salinity may not show themselves for some years and then perhaps only to those irrigators in the lower parts of an irrigation area.

The quantity of water to be applied and the frequency of application depend on soil, plant, meteorological and economic conditions. The usual object in an irrigation is to recharge the soil to field capacity throughout the zone from which roots withdraw the water. Then, after the soil has been dried by evapotranspiration to some allowable limit, another application will be needed. The amount to be applied can be estimated from the difference between the water content of the dried soil and its field capacity. Alternatively the excess of evapotranspiration over precipitation is calculated from meteorological data for the intervening period, as discussed in Chapter 12. This excess represents the water deficit in the soil that has to be made good by irrigation. The suction to which the soil should be allowed to dry before being irrigated again varies with the crop as well as with the economics of the project including the cost of water and labour. It may be of the order of 10 m in the case of vegetables or 100 m for a grain crop. The effect of this limiting suction on water uptake and growth is discussed in Chapter 11.

Local irrigation schedules often become established from simple observations of changing soil water, plant and weather conditions. However, the timing of an application can be assessed more critically when suction is measured by tensiometers or gypsum blocks, water content is measured by the neutron moisture meter or water deficit is assessed by monitoring the evapotranspiration since the last application. Evapotranspiration is widely used and meteorological services report it daily for the timing of irrigations in many regions. With an occasional check on water content to avoid cumulative errors, it is satisfactory for estimating water deficits. Advantages and limitations of the various methods for scheduling irrigation are discussed by Campbell and Campbell (1982). Plant, soil and meteorological conditions that affect the time that can be allowed between irrigations are given in Table 10.2 which is selected from a more detailed account by Hagan (Kovda *et al.*, 1973, p. 247).

Table 10.2. *Effect of plant, soil and meteorological conditions on frequency of irrigation*

	Conditions that tend to shorten the period between irrigations
Plant	Sparse rooting. Vegetative parts of plant are to be harvested.
Soil	Small amount of water storable in the root zone within the allowable suction limit. High content of dissolved salts in soil water.
Meteorology	High rate of evaporation during the period. No rain during the period.

Land is irrigated by running the water down furrows or over the soil surface between parallel check banks, by flooding and ponding it between contour banks, by sprinklers, or by dripping water from a distribution system of plastic tubes. Flooding systems are suitable for soils of low permeability and are commonly used for rice. Furrow and check bank systems are widely used (for example for grape vines and pastures, respectively) but they are not well suited to soil of high permeability because too large a part of the applied water then enters the upper part of the run. Sprinklers are used for many different crops on soils of high or medium permeability. They distribute water reasonably uniformly over soils of uneven terrain and differing texture. Systems have been developed that enable large areas to be traversed mechanically by sprinklers.

Drip or trickle irrigation does not expose as much wet soil to direct evaporation as other methods do and it supplies water continuously at a high matric potential to the zone where the roots are. Because the whole surface area is not wetted, there is a tendency for salt to accumulate in the surface soil between the dripped zones and this can cause damage if washed down into the root zone by rain (Bernstein and Francois, 1973). This happens also to a lesser degree in furrow irrigation. Further details of the principles and expanding application of drip irrigation are given by Bresler (1977).

10.6 Soil salinity and its control

The soil solution

Soil water contains dissolved salts which may be present in sufficient concentration to restrict water uptake by plants. Work has to be done in extracting water from the solution as well as from the soil matrix, and the solute (or osmotic) potential acts additively to the matric potential in this. Thus the soil water potential, as given by the sum of these two, influences uptake, transpiration and growth. The effect on growth is illustrated in Fig. 10.3 which shows that the total suction, representing the sum of matric suction and solute suction (or osmotic pressure) was closely related to the growth of beans.

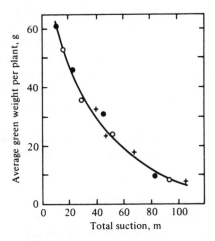

Fig. 10.3. Relation of plant growth (beans) to the average total suction of the soil water (Wadleigh and Ayres, 1945). ●, low matric suction; ○, medium matric suction; +, high matric suction.

There is however a possible qualification to be made concerning the relative effects of the matric and osmotic components of soil water potential on plant growth. A decrease in matric potential with drying is accompanied by a decrease in the hydraulic conductivity of the soil and more particularly there is an increase in the resistance offered to the flow of water at the contact of roots with soil and within the roots themselves. This imposes an additional restriction on water uptake by plants. Thus while the matric and osmotic potentials are additive in their effects on growth and transpiration, the effect per unit potential may be greater for matric potential than it is for osmotic potential. This is not evident in the three sets of data in Fig. 10.3 which fall uniformly on a single curve rather than separating into three. However evidence of a greater effect of matric potential on plant growth is given by Gingrich and Russell (1957) and more recently by Parra and Romero (1980) and others (see Meiri, 1984). Shalhevet and Hsiao (1986) suggest that in addition to any flow restriction that accompanies a lowering of the matric potential, incomplete osmotic adjustment within the plant under water stress might also be a cause. The role of osmotic adjustment within plant tissues will be taken up in Section 12.1.

The effects of solutes on growth are not confined to osmotic effects. Some of the ions in soil solutions can have specific effects on plants and so retard plant growth independently of the osmotic pressure of the soil solution. Bernstein (1975) has reviewed this aspect.

The soil water potential, representing the sum of matric and osmotic (or solute) potential, can be determined directly by the psychrometric methods given in Section 3.3. The freezing point depression of water in soil samples has also been used for this purpose but the method is not satisfactory for routine use for reasons set out by Holmes *et al.* (1967). It remains now for us to deal with osmotic potential and its measurement, recalling from Section 2.2 that the osmotic potential expressed on a volume basis is equal but of opposite sign to the osmotic pressure or solute suction.

The dependence of osmotic pressure, π, on concentration, first demonstrated by Pfeffer in 1877, is given in van't Hoff's equation

$$\pi V = nRT$$

where n is the number of moles of a solute in a volume, V, of solution, R is the gas constant ($8.314 \, J \, K^{-1} \, mol^{-1}$), and T is the temperature in kelvin. This equation closely resembles the gas law. The kinetics of the solute molecules and ions distributed through a volume of solution is analogous to that of the molecules comprising a volume of gas. From it we may calculate that a solution with a concentration of 1 mole per litre (or 1 mol per $10^{-3} \, m^3$) would have an osmotic pressure given by $\pi \times 10^{-3} = 1 \times 8.314 \times 273$ at 0 °C, so that $\pi = 2.27 \times 10^6$ Pa. This equation applies for an ideal solution of a non-dissociating substance. For a dilute solution of a salt such as sodium chloride that dissociates into two species of ions each of which contributes to osmotic pressure, π is nearly twice as large. Because of uncertainties about the composition of a soil solution and the degree of dissociation of its solutes, the van't Hoff equation does not provide an accurate means for determining π for soil water.

Osmotic pressure of a solution can be measured directly by finding the excess pressure reached when the solution and the water are separated by a membrane through which water but not solutes can pass. Because of difficulties with semi-permeable membranes, they are not suited to routine work. Consequently use is made of depression of the vapour pressure and of the freezing point of water in a solution and the electrical conductivity of solutions. All of these properties depend, as osmotic pressure does, on the concentration and degree of dissociation of the solutes. Osmotic pressure, π, and osmotic potential, Π, are obtained from the vapour pressure of the solution using the psychrometric methods described in Chapter 3 and calculations similar to the examples given with Equation (2.6) for matric potential. With potential expressed per unit volume of water, this becomes

$$\Pi = -\pi = RTV_m^{-1} \ln (e/e_0), \tag{10.3}$$

where R is the gas constant, T is the temperature in kelvin, V_m is the molar volume (the volume of 1 mole of liquid water), e_0 is the vapour pressure of water and e is that of the solution. $V_m \rho$ is equal to M as used in Equation (2.6) where ρ is the density of water. It may be noted that, when osmotic potential, Π, is expressed on the volume basis as it is here, it has the same dimensions as osmotic pressure, π, but is of opposite sign.

The freezing point depression of an ideal solution with a concentration of 1 mole per litre is found by experiment and theory to be 1.86 °C. Also the osmotic pressure of this solution is 2.27×10^6 Pa at 0 °C as calculated above. If the osmotic pressure of some other solution needs to be known, it can be calculated proportionally from its freezing point depression, ΔT, by means of $\pi/(2.27 \times 10^6) = \Delta T/1.86$ or $\pi = 1.22 \times 10^6 \Delta T$ Pa. For solutions extracted from soils and plants, this is used by Crafts, Currier and Stocking (1949) and US Salinity Laboratory Staff (1954) in a modified form equivalent to

$$\pi = 1.221 \times 10^6 \Delta T - 2.1 \times 10^3 (\Delta T)^2 \text{ Pa}. \tag{10.4}$$

This is for a solution at 0 °C or 273 K. For some other temperature T the osmotic pressure is given by $\pi_0 T/273$.

In the third type of method, use is made of an empirical relation established by the US Salinity Laboratory Staff (1954) between the osmotic pressure and the electrical conductivity of river waters and solutions extracted from soils in the western states of USA. This is

$$\pi = 3.6 \times 10^5 \gamma \text{ Pa}, \tag{10.5}$$

where γ is the electrical conductivity at 25 °C in siemens per metre (S m^{-1}). It is necessary to note that in irrigation practice the conventional symbol is EC and values are expressed in units of dS m^{-1}.

The salinity of soils can be represented by the amount of soluble salts in a given mass of dry soil or indirectly by the electrical conductivity of a solution extracted from the soil. Methods are described by US Salinity Laboratory Staff (1954), Beatty and Loveday (1974), and Bresler, McNeal and Carter (1982). For the electrical conductivity measurement, the solution can be taken from saturated soil by suction. The electrical conductivity of the saturation extract, EC_e, in units of dS m^{-1}, is widely used as a measure of soil salinity. Guides have been established showing how yield decreases with increasing salinity after a threshold value of EC_e has been reached that varies with the type of plant. At 4 dS m^{-1} or greater soil can be considered to be saline and the yields of most plants are restricted between 4 and 8. Sensitive plants (e.g. beans and carrots) may be affected between 2 and 4, while some salt tolerant crops (e.g. barley and cotton) may yield satisfactorily between 8 and 16 dS m^{-1}. Tables and diagrams are available showing how the yields of many different plants are affected by salinity (e.g. Bresler, McNeal and Carter, 1982). In terms of salt content, the level of salinity represented above by 4 dS m^{-1} corresponds approximately to 1.5×10^{-3} g total dissolved solids per g of dry soil in soils of fine texture, and less (down to about half of that) in soils of coarser texture.

For the monitoring of salinity in irrigated land, sensors imbedded within porous ceramics can provide continuous measurements of the electrical conductivity of the soil solution *in situ* for soil that is fairly wet. Suction sampling is also used to withdraw solutions from soil in the field by means of porous ceramic cups when the soil is wet enough. These *in situ* methods enable changes in salinity with time to be followed. When the salinity of a large area is to be assessed soil samples are taken and aqueous extracts are obtained from them for analysis or for the measurement of electrical conductivity. It is possible also to measure the electrical conductivity in the field by a four-electrode method, by electromagnetic induction or by time domain reflectometry. These methods provide less accurate measurements than the sampling methods but they represent larger sampling volumes which is an advantage for broad surveys. A full account of field methods is given by Rhoades (1984).

Causes of salinity

Salts originate from weathering processes in soil and accessions brought to the soil in rain, dust, surface water or ground water. When there is sufficient leaching of the soil, as in humid localities, they do not constitute a problem. In soils of arid and semi-arid localities, a salt profile is developed when drainage is absent or inadequate. The content of soluble salts may be relatively low near the surface, and for a given environment this zone extends more deeply the sandier the soil because with a smaller field capacity to be satisfied in sandy soils the water from an effective fall of rain leaches the salt down further. If the roots of vegetation dry the subsoil between wet periods, the salt remains and the profile persists with only minor seasonal variations. Changes in vegetative cover brought about in dryland agriculture or increased water use in irrigation agriculture can upset this equilibrium, the first gradually over a long period and the second rapidly in the first year of irrigation.

As discussed in Section 6.2, the clearing of a natural perennial vegetation of trees and shrubs and the substitution of crops and pastures that use water for only a part instead of for the whole year can so reduce the annual transpiration that water accumulates in the subsoil and ultimately seeps down slope. Fallowing to conserve water for cereal crops contributes to the seepage. The shallow water tables that develop as a consequence on slopes and in valleys allow water to evaporate continuously from the surface of the soil which then becomes saline in regions of relatively low rainfall. Preventive action includes the use of deep rooting perennials such as lucerne to dry the subsoils of intake areas over occasional periods of a few years (Rowan, 1971). Doering and Sandoval (1976) have found interceptor drains to be effective in saline seeps on farms in the northern Great Plains of USA. Early detection of new seeps has been urged by Halvorson and Rhoades (1976) who have used the four-electrode conductivity technique for measuring salinity rapidly in the field without having to take soil samples.

When irrigation water is introduced to semi-arid or arid areas, the soluble salts are rapidly redistributed. In general the salt content is reduced greatly within the top metre of soil as in the example in Table 10.3 for a saline soil in southern Kazakhstan, USSR, that was leached with four different amounts of water applied in irrigations of 0.4 m several days apart (Kalinin, 1969). Chlorides were readily removed and the salts, $CaSO_4$ and $Ca(HCO_3)_2$, that persisted and are represented in the total dissolved solids of Table 10.3, were due to gypsum and lime in the soils.

Table 10.3. *The leaching of salt from saline soil* (*after Kalinin, 1969*)

Amount of water used for leaching m	Salt removed from top metre of soil as fraction of initial content	
	Total dissolved solids	Chlorides
0.4–0.5	0.30	0.84
0.7–1.0	0.55	0.89
1.1–1.4	0.70	0.96
1.5–2	0.75	0.96

Leaching may, however, be followed by secondary salinization of soil in localities where a water table raised by seepage, comes sufficiently close to the surface to stimulate evaporation of water directly from the soil, thus increasing its salinity. If natural drainage through the soil is not able to prevent this, artificial underground drains have to be introduced to carry drainage water and dissolved salts away. Salinity is certain to be a problem if neither natural nor artificial drainage is satisfactory, because otherwise salts brought in with the irrigation water will tend to accumulate.

Salt balance in irrigated land

The water balance of irrigated land can be expressed for a given period such as a year as

$$I + P = E + D, \tag{10.6}$$

where I and P are, respectively, the amounts of water (as depths) gained in irrigation and precipitation and E and D are those lost in the same period in evaporation plus transpiration (evapotranspiration) and in drainage below the root zone. Gains or losses from surface drainage and changes in soil storage are here neglected. When each of the items in Equation (10.6) is multiplied by the concentration C of salts that it contains, we obtain the salt balance for the period as

$$C_I I + C_P P = C_E E + C_D D.$$

If C_P and C_E are taken as zero, the leaching requirement (US Salinity Laboratory, Staff, 1954) is obtained from

$$D/I = C_I/C_D = \gamma_I/\gamma_D, \tag{10.7}$$

where γ_I and γ_D are, respectively, the electrical conductivity of irrigation and drainage waters. The leaching requirement is that fraction of the irrigation water, D/I, that must drain below the root zone for C_D to have a specified value at the end of the period. Thus if the maximum concentration to be allowed in the drainage water is five times that of the irrigation water, one fifth of the irrigation water must drain through. Equations (10.6) and (10.7) may be combined to give

$$I = (E - P)C_D/(C_D - C_I), \tag{10.8}$$

which defines the amount of irrigation water of concentration, C_I, that is to be applied if the drainage water is to be held at a specified concentration, C_D. The equation shows also how the amount required is reduced by precipitation.

The salt concentration of drainage water that must not be exceeded in applying Equation (10.8) is judged in relation to the tolerance of the particular crop to salinity. Bernstein (1975) quotes maximum electrical conductivity to be allowed in drainage waters as 40–45 dS m^{-1} at 25 °C for tolerant crops, 32–35 for moderately tolerant crops and 14–16 for sensitive crops, if normal yields are to be obtained. It will be noted that these conductivities are higher than the maxima quoted above for the saturation extracts controlling yields. This is partly because salinity increases with depth within the root zone where water without all its solutes is being continually withdrawn by plants. Hence the roots may be in solutions that range in concentration from that of the irrigation water near the surface to that of the drainage water at the bottom of the zone. The greater the leaching fraction D/I, the lower will be the salinity of the drainage water for a given quality of irrigation water. Hence salinity profiles are relatively uniform when the leaching fraction is large (near one) but salinity increases greatly with depth as it approaches zero. Salinity profiles showing these characteristics have been published by Bower, Ogata and Tucker (1969) from experimental results and by Raats (1975) from theoretical analysis (Fig. 10.4).

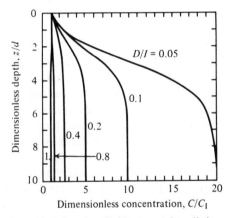

Fig. 10.4. Effect of leaching fraction D/I, on steady salinity profile of soil under a plant cover that is irrigated with water of concentration C_I. C is the concentration at depth z and d may be interpreted as an equivalent rooting depth (Raats, 1975).

The salt concentration of water in the soil increases with time following an irrigation as water is withdrawn for transpiration and in evaporation. But, as pointed out by Yaron *et al.* (1972), this withdrawal affects the matric more than the osmotic component of the soil water suction in relation to the uptake of water by plants. While matric suction can increase by two orders of magnitude in the drying of soil from field capacity to permanent wilting point, the osmotic pressure increases only two- or three-fold due to the increase in salt concentration.

10.7 Miscible displacement

So far we have dealt with salts as though their movement can be identified with the macroscopic movement of the soil water. Such an assumption neglects the natural dispersion of substances of all kinds dissolved in the water when it moves through the soil pore spaces. The solute dispersion about an interface between a resident and a different, displacing solution that are miscible with each other, is caused by two physical processes. These are molecular diffusion consequent upon the thermal agitation of all the molecular and ionic entities present and mechanical dispersion caused by fluid velocities that are different from place to place in the porous medium when the solutions are in motion.

Mechanical dispersion is more often termed hydrodynamic dispersion. In the derivation of the law of Hagen–Poiseuille for laminar flow through a small-bore tube (see Childs, 1969, p. 195), the velocity of the fluid, v, along any streamline as a function of its distance, r, from the axis of the tube appears as the expression

$$v = g\rho\Delta\phi(R^2 - r^2)/4\eta l, \qquad (10.9)$$

where R is the radius of the tube $(0 < r < R)$, $\Delta\phi$ is the hydraulic potential difference between the ends of the tube of length l, ρ is the density of the fluid and η is its dynamical viscosity. The ratio $\eta/\rho = \nu$ is sometimes written explicitly and referred to as the kinematic viscosity. Inspection of Equation (10.9) shows that fluid moving along the axis of the tube where the velocity is largest would eventually move far ahead of the fluid that began its flow near the wall. A plane displacement front entering the inflow end of the tube begins to assume the shape of an ellipsoid, whose apex would advance along the axis at a rate faster than the interface at any intermediate position between the walls and the centre line. The mean concentration of solute assessed at any cross-section of the tube upstream of the apex of the ellipsoid would obviously depend upon the relative, sectional areas occupied by the originally resident fluid and the fluid displacing it, which flows in a central circular area surrounded by an outer annulus of the original fluid that has not yet been displaced. The distribution of mean solute concentration at any cross-section along the tube is therefore suitably referred to as hydrodynamic dispersion.

Taylor (1953) presented the mathematical description of such dispersion, in a small-bore tube, to which he added the refinement of allowing for the molecular diffusion across the ellipsoidal surface of the actual interface, which theoretically should become coincident with the walls of the tube only as time of displacement approaches infinity. These ideas become very helpful when they are taken into the consideration of dispersion of solutes in a fluid flowing through a porous medium. Within any one pore, flow velocity is least near the walls. It is also less in small than in large pores. The overall effect of the mean velocity of the liquid on dispersion is given empirically (Biggar and Nielsen, 1976) by

$$D = D_m + \alpha \bar{v}^n, \tag{10.10}$$

where D is the dispersion coefficient for flow in the porous medium, D_m is the coefficient of molecular diffusivity for the solute, $\bar{v}$ is the actual mean velocity $(= v'/\theta$, where θ is the volumetric water content, v' is the Darcian velocity) and n is an empirical exponent whose value is often close to unity. The proportionality constant, α, not necessarily constant, is referred to as the dispersivity, with dimensions L, since a diffusion coefficient always has dimensions L^2T^{-1}, independent of whatever entity is diffusing. Note that the dispersion coefficient, D, of Equation (10.10) is appropriate for flow in the porous medium. Denoting by D_0 the molecular diffusion coefficient of the solute in bulk aqueous solution, we have $D_m = b\theta D_0$, where b is a tortuosity factor for flow in a porous medium and θ is the volumetric water content.

The dispersion coefficient has directional properties to the extent that the hydrodynamic dispersion is more vigorous in the direction of the streamlines than in directions normal to the streamlines. The mathematical description of the phenomenon is therefore rendered more difficult because the use of tensor calculus is required for a comprehensive, three-dimensional statement of the equations. If we restrict our treatment to one-dimensional considerations, the motion of solute in the flowing solution can be described by the following equations. Representing the flux of solute transported in the positive x-direction across a section normal to that direction by q_s, we have

$$q_s = \theta C \bar{v} \tag{10.11}$$

where the concentration, C, is expressed as $kg\ m^{-3}$ of solution, $\bar{v}$ is the mean velocity of the solution in the pore space and θ is the fraction of the total-soil cross-sectional area through which the soil solution flows $(m^2\ m^{-2})$, i.e., the volumetric water content. The convection of solute without dispersion is given by Equation (10.11).

The additional flux of solute contributed by dispersion is given by

$$q_s = -D\frac{d(\theta C)}{dx}, \tag{10.12}$$

where D is the dispersion coefficient as given in Equation (10.10). It is permissible to add Equations (10.11) and (10.12) to give the total flux, thus

$$q_s = \theta C \bar{v} - D \frac{d(\theta C)}{dx}. \tag{10.13}$$

The divergence of the flux is given by (see Appendix C)

$$-\frac{\partial q_s}{\partial x} = \frac{\partial(\theta C)}{\partial t} - \text{sources} + \text{sinks}. \tag{10.14}$$

The expository virtue of the mathematical development is improved if we assume that there are no sources or sinks of the solute material in the region governed by the equations, so Equations (10.14) and (10.13) in combination become

$$-\frac{\partial(\theta C \bar{v})}{\partial x} + \frac{\partial}{\partial x}\left(D \frac{\partial(\theta C)}{\partial x}\right) = \frac{\partial(\theta C)}{\partial t}. \tag{10.15}$$

Simplification of Equation (10.15) allows it to be written as

$$D \frac{\partial^2 C}{\partial x^2} - \bar{v} \frac{\partial C}{\partial x} = \frac{\partial C}{\partial t}, \tag{10.16}$$

but the simplifications impose restrictions upon its use. Firstly, dividing Equation (10.15) through by θ implies that the water content of the soil is constant in time and space. This limits the application to saturated media, in which $\theta = \varepsilon$, the total porosity. Then, the dispersion coefficient, D, must be assumed to be constant as then must the mean velocity, $\bar{v}$. That is, Equation (10.16) is appropriate for homogeneous and isotropic soils, through which the fluids are made to flow at a steady rate. These are the physical conditions of many laboratory experiments.

Consider what happens when a resident solution in the pore spaces of a soil column begins to be displaced by a solution with a different concentration. The interface between the original and the displacing fluid, at the inflow end of the column, is initially planar and abrupt. But the fluids pass more quickly through the larger pore spaces and their interconnecting passage ways so that the interface becomes more and more diffuse as time goes on, though it can be regarded as planar still, on the macroscopic scale. If the solutes did not disperse at all, the interface would remain abrupt and move through the column with velocity $\bar{v} = v'/\theta$. It is convenient mathematically to substitute for the coordinate system fixed in laboratory space, for which Equation (10.16) is appropriate, a system with a moving origin, namely

$$\xi = x - \bar{v}t. \tag{10.17}$$

Then Equation (10.16) becomes

$$D \frac{\partial^2 C}{\partial \xi^2} = \frac{\partial C}{\partial t}. \tag{10.18}$$

There are some subtleties in this approach, ignored in this treatment, which have been well described by Smiles and Bond (1982) and Bird *et al.* (1960). It is sufficient here to observe that ξ is a distance measured from that position which the interface would occupy if it were still abrupt and had been convected with velocity $\bar{v}$. With the initial and boundary conditions,

$$t = 0, \quad x > 0, \quad C = C_n$$

$$t > 0, \quad x = 0, \quad C = C_0 \qquad\qquad (10.19)$$

$$t > 0, \quad x \to \infty, \quad C = C_n$$

the solution of Equation (10.18) is

$$\frac{C - C_n}{C_0 - C_n} = \frac{1}{2}\left(\operatorname{erfc}\frac{\xi}{(4Dt)^{1/2}}\right). \qquad\qquad (10.20)$$

The error function complement (erfc) was defined in Chapter 5.1 in connection with the horizontal infiltration of water into dry soil.

Fig. 10.5 displays the dependence of the normalized concentration, C/C_0, upon the time of infiltration at a fixed location in the column (curve A) and the distribution of the concentration of solutes along the column at a fixed time (curve B). The use of C/C_0 instead of $(C - C_n)/(C_0 - C_n)$ is not only convenient but also recognizes that many experiments have been performed by the displacement of pure water ($C_n = 0$) by an electrolyte solution which is easily monitored by its electrical conductivity.

The data points for the construction of Fig. 10.5 were calculated using a dispersion coefficient of 1.6×10^{-9} m^2 s^{-1}. Biggar and Nielsen (1976) suggested such a value as falling near the middle of the range of values for a variable field soil. The mean physical velocity of the fluid flow was chosen to be 9.3×10^{-8} m s^{-1} (approximately 8 mm day^{-1} or an infiltration rate of $\theta \times 8$ mm day^{-1}) so that, after the chosen time of 1.075×10^7 s (approximately 124 days) a volume of solution exactly equal to the volume of pore space in a one-metre segment of the column would have passed through (one pore-volume). The slight asymmetry of curve A about the vertical line $t = 124$ days is real and is caused by the solute having a range of times during which dispersion is able to occur.

Departures from the idealized behaviour exemplified in Fig. 10.5 are frequently observed. Fig. 10.6 shows two departures from the simple break-through curve, A of Fig. 10.5. Curve C would be appropriate to infiltration and dispersion of an electrolyte solution through a well-aggregated clay bed, when the anion is used to detect the displacing solution. Exclusion of the anion from the region of the electrical double-layers of the clay allows a more rapid transit of tracers such as Cl$^-$ or

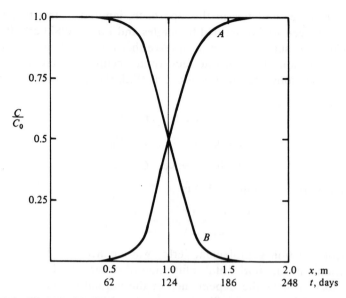

Fig. 10.5. The miscible displacement of pure water by a fluid with solute concentration, C_0, through a column of semi-infinite extent. Curve A shows C/C_0 calculated at a fixed distance (1 m) from the entry end v. time. Curve B shows C/C_0 calculated at a fixed time v. distance along the column. (See the text for further details.)

Br^- which are frequently used. Or the exclusion could be mechanical, caused by a small hydraulic conductivity of the clay peds. The two processes probably operate together in natural systems. Curve D is appropriate to cation absorption, which causes a delay in appearance of the tracer, together with a retention in the soil of a significant quantity of the amount applied at the inflow end.

White (1985) reviewed experimental evidence as well as theoretical arguments for the influence of macropores upon the dispersion of dissolved and suspended matter in the soil water. Generally, the rapid passage of fluid through channels that are continuous flattens the breakthrough curve so that the displacing fluid begins to appear much before it has actually displaced one pore-volume of the original solution. In consequence, schemes for the protection of the waters of the environment against contamination by biocides, radioactive substances, etc. need to be designed with knowledge of the real properties of the soils and subsoil sediments and fractured rocks.

Salt can be leached from saline soil with less consumption of water if the soil is unsaturated. When the largest pores are empty the water moves more slowly through the soil thus allowing more salt to be transferred

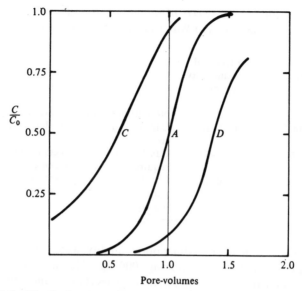

Fig. 10.6. Miscible displacement curves. Curve *A* is a simple break-through curve, such as *A* of Fig. 10.5. Curve *C* is for an anion tracer, showing the effect of anion exclusion and rapid transit between clay aggregates. Curve *D* is for a cation tracer, showing the effect of delay caused by cation adsorption or exchange. (After Rose, 1977.)

by diffusion from poorly conducting pores into the moving water. Hence allowing rain or sprinkler water to enter at a rate less than the potential rate for the soil when flooded is more economical of water than continuous ponding to obtain a given amount of leaching. Intermittent ponding with drainage is a practical management to achieve water economy (Biggar and Nielsen, 1967; Nielsen *et al.*, 1972, p. 125).

   Miscible displacement and the many aspects of the processes have been reviewed in books by Bresler *et al.* (1982) and Scheidegger (1974). Rose (1977) presented a unifying treatment of hydrodynamic dispersion that helps to resolve some of the contradictions apparent in the large literature on the subject.

10.8 Quality of irrigation water

The quality of irrigation water depends on the concentration and composition of the salts dissolved in it. Concentration limits set out in Table 10.4 provide an arbitrary guide to the first of these two aspects of quality. They apply in a broad way to arid and semi-arid regions, but higher limits are applicable in more humid areas where irrigation water is used as a supplement to rain. Most lands that are regularly irrigated use water

Table 10.4. *Guide to the effects of salinity of irrigation water on crops in arid and semi-arid regions* (*Water Quality Criteria Committee, 1972*)

Total dissolved solids mg l^{-1}	Electrical conductivity dS m^{-1}	Effects
<500	<0.75	No detrimental effects are usually noted.
500–1000	0.75–1.5	Can have detrimental effects on sensitive crops.
1000–2000	1.5–3.0	Can have adverse effects on many crops, requires careful management.
2000–5000	3.0–7.5	Can be used for tolerant plants on permeable soils with careful management.

that contains <2000 mg l^{-1}. But poorer quality water finds some uses. For example, groundwater containing 5000 mg l^{-1} is used to irrigate wheat in Western Rajasthan, India, in alternate years (Gupta and Abichandani, 1968). On 1 per cent of the land irrigated in Italy, brackish water is used with a concentration up to 5000 mg l^{-1} according to Cavazza (1969). Good drainage and occasional spelling of the land to allow rains to remove the accumulated salt are necessary when this is done.

In using brackish water, frequent and ample applications are usually recommended to prevent the soil water potential decreasing unduly when its osmotic component is already low (Yaron et al., 1972). However Shalhevet (1984) has concluded from the available evidence that the relative increase in yield caused by increasing the frequency of irrigation is much the same at any level of water quality so that there is no unusual benefit in the case of brackish water.

Even saline water may have some use. Water with concentration of sea water (35 000 mg l^{-1}) has been used together with gypsum in the early stage of reclamation of sodic soil to increase its flocculation and hydraulic conductivity and so allow Ca^{2+} to be brought into the soil to replace Na^+. Reeve and Bower (1960) proposed a system of progressive dilution to remove both Na^+ and soluble salts, and this has been successfully used in laboratory and field trials referred to by Roos, Awadalla and Khalaf (1976).

The second aspect of quality – the composition of the dissolved material – includes especially the concentration of Na^+ in the water. If this is unduly high in comparison with the concentration of the divalent ions Ca^{2+} and Mg^{2+} the irrigation water will cause the soil to become sodic. The sodium adsorption ratio (SAR) of water (U.S. Salinity Laboratory Staff, 1954; Sposito and Mattigod, 1977; Bresler, McNeal and Carter,

1982) is used to predict the effect of these ions on the exchange complex of irrigated soil. It is calculated from

$$SAR = Na^+ (Ca^{2+} + Mg^{2+})^{-1/2}, \qquad (10.21)$$

where Na^+, Ca^{2+} and Mg^{2+} represent concentrations of the cations in millimoles per litre. A nomogram is available for estimating the exchangeable sodium percentage (ESP) that will be in equilibrium with water of a given composition (U.S. Salinity Laboratory Staff, 1954; Wilcox and Durum, 1967). Values of SAR do not differ greatly from the resulting ESP so that, for example, a soil with ESP of about 6 will result from using water with SAR of 5. Water with a high SAR can be used safely only on sands whose structure and hydraulic conductivity are little affected by ESP.

10.9 Use and disposal of wastes on soil

Farm yard manure from domesticated animals has always been valued for improving the fertility and physical properties of soils, and many other waste materials are applied to soils. With rapidly increasing human population, increasing urbanization, development of industrial plant and the establishment of large feed lots for animals, there has been an increasing need to dispose of wastes from centres where they accumulate. Schemes for recycling wastes from cities through soil have existed for many years and interest in them has grown because of the serious pollution of streams and lakes fed directly by effluents. Since wastes can provide water for irrigation, organic matter to improve physical properties of the soil, and useful additions of nitrogen and phosphorus for crops, they can have a positive value in agriculture.

The composition of wastes varies greatly with their origins. Those from agriculture and food processing plants are discussed fully by Loehr (1974). Effluents from cities differ according to the relative contribution from domestic, commercial and industrial areas, and from storm water. For use in irrigation, waste water is often of marginal quality. For example, Pound and Crites (1973) found that the effluents from a number of sewage treatment plants located mainly in southern California, USA, had an average content of about 1000 mg l^{-1} of total dissolved solids and a sodium adsorption ratio of 5. On the basis of Table 10.4, this is not good quality water for irrigation of arid lands. In humid regions, dissolved salts are not of such concern because of the dilution and leaching by high quality water that comes to the land as rain. Here the problem is more one of low infiltration rate of moist soils. This is aggravated by clogging of pores by organic matter, by the micro-organisms that degrade it and by their gaseous products. According to Erickson (1973) working in Michigan, USA, medium and fine textured soils used in waste water disposal may need to dry periodically to allow the infiltration rate to recover. Sandy soils are most suited hydraulically to water disposal in humid climates, but if they are too coarse the chemical and biological renovation of water passing too quickly through may not be as satisfactory as in medium or fine textured soil. Freezing conditions can make it impossible to dispose of liquid wastes in any type of soil. Under suitable conditions, 30 or 40 mm of water can be disposed of per week onto as large an area as is needed of grassland, crop or forest by sprinkler, furrow or flood irrigation. Alternatively, it can be applied at a high rate of 3 or 4 m per week to a seepage basin through which the water infiltrates to the groundwater. In this type of system, described by Bouwer (1973) in Arizona, USA, drying periods of two or three weeks duration are necessary periodically to restore the infiltration rate.

In addition to the problems of dissolved salts and infiltration behaviour, waste water presents other chemical and biological problems. There are possible nutritional hazards for man and other animals from high nitrate concentrations in groundwater and from metal ions that accumulate in the soil and are taken up by plants growing on it. Human and animal pathogens in sewage effluents are always of concern and raw sewage is usually treated to reduce the amount of organic matter before being applied to land so as to reduce its population of micro-organisms. Surface run-off to streams and the use of crops like lettuce have to be avoided to prevent contamination of water or food used by man. However, farms, such as that at Werribee for the city of Melbourne, Australia, with a present population of over two million, show that irrigation with waste water can be made to operate successfully over many years (Kirby, 1968). The Werribee farm of about 10 000 hectares, about half of which is irrigated for grazing animals, disposes of one-third of Melbourne's waste water consisting of 74 per cent domestic wastes, 17 per cent industrial wastes and 9 per cent surface and groundwater. It was established in 1897 on an area receiving an average annual rainfall of 500 mm.

10.10 Drainage

More than 100 million hectares of land have been drained throughout the world. Much of this is in humid temperate areas like the eastern USA and Europe but artificial drainage is also necessary in many irrigated lands of arid and semi-arid areas to prevent salinization. The Netherlands are a good example of an intensively drained humid area with more than half of the 2.5 million hectares of cultivated land drained.

Theory and methods of draining are dealt with in Chapter 6 and by van Schilfgaarde (1974a). Drainage is undertaken in humid areas to reduce the water content of wet soils and so increase their aeration, temperature or strength. Poor aeration is the principal disadvantage of a soil that is too wet. Wet soils are characterized by mottling (or 'gleying') due to low oxygen content and the consequent reduction of iron, manganese and other oxides. Gases diffuse through water much more slowly than through air, the ratio between the diffusion coefficients being about 10^{-4}. Hence exchange of oxygen and carbon dioxide between plant roots and air is limited in wet soil and all but a few specialized plants such as rice suffer. The microbial population is similarly affected so that reducing rather than oxidizing organisms dominate. Oxidation of organic materials including the formation of soluble nitrates from organic nitrogen (nitrification) is consequently inhibited. Effects of water content on aeration and temperature are discussed in Chapter 11 and effects on strength in Chapter 9.

Irrigated lands in non-humid areas may need to be drained to remove excess salts and to lower water tables so that salts are not concentrated near the surface by water moving freely upwards in wet soil and evaporating there. If water tables are kept from rising much above 2 m below ground level, the rate at which water moves to the surface is usually small enough for salinity to be controlled (Fig. 10.7). When tile-drains are used to do this in irrigated land, they are usually placed at about that depth, provided the subsoil is sufficiently permeable. If the salt content of the groundwater is high or the period between irrigations is long, the effects of evaporation from a water table will be aggravated. Accordingly critical depths to water tables in the Hungarian lowland were found by Szabolcs, Darab and Varallyay (1972) to range from 2 to 4.5 m, according to water management practices.

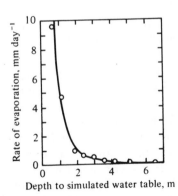

Depth to simulated water table, m

Fig. 10.7. Theoretical and experimental rates of evaporation of water from a column of sandy loam as affected by depth to a simulated water table (Gardner and Fireman, 1958).

In humid regions where salinity is not a problem, drains are usually at a much more shallow depth of about 1 m. The optimum depth depends on seasonal conditions as well as type of soil and crop. The object is to reduce water content sufficiently to allow aeration of the root zone and handling of the soil, without depriving the crop of a reserve of subsoil water during dry periods. The rate at which water can be supplied to the root zone of a clay soil from a water table is illustrated in Fig. 10.8. Since in many cases the rate is inadequate for crops, drainage ditches and tile-drains are sometimes used to supply water underground to the root zone of permeable soils by sub-irrigation (Criddle and Kalisvaart, 1967). The supply is augmented by raising the water table into the root zone. Raats and Gardner (1974) have reviewed work on vertical flow of water from a water table to an evaporating surface or a root zone.

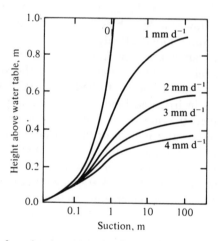

Fig. 10.8. Profiles of suction in a clay soil for various rates of upward water flux from the water table (Wind, 1955).

In deep sands it would often be an advantage to retard the rate at which water drains beyond the root zone. To do this, Erickson, Hansen, and Smucker (1968) have inserted asphalt barriers at a depth of about 70 cm in sandy soils in Michigan, USA.

Water draining from irrigated areas and from land used for liquid waste disposal can lower the quality of a body of water into which it may flow. Although soil acts as a sink for many substances including phosphate fertilizers, heavy metals, pesticides and organic wastes from animals, towns and factories, most soluble salts come through readily in drainage water. Because water is withdrawn in evapotranspiration, these are at a higher concentration than in the water applied to the land. That being so, an increase in salinity of the river water downstream results when irrigation is practised along its course (Table 10.5). Sodium ion may also be at a higher concentration in drainage water than in applied water. Surface drainage poses additional problems. If water running off a waste disposal area reaches a water supply source before organic wastes have been fully degraded by micro-organisms, it can be hazardous to the health of man and animals. Also surface and subsoil drainage if charged with plant nutrients from fertilizers, sewage or animal wastes, can cause unwanted growth of aquatic plants by the eutrophication of water bodies into which it flows.

Pollution of water supplies in most of these ways can be guarded against but it is often difficult to prevent dissolved salts from draining back into the water supply. The most effective protection is by using evaporation pans or salt lakes to accept any drainage water that is too saline for further use. The Salton Sea performs this function for irrigated lands of the Imperial and Coachella Valleys of southern California, USA. Evaporation pans are sometimes established at a level which is only reached and flushed by a river at times of exceptional flood. Irrigation management can help to control the amount of salt going into streams and groundwater from drains and seeps. Ways of doing this are reviewed by Bower (1974). In particular, the efficient conveyance and application of irrigation water as discussed in Section 10.5 can reduce the amount of water seeping through unnecessarily and collecting connate salt from underlying strata on the way. Some reduction in the amount of salt draining through from the irrigated soil can also be brought about by keeping the leaching fraction to the minimum allowable for satisfactory plant growth. When this is done, the salt concentration of the drainage water may be great enough to cause the precipitation of $CaCO_3$ and $CaSO_4 \cdot 2H_2O$ and so lower the salt burden. As evidence of this, Rhoades *et al.* (1974) showed that the amount of applied salt returned in drainage water from lysimeters irrigated with waters of different composition decreased as the leaching fraction, D/I, decreased from 0.3 to 0.1.

Table 10.5. *The mean salinity of the Murray River, Australia, in 1966–7 (from River Murray Commission, 1970)*

Distance up river from ocean, km	2200	1640	1250	910	525	89
Total dissolved solids, mg l^{-1}	37	65	138	151	324	391

11

The physical environment of roots

11.1 Introduction

The living plants and animals that inhabit soil depend upon a supply of water and nutrients, an exchange of gases between the soil and the atmosphere, and the maintenance of suitable soil temperatures. Also, if roots are to extend and expand and if organisms are to move through soil, their activities will be affected by the structure and strength of the soil. With the exception of water and nutrient supply which will be covered in Chapter 12, these physical aspects of the environment of roots and soil organisms form the subject of the present chapter.

Among the transport processes to be discussed in this and the next chapter are the diffusion of solute molecules in soil water, diffusion of gas and vapour molecules in soil air, and the conduction of heat by soils. These three processes are basically similar since each depends on the random motion of molecules rather than on mass flow. For example, in the case of a solution the solute molecules (or ions) are in thermal motion in random directions so that some cross from each side of any plane normal to the concentration gradient. However, more molecules will cross the plane in a given time from the side with the higher concentration into that with the lower than in the reverse direction. Consequently, diffusion will occur in the direction of decreasing concentration, and the flux is found by experiment to be proportional to the concentration gradient. The circumstances are similar for the diffusion of one gas through another. This takes place because of the thermal motion of the gas molecules and, as in the case of solute molecules, net movement across a plane in the direction of decreasing concentration is proportional to the gradient. In the case of heat conduction, the thermal motion of the molecules of the conducting medium increases with temperature. Those molecules at a high temperature contribute some of their thermal energy to others with which they collide that are at a lower temperature. The movement of heat is therefore in the direction of decreasing temperature and is found to be proportional to the temperature gradient.

11.2 Resistance to penetration

Roots provide water, nutrients and anchorage to plants. Although they may occupy only about 1 per cent or less of the soil volume, their penetration into the soil can be thorough. Rooting density decreases with depth in a manner that depends on the type and age of the plant and the soil conditions including water distribution, structure and strength. Examples for different plants are given in Table 11.1 using some of the data collected by Milthorpe and Moorby (1974) from various sources.

Table 11.1. *Rooting density of established plants in the field (from Milthorpe and Moorby, 1974)*

Type of plant	Depth cm	Rooting density cm cm^{-3} or 10^4 m m^{-3}
Grasses	0–10	30–50
Cereals (oats, rye, wheat)	0–15	5–25
	25–50	4
	75–100	2
Trees (*Pinus radiata*)	0–8	2
	25–45	0.8
	91–106	0.4

The ability of plants to exploit the water and nutrients contained in a soil horizon depends largely on the concentration of roots there. For example, deep rooting makes plants less vulnerable to periods of drought. Roots can grow from 1 to 50 mm per day depending on the type of plant and the soil environment of structure, strength, temperature, aeration and water content. Growth occurs in a region extending about 1 cm back from the root tip. The root's diameter ranges from about 0.05 to 0.5 mm in annual plants but its influence is extended further out by root hairs which are protrusions about 1 mm long from epidermal cells. Milthorpe and Moorby (1979), Russell (1977) and Kramer (1983) provide further details.

In Chapter 8 it was shown how roots help to develop the mellow structure found in surface soils under pasture. In turn, a structure of that kind, with satisfactory porosity, pore sizes, root channels, and with soft but water-stable crumbs, provides an environment in which roots grow without meeting much resistance. But as shown in Fig. 11.1, the rate of growth of roots decreases progressively as soil strength increases.

The basic properties affecting rate of elongation of roots in the samples represented in Fig. 11.2 were shown by Eavis (1972) to be aeration and resistance to penetration. When the wet, poorly aerated, samples at high

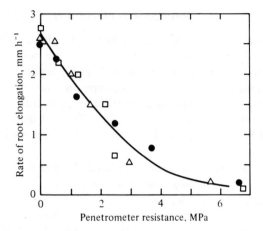

Fig. 11.1. Effect of soil strength (penetrometer resistance) on rate of elongation of a peanut root in a loamy sand at three water contents: 0.07 (●), 0.055 (△), and 0.038 g cm^{-3} (□) (Taylor and Ratliff, 1969).

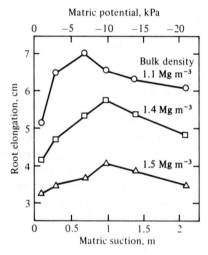

Fig. 11.2. Elongation of roots of pea seedlings in 48 h in a sandy loam soil packed at the three bulk densities shown (Eavis, 1972).

matric potentials are excluded, his data show that the rate of root growth is reduced by increasing dryness and bulk density both of which tend to increase soil strength. In all soils, but especially in those of single grain or massive structure, soil strength affects root growth because the soil must be deformed to create a root channel. Roots cannot extend into rigid pores if their diameter is smaller than about 0.2 mm according to

Wiersum (1957). Similarly Scholefield and Hall (1985) found that 0.3 mm was the lower limit for the growth of rye grass in rigid material. In growing, the root exerts an axial pressure at its tip and it expands cylindrically behind the tip. An observed consequence of this is an increase in bulk density of soil bordering the root channel (Greacen, Barley and Farrell, 1968). Roots that are well supplied with oxygen can exert a pressure up to about 1 MPa on the soil (Taylor, 1974) by virtue of the hydrostatic pressure (or turgor) that can exist in an enlarging cell.

Neglecting gravity, the relation between the total potential of water in the soil and in the nearby root cell at equilibrium is given by

$$\psi_s + \Pi_s = P_c + \Pi_c,$$
(11.1)

where ψ_s is the matric potential and Π_s the osmotic potential of the soil water, P_c is the pressure potential and Π_c the osmotic potential in the cell with $P_c > 0$ and ψ_s, Π_s and Π_c each < 0. With potential expressed on a volume basis, P_c has the dimensions of pressure. It follows from Equation (11.1) that the cell can regulate the hydrostatic pressure of its contents by adjusting its osmotic potential. If this is decreased by increasing the solute content, P_c can be increased to obtain greater turgor for a given water potential in the soil. It then has greater ability to enlarge against resistant soil. The resulting turgor pressure in the cell, P_c, is opposed by the pressure, W, exerted by the stretched wall and the external pressure of the soil so that

$$P_c + W + \sigma_N = 0.$$
(11.2)

Greacen and Oh (1972) obtained P_c from the measured values of the other quantities in Equation (11.1) and they derived the normal stress of the soil, σ_N, from penetrometer readings. They thus obtained W from Equation (11.2) and used this in an equation that they set up for root growth,

$$G = E(W - W_{min}),$$
(11.3)

where G is the rate of growth of the root, E is a coefficient for the extensibility of the cell wall, and W_{min} is a minimum value of W below which the cell wall will not extend. Equation (11.3) is similar to one used for above-ground plant tissues to show the relation between growth and turgor pressure. The linear relation between G and W represented in Equation (11.3) was demonstrated experimentally for pea radicles growing in a loam soil of varied strength and water content. This hypothesis of osmotic adjustment for cell growth goes back to the pioneering work of W. Pfeffer in 1893. Weaknesses in it and alternatives to it are discussed by Russell and Goss (1974) and Barley (1976).

The penetrometers that are used to obtain a relative measure of the resistance offered by soil to the penetration of roots are usually fine probes with a conical point. The force required to penetrate the soil per unit area of the probe is referred to as penetrometer pressure or resistance. The forces acting on the probe and on a growing root are compared and fully discussed by Greacen (1986). Fig. 11.1 illustrates the type of relation that exists between penetrometer pressure and growth of roots. However a penetrometer has to exert greater pressure than a root tip in penetrating a soil. One reason for this is that, unlike a root, a penetrometer cannot diverge from the direct line of advance when a resistant aggregate is in the way. In an attempt to reduce this defect of penetrometers, Grant *et al.* (1985) used a micropenetrometer that detects variations over a distance of 1 mm. By this means the frequency of zones of high and low resistance can be explored. As an example of the root's exploitation of paths not available to the penetrometer, we may cite Ehlers *et al.* (1983) who found that while root growth was severely limited at a penetrometer pressure of 3.6 MPa in conventionally tilled soil, the corresponding limit in untilled soil was higher at about 5 MPa. There the roots avoided resistant barriers by using continuous channels left by earthworms and decayed roots that were not preserved in the tilled soil.

It has been found that the ability of a root to penetrate an aggregate decreases with increasing angle of incidence (Gardner and Danielsen, 1964; Greacen, Barley and Farrell, 1968), and that large aggregates are more resistant to penetration than are small component aggregates when they have been separated (Grant *et al.*, 1985). If a root meets a small impenetrable flat barrier, it grows across it and continues on at an angle of deflection that varies with the angle of incidence, both being measured with reference to the perpendicular to the surface of the barrier (Bandara and Fritton, 1986). In the case of a very extensive barrier such as a hard pan or plough sole, the roots may be deflected horizontally and the plant will suffer from being unable to go deeper.

Dexter (1986*a, b, c*) has made a comprehensive study of root behaviour at the interface between a tilled seed-bed and a compacted subsoil. He found that penetration of the subsoil decreased exponentially with increasing strength of the subsoil. Penetration was better when the base of the seed-bed consisted of fine rather than coarse aggregates with less freedom for horizontal deflection of the root at the interface. With several fine seminal roots, a wheat seedling did better than a pea with its single root in penetrating the compacted subsoil. As a consequence, the limiting subsoil strength determined by penetrometer pressure was much greater for the penetration of wheat (3 MPa) than it was for peas (0.4 MPa). Dexter also conducted experiments with roots that had been deflected horizontally at the interface of the seed-bed with an artificial subsoil that was impenetrable except for cracks 0.5 to 3 mm wide. He found that the proportion of roots entering cracks decreased with decreasing width of crack and with increasing angle of approach between the root and the crack. Similarly with cylindrical holes instead of cracks in the artificial subsoil he found that the proportion of roots going into the holes decreased with decreasing size of hole. There was no preferential growth of the root towards the holes in the well-aerated system. In these experiments wheat with more seminal roots than peas had the advantage in gaining entry into cracks.

Management favouring root penetration includes tilling the surface soil to reduce strength and improve porosity together with avoidance of compaction by implements and traffic. Subsoil tillage to break up pans in the subsoil may sometimes be necessary also. These and other ways of improving the environment of roots are discussed in detail in a book edited by Arkin and Taylor (1981).

Crusts that form on the surface of bare soil present a special problem of penetration for seedlings attempting to emerge following germination. McIntyre (1958*b*) showed that the breakdown of aggregates by raindrops is accompanied by a resorting of the material to form a thin skin that acts as a seal against entry of water or air (Section 9.9). As this dries it increases in strength and acts as a crust that opposes and may prevent seedlings from emerging. The strength of the crust increases with increasing clay content, bulk density of the top soil and exchangeable Na^+ and it decreases with increasing organic matter content, size of the aggregates prior to wetting and water content. There is also an effect of time due to the age-hardening that can occur in a clay material whose structure has been severely disturbed (Yong and Warkentin, 1975, p. 68). Management measures that will reduce the degree of crusting are covered in Section 8.9 and in a review of the literature on crusts by Awadhwal and Thierstein (1985).

11.3 Aeration

Aeration of the upper part of the soil profile is necessary for the growth of most terrestrial plants. In particular, oxygen is used in respiration to provide the energy for roots to extend, expand, and take up nutrient ions selectively. Carbon dioxide is a product of this respiration. Although some plants such as rice are able to exchange gases between roots and the air above ground through plant tissues, most depend on the transport through the soil of oxygen from and of carbon dioxide to the external air. Bacteria, fungi and other micro-organisms inhabiting the soil have their own requirements also so that gas exchange goes on in the absence of plants (Table 11.2). Those that break down organic matter aerobically

Table 11.2.　*Respiration, $l\,m^{-2}\,day^{-1}$ in uncropped soil and in soil carrying a crop of kale at Rothamsted Experimental Station, England (Currie, 1970)*

Soil temperature at 30 cm	17 °C (July)		3 °C January)	
Treatment	Cropped	Uncropped	Cropped	Uncropped
O_2 demand	16.6	8.1	1.4	0.5
CO_2 evolution	17.4	8.0	1.5	0.6

are responsible for a great deal of the oxygen consumption in cropped as well as uncropped soil. Their association with root systems in the rhizosphere, already noted in Chapter 8, is responsible for the breakdown of much root material derived directly from the living plant (Sauerbeck and Johnen, 1976). Respiration of roots and micro-organisms brings the oxygen content of soil air below that of the atmosphere (20.96 per cent) and brings the carbon dioxide content above the 0.03 per cent contained by the atmosphere. These complementary trends become more marked with increasing depth. Other substances whose transport in the gas phase can be important in soil include nitrogen, ammonia, water vapour, and fumigants.

In wet soils, where the volume fraction, ε_a, of soil that contains air falls below about 0.1, plant growth may be affected by low oxygen concentration. This is not, however, a general criterion for poor aeration. Wesseling (1974) quotes a range of values given in the literature from 0.005 to 0.2 over which ε_a has been found to become critical. When the soil is very wet, the resulting anaerobic conditions encourage micro-organisms that reduce nitrates and produce substances toxic to plants instead of the products carbon dioxide and nitrates that occur with aerobic micro-organisms. Mottled colours characterize soil horizons that have been subjected periodically to anaerobic conditions. Ferrous iron is transported from sites in which oxygen is deficient to sites that on drying become better supplied with oxygen. There the iron accumulates on being precipitated as rust coloured ferric hydroxide (Blume, 1968).

Gases and vapours are transported in soil air by convection and diffusion, the latter being the main mechanism. The streaming of air by convection can be caused by changes in water content as when infiltrating water displaces air, and by pressure variations in the soil air resulting from wind eddies, changes in atmospheric pressure and changes in soil temperature. In a review of the work of Buckingham, Romell and others, Keen (1931, p. 336) concluded that convection from all causes was small in comparison with diffusion. However, the effect of wind eddies at the soil surface has been re-examined by Farrell *et al.* (1966). Their analysis showed that the effective diffusion (or dispersion) coefficient resulting from the pressure fluctuations produced in soil air by wind could be up to 100 times the molecular diffusion coefficient in a coarse mulch. Surface air could penetrate several centimetres down in this way. They believed that this explained reported effects of wind in enhancing the rate of evaporation from mulched soils during the later stages of drying.

Diffusion of gases or vapours occurs when a concentration gradient exists, in accordance with Fick's law

$$q = -D_0 \, \partial C / \partial x, \qquad (11.4)$$

where the flux, q, represents the rate of transfer of a gas per unit area of cross-section, C is its concentration and x is the space co-ordinate normal to the section. Since the partial pressure, p, of the gas in the mixture of gases is proportional to C, Equation (11.4) can also be given as

$$q = -D_0' \, \partial p / \partial x. \qquad (11.5)$$

It is to be noted that, although a partial pressure gradient results in diffusion, a total pressure gradient for the mixtures of gases would cause mass flow in accordance with Darcy's law. In soils the total pressure of soil air (i.e. the sum of the partial pressures of all the components) is usually that of the atmosphere above the surface but the partial pressure of O_2 can be much lower and that of CO_2 much higher than in the atmosphere. In these equations D_0 or D_0' is the diffusion coefficient for the gas. For the diffusion of O_2 and CO_2 through air, D_0 has the values given in Table 11.3 according to sources quoted by Grable (1966).

Table 11.3. *Diffusion coefficients* $(m^2 \, s^{-1})$ *for oxygen and carbon dioxide in water and in air at* 25 °C (*Grable*, 1966)

	Oxygen	Carbon dioxide
Air	2.26×10^{-5}	1.81×10^{-5}
Water	2.60×10^{-9}	2.04×10^{-9}

The diffusion coefficient for a gas in soil can be measured using methods described by Rolston (1986) or it can be estimated less reliably from the air content of the soil. Equation (11.4) represents the diffusion of one gas through another specified gas such as air unimpeded by the presence of a solid phase. Soil particles reduce the effective cross-sectional area available for diffusion of a gas in soil to that fraction of the soil that is occupied by air. This is the same as the air-filled porosity, ε_a, which is the fraction of the bulk volume that contains air. The particles also impede the transfer by changing the pathway for flow into a tortuous and poorly connected system that reduces the flux by an impedance factor, b. The flux, q_s, of a gas through soil by diffusion is thus

$$q_s = -D_0 b \varepsilon_a \, \partial C / \partial x. \tag{11.6}$$

If D is the diffusion coefficient for a gas diffusing through the soil, then

$$D/D_0 = b \varepsilon_a. \tag{11.7}$$

Penman (1940) adopted a constant value 0.66 for b because this accorded approximately with his diffusion data and also with the inverse of Carman's value, $\sqrt{2}$, for tortuosity which was generally considered at the time to be a constant (see Section 8.6). However, this makes no allowance for improved communication between pores as ε_a increases. Subsequent work especially in the field of petroleum research showed that impedance decreases as porosity increases (Wyllie and Gregory, 1955; Marshall, 1958) and accordingly Marshall (1959b) proposed $\varepsilon_a^{0.5}$ for the factor b in Equation (11.7). Millington (1959) also deduced $b = \varepsilon_a^{1/3}$. With these substitutions, D/D_0 becomes equal to $\varepsilon_a^{1.5}$ and $\varepsilon_a^{4/3}$, respectively. It may be noted that Buckingham (1904) had earlier assumed that D was proportional to ε_a^m and he found that this fitted his diffusion data fairly well with $m = 2$. According to Wyllie and Gregory (1953) and Currie (1960, 1970), m varies with particle shape from about 1.5 for sand to 10 for plate-shaped particles of mica. Considerations of shape stem from the theory of electrical conductivity as applied for example by de Vries (1950) and Wyllie and Gregory (1953).

These equations make no special provision for effects of water on diffusion other than by reducing the air-filled porosity. Millington (1959) recognized that water impedes diffusion further by closing or reducing the necks connecting adjoining pores. This effect of water is included in his equation for moist materials which can be expressed as

$$D/D_0 = \varepsilon_a^{10/3} \varepsilon^{-2} \tag{11.8}$$

where ε is the porosity. Currie (1961, 1970) set up the equation

$$D/D_0 = (\varepsilon_a/\varepsilon)^4 \varepsilon^m \tag{11.9}$$

for moist soil in which, as above, $m \simeq 1.5$ in sands. He also showed that this type of equation was suited only to non-aggregated material. For beds of aggregates with varied compaction and water content, Currie (1984) concluded that no single relationship between D/D_0 and ε_a would fit the results even when only the one soil was involved in the measurements. In a bed of aggregates, water first reduces the air-filled pore space within dry aggregates without greatly affecting the continuity of pore space between them. Other equations with more empirical parameters are discussed by Troeh, Jabro and Kirkham (1982) and Glinski and Stepniewski (1985).

Fick's law, as given in Equation (11.4), applies to steady state conditions of diffusion. For the non-steady state we combine this with the continuity equation (see Appendix C):

$$-\partial q/\partial x = \partial C/\partial t, \tag{11.10}$$

which relates the difference between the amount of a gas entering and leaving a volume element in unit time with the change in the amount of the gas in the volume element in unit time. This gives the general diffusion equation

$$\partial C/\partial t = D_0 \, \partial^2 C/\partial x^2, \tag{11.11}$$

if the diffusion coefficient is constant. In three dimensions, this is expressed for an isotropic medium as

$$\partial C/\partial t = D_0 \nabla^2 C. \tag{11.12}$$

With D_0 replaced by the coefficient of diffusion in soil, D, and a term, S, introduced in the continuity equation to cover for example a source of carbon dioxide or a sink of oxygen in soil arising from the activities of roots and micro-organisms, Equation (11.11) becomes

$$\partial C/\partial t = D \, \partial^2 C/\partial x^2 + S. \tag{11.13}$$

S is the production rate per unit volume and has positive values for carbon dioxide and negative for oxygen. For steady state diffusion, for which $\partial C/\partial t = 0$, Equation (11.13) becomes

$$\partial^2 C/\partial x^2 + S/D = 0. \tag{11.14}$$

Solutions of this equation were presented by van Bavel (1951). For the case of a uniform production rate of carbon dioxide, an assumed boundary impermeable to the passage of carbon dioxide at the water table (depth L) and the concentration, C_0, of carbon dioxide in the free atmosphere at the soil surface, the boundary conditions are

$$0 \leqslant x \leqslant L, \quad S \text{ is the production rate;}$$

$$x = 0. \quad C = C_0 \, ;$$

$$x = L, \quad dC/dx = 0.$$

For such conditions, the solution to Equation (11.14) is

$$C = C_0 + (S/D)(Lx - (x^2/2)). \tag{11.15}$$

Kirkham and Powers (1972) give details of solutions of Equations (11.13) and (11.14).

Variation of the concentration of carbon dioxide with depth, as calculated from Equation (11.15) by van Bavel (1951) for air-filled porosities 0.1 and 0.01 is illustrated in Fig. 11.3 taking $S = 4.47 \times 10^{-7}$ kg m^{-3} s^{-1} and obtaining D from Equation (11.7) with $b = 0.66$. Equation (11.15) shows that the increase in carbon dioxide and the reduction in oxygen, as given by $C - C_0$, varies directly with S and inversely with D. Clearly both affect the aeration status of a soil. The effect of D can be interpreted from Fig. 11.3 because D depends on air-filled porosity. The effect of S is illustrated in Fig. 11.4 for the oxygen profiles in a soil as calculated by Currie (1962) for two rates of respiration. It will be appreciated that S is negative in Fig. 11.4, corresponding to consumption of oxygen, and positive in Fig. 11.3, corresponding to production of carbon dioxide. Both effects are further illustrated by van Bavel and Currie.

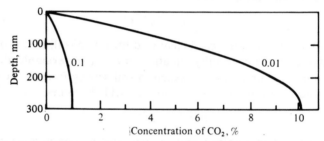

Fig. 11.3. Carbon dioxide profiles in soil of air-filled porosity 0.1 and 0.01 to a depth of 300 mm when the rate of CO_2 production is taken to be uniformly 4.47×10^{-7} kg m^{-3} s^{-1}. (After van Bavel, 1951.)

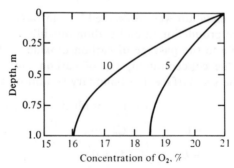

Fig. 11.4. Oxygen profiles in soil resulting from uniform activity to a depth of 1 m with respiration rates of 10 and 5 l m^{-2} day^{-1}. D/D_0 was taken to be uniformly 0.0625. (Currie, 1962.)

Since, in wet soils, roots and micro-organisms are separated from the soil air by water films, part of the path for diffusion of oxygen and carbon dioxide is through water. The motion of the diffusing molecules is greatly reduced in water and the diffusion coefficient is lower than in air by four orders of magnitude (Table 11.3). Hence diffusion coefficients for the gas phase are not necessarily critical for plant growth. In order to measure the rate at which oxygen can be supplied to a root in its environment of water and air, Lemon and Erickson (1952) used a platinum microelectrode inserted in the soil. The current consumed in reducing oxygen electrochemically was measured to determine the oxygen diffusion rate. The method has been found to be unreliable in unsaturated soil (McIntyre, 1970) because of uncertainty about the continuity and thickness of water films around the electrode. It continues in use as a relative rather than as an intrinsic measure of oxygen diffusion rate. Glinski and Stepniewski (1985) consider that, despite its limitations, it is the best guide at present to oxygen availability to roots. They provide descriptions of this and other methods for measuring soil aeration as part of a full account of aeration and its role for plants.

Greenwood (1975) has also reviewed the use and limitations of means available for assessing aeration. These include measuring the composition of soil air, air-filled porosity, diffusion coefficient, and oxygen diffusion rate. Although none provides a critical means for assessing aeration in relation to plant growth, Greenwood considers that soils with $\varepsilon_a > 0.1$ are likely to be adequately aerated and that an oxygen content < 5 per cent in the soil air is likely to affect plant growth. However, soils with a higher content may nevertheless have zones that are deficient. Criteria are difficult to establish because of differences in oxygen content inside and outside large aggregates, the effect of duration of anaerobic conditions, and the varying activities of plants and micro-organisms in different seasons and different soils. For example, wheat that is in water-logged soil in winter may suffer no damage because oxygen requirements at that time are small (Currie, 1975).

The main avenue for managing aeration lies in the control of water table depth by drainage. Structure control has relevance also since anaerobic conditions are encouraged by a surface crust or a massive structure into which air cannot enter while the matric potential is high. In this connection Smith (1980) has developed a model of anaerobic zones in aggregated soil. Also Willoughby and Willatt (1981) suggest from experiments with plants that aggregates larger than 6 mm might not be fully exploited by roots when the supply of oxygen is reduced due to the microbial decomposition of organic matter in wet soil. Management aimed at reducing gaseous diffusion is also of some interest as it concerns the burying of materials that release hazardous gases (Clifford and Hillel,

1986). Clay that has been compacted to reduce the number of large pores can be fairly effective even when dry if much of the pore space is in pores smaller than 0.1 μm in diameter. They are then smaller than the mean free path of the gas molecules and under these circumstances a reduced rate of diffusion (called Knudsen diffusion) occurs in them.

11.4 Soil temperature

Thermal regime of the soil surface

The temperature of the soil surface responds closely to the radiant energy budget. Daytime heating is by short-wave radiation (mostly shorter than 2 μm) from the sun and sky which has a spectrum at the earth's surface that would be appropriate to a radiator at about 6000 K. Night-time cooling results from the loss of energy emitted from the surface of the soil as long-wave radiation (mostly longer than 4 μm) with a spectrum appropriate to a radiator at about 280 K. The balance between incoming and outgoing energy at the soil surface includes also other items as shown in the energy balance equation

$$(1 - \alpha)R_s = R_l + G + H + LE, \qquad (11.16)$$

where R_s is the incoming short-wave radiation, a fraction α of which is reflected from the surface and so does not affect soil temperature, R_l is the long-wave radiation (outgoing minus incoming from the atmosphere), G and H represent heat flow into the ground and air respectively, and LE is the latent heat consumed by the evaporation flux E, L being the specific latent heat of vaporization of water.

The temperature regime of the soil surface has two cyclical periods, the diurnal and annual. Daytime heating and night-time cooling are responsible for the diurnal period, and the annual period results from the variation in short-wave radiation, R_s, throughout the year. This varies little at the equator but at a latitude of 50° for example the summer to winter ratio is about 5 for 30 days centred on the solstices (Budyko, 1974, p. 46). Following the winter solstice, on 22 December in the northern hemisphere and 22 June in the southern hemisphere, short-wave radiation increases and soon begins to provide an energy surplus sufficient for warming of the surface to commence. Warming then continues for the next six months until incoming short-wave radiation begins to decrease. The reverse then holds for the half year summer to winter solstices.

Conduction of heat

The manner in which heat flows through the soil and the amplitude and phase of the temperature waves below the surface are of considerable importance in plant cultural practices and in engineering uses of soils.

The conduction of heat through a solid body is described by the Fourier equation

$$Q = -\lambda \, dT/dz, \qquad (11.17)$$

where Q is the flux density of heat which is the rate of heat transfer per unit area. When Q is in units of $J\,m^{-2}\,s^{-1}$, the gradient of the temperature, dT/dz, is expressed as $K\,m^{-1}$, and therefore the thermal conductivity, λ, has the units $J\,m^{-1}\,s^{-1}\,K^{-1}$ or $W\,m^{-1}\,K^{-1}$. The formal similarity of Equation (11.17) to the Darcy equation (4.5) may be noted, and the negative sign is necessary since the heat flux is in the direction of decreasing temperature.

Equation (11.17) holds instantaneously, and it may be used to obtain the thermal conductivity by laboratory experiment, when the temperature gradient can be arranged to remain steady. It is the equation that describes steady state heat conduction. In the field, the soil temperatures and the temperature gradients do not remain steady for long. The equation of conservation of heat, the ubiquitous continuity equation, is therefore used together with the Fourier equation to describe the non-steady temperature regime. The continuity equation is obtained by equating the differences between the amount of heat entering and leaving a volume element of the material in unit time with the change in heat content of the element in unit time (Appendix C). It is given for one-dimensional flow by

$$-\partial Q/\partial z = \rho c \, \partial T/\partial t, \qquad (11.18)$$

where c is the specific heat capacity of the soil, ρ is the soil bulk density and t is time. Upon combining Equations (11.17) and (11.18) we have

$$\partial T/\partial t = (\lambda/\rho c) \, \partial^2 T/\partial z^2, \qquad (11.19)$$

which is the general equation for the conduction of heat through a body with thermal conductivity λ and volumetric heat capacity ρc. We may note the similarity of Equation (11.19) to Equation (4.32) and observe that the thermal diffusivity $\kappa = \lambda/\rho c$ is the parameter that serves for the thermal flux the same role that the moisture diffusivity serves for water flux.

The heterogeneous nature of soil with varying proportions of air, water, organic and mineral matter produces a range of thermal conductivities and heat capacities. The relevant properties of important soil constituents are given in Table 11.4 which is based principally on data assembled by de Vries (1963). The volumetric heat capacity, ρc, of the whole soil may be estimated from the properties of the constituents by taking the weighted mean

$$\rho c = \sum \theta_i (\rho c)_i. \qquad (11.20)$$

Table 11.4. *Thermal properties of soil constituents (after de Vries, 1963)*

	Specific heat capacity kJ kg^{-1} K^{-1}	Density kg m^{-3}	Volumetric heat capacity kJ m^{-3} K^{-1}	Thermal conductivity W m^{-1} K^{-1}
Air (20 °C)	1	1.2	1.2	0.025
Water	4.2	1.0×10^3	4.2×10^3	0.6
Ice (0 °C)	2.1	0.9×10^3	1.9×10^3	2.2
Quartz	0.8	2.7×10^3	2×10^3	8.8
Clay minerals	0.8	2.7×10^3	2×10^3	2.9
Soil organic matter	2.5	1.1×10^3	2.7×10^3	0.25

where θ_i is the content of constituent i expressed as a volumetric fraction of the soil. As an example, a soil possessing the composition θ (air) = 0.10, θ (water) = 0.35, θ (organic matter) = 0.02 and θ (clay materials) = 0.53 has a volumetric heat capacity of about 2.6×10^3 kJ m^{-3} K^{-1}, of which the soil water and clay mineral matter contribute 58 per cent and 40 per cent, respectively.

The thermal conductivity of the soil is greatly affected by its constituents. In particular, air is a poor conductor (Table 11.4) and in soil it reduces the effectiveness of the solid and liquid phases. Of the three constituent phases, the solid phase has the highest conductivity. Therefore conductivity is increased as bulk density increases, as is illustrated in Fig. 11.5 for a loam soil. An increase in bulk density reduces the air content and brings the solid particles more closely into contact. Water content has a marked effect because when water replaces air it provides bridges between particles that greatly increase the conductivity of the soil.

Thermal conductivity can be estimated approximately from bulk density and water content by means of charts prepared empirically by Kersten (1949) for soils of different texture in the frozen and unfrozen state. A more basic relation between the thermal conductivity of a soil and that of its constituents (air, water, and particles) has been derived by de Vries (1963) from an equation of Burger for the electrical conductivity of a medium composed of ellipsoidal particles. De Vries has checked calculated values of thermal conductivity of soils against measured values satisfactorily using his relation, and Penner (1970) has done so for frozen soils. Kimball *et al.* (1976*b*) concluded from their own and other comparisons that calibration is necessary for each particular soil if reliable calculations are to be made by means of this relation. However, they successfully applied it to obtain λ for use in Equation (11.17) when computing the heat flux from the temperature gradient in soil (Kimball *et al.* 1976*a*). Their paper provides references to other methods of

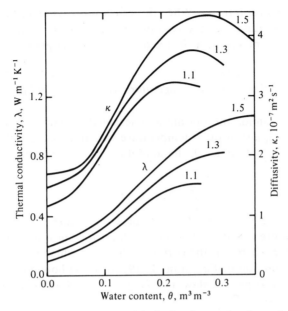

Fig. 11.5. Effect of water content and bulk density on the thermal conductivity (λ) and diffusivity (κ) of a loam soil. The bulk density is shown for each curve in Mg m^{-3}. (Adapted from Kolyasev and Gupalo, 1958.)

determining heat flux. The measurement of thermal conductivity is complicated by the water present in the soil. As shown in Section 4.7, water moves under the influence of a gradient in temperature thus causing a transfer of sensible heat in the liquid phase and of latent heat in the vapour phase. Although methods are devised to minimize this, most published values, from those of H. E. Patten in 1909 onwards, include a contribution from this source as part of an apparent thermal conductivity. De Vries (1963) has provided a basis for estimating the contribution to heat transfer made by the transfer of water vapour. Moench and Evans (1970) have applied this to obtain 'real' thermal conductivity from their measured values. They found that real conductivity of a sandy loam soil increased rapidly with increasing water content due to bridging until the soil reached a degree of saturation of about 25 per cent. After that the increase was no greater than that accountable to the conductivity of the water which was greater than that of the air it replaced.

The effects of water content and bulk density on thermal diffusivity are shown in Fig. 11.5. Because of the inverse relation with volumetric heat capacity, diffusivity does not continue to increase with water content as does thermal conductivity. The maxima shown by these and most other published curves are probably due to the included effect of vapour movement on heat transfer. When this effect was excluded, Moench and Evans (1970) found that the resulting real diffusivity of their soil was relatively constant when the degree of saturation was greater than 30 per cent.

Diurnal and annual temperature waves in the soil profile

In order to examine the diurnal and seasonal penetration of heat into the soil profile, a sinusoidal temperature variation at the soil surface is assumed such that, at $z = 0$, the amplitude of the temperature wave is A_0, i.e.

$$T = T_a + A_0 \sin \omega t, \tag{11.21}$$

and $T \to T_a$ as $z \to \infty$. Here T is the soil temperature, T_a is the average soil temperature, t is time, z is the depth positive downwards, and ω is equal to $2\pi/\tau$ where τ is the period of the cycle which may be a day or a year. The solution of Equation (11.19) for these boundary conditions can be shown to be (Carslaw and Jaeger, 1959; Kirkham and Powers, 1972)

$$T(z, t) = T_a + A_0 \exp -\{(\omega/2\kappa)^{1/2}z\} \sin \{\omega t - (\omega/2\kappa)^{1/2}z\}, \tag{11.22}$$

if the thermal diffusivity, κ, is assumed to be constant. Equation (11.22) states that the amplitude of the temperature fluctuation decreases exponentially with depth below the soil surface. It should be remembered that the sinusoidal function of time is only an approximate specification of the way the soil surface temperature varies throughout the year. Nevertheless, and particularly for the seasonal temperature regime, it is acceptable.

The amplitude would decrease to about 5 per cent of the surface amplitude where the exponent is -3. Therefore, for a daily period, for which $2\omega = \pi/\tau = 0.73 \times 10^{-4}\,\mathrm{s}^{-1}$ and adopting $\kappa = 0.4 \times 10^{-6}\,\mathrm{m}^2\,\mathrm{s}^{-1}$, we find that the depth at which this happens would be approximately 0.3 m. Similarly, the depth at which the seasonal amplitude of the temperature variation would be damped to 5 per cent of the surface amplitude would be approximately 6.5 m.

Besides predicting the increased damping of amplitude with depth, the equation predicts the increased delay in phase of the temperature wave with depth diurnally and annually. These predictions are well satisfied by actual observations. For example, West (1952) found that fluctuations in soil temperatures observed over a period of eight years at Griffith, Australia (latitude 34°S, longitude 146°E) fitted Equation (11.22) closely. Figs. 11.6 and 11.7 taken from his data illustrate the nature of the amplitude and phase changes with depth.

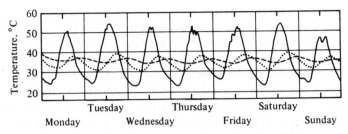

Fig. 11.6. Diurnal temperature wave in soil at three depths during a week in midsummer at Griffith, Australia (West, 1952). ———, 25 mm depth; . . . . , 150 mm; – – – , 300 mm.

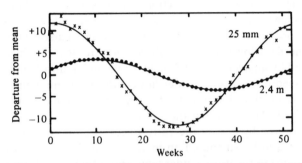

Fig. 11.7. Annual temperature wave in soil at two depths for 52 weeks commencing January 1. Data points are averages formed from 8 years of records at Griffith, Australia (West, 1952).

A comprehensive collection and review of soil temperature records for many localities throughout the world has been made by Chang (1958).

Management of soil temperature

Soil temperature greatly affects the plants and micro-organisms living in the soil. The rate at which organic matter is decomposed by micro-organisms increases as the soil temperature increases from about 5 °C up to about 30 or 40 °C (Nyhan, 1976). The optimum soil temperature for the growth of plants varies widely with the species over the range of about 20° to 30 °C as shown in data collected together by Voorhees, Allmaras and Johnson (1981). The growth rate of micro-organisms and plants may increase two- or three-fold with a 10 °C increase in soil temperature up to the optimum. Soil management for the control of soil temperature is therefore attempted under a variety of conditions ranging from those in which low temperature restricts germination and early growth of plants in a cold winter or spring to those in which high temperature is unfavourable to germination and root growth in hot summers.

Plastic covers over soil can increase soil temperature by reducing evaporation and night-time losses of heat to the atmosphere. Revut (1973) found that transparent polythene film over the surface increased the soil temperature by 4 to 6 °C to a depth of 200 mm on clear days in the north of the USSR. Black film helped on moist soil but not as greatly as did transparent film, which transmits short-wave radiation to the soil. Films are used in strawberry and vegetable growing but are too expensive for most other crops. Mulches of straw or other residues also affect soil temperature. While these have favourable effects on water retention and erosion control they can also unfavourably delay the warming of soil in early spring because they reduce the amount of short-wave radiation received due to shading and reflection. This has been found to constitute a disadvantage for spring seeding of straw-mulched soils in maize growing areas of North America. Allmaras, Burrows and Larson (1964) found that in nine field locations the soil temperature at a depth of 100 mm was reduced by an average of 1.2 °C by straw mulch during the six weeks following seeding. However under hot conditions there is an advantage in cooling the soil with a mulch. Working in the humid tropics of west Africa, Lal (1974) found that the soil temperature at a depth of 50 mm could be lowered by 8 °C at the early stages of crop growth by a mulch of rice straw and he attributed an increased yield of maize primarily to the decrease in temperature. Under semi-arid tropical conditions, a mulch can help further by reducing the rate of evaporative loss of water from the soil and thus causing also some delay in the hardening of a surface crust during the critical stage of seedling emergence. McCown, Jones and Peake (1985) found that a mulch of chemically killed pasture increased the yield of maize by over 20 per cent in northern Australia. Ross, Williams and McCown (1985) estimated from an analysis of the radiation balance and convective heat loss that such a mulch in the semi-arid tropics could reduce the soil temperature by as much as 20 °C. Other covers that affect temperature include black materials such as soot or bitumin that decrease reflection of short-wave radiation, or white materials such as chalk that increase it. It may be noted also that, as vegetative mulches darken with age, reflection decreases.

Drainage can help the soil to warm faster in the spring according to Feddes (1972) who found that soil near the surface was 1 or 2 °C higher in plots with lowered groundwater tables in the Netherlands. However Scotter and Horne (1985) found no effect of drainage on soil temperature under pasture in a fine textured soil in New Zealand and they questioned whether any general effect of drainage on temperature should be expected.

Tillage affects the soil temperature in various ways. The removal of vegetative cover will increase the amplitude of the diurnal wave, a decrease in bulk density will reduce the conductivity, and roughness can affect temperature distribution in the surface soil. Burrows (1963) found that the soil temperature at 100 mm depth in ridges with east-west orientation was greater than under a horizontal soil surface. In accordance with the well-known effects of slope and aspect, the temperature is greatest on the south side of the ridges (in the northern hemisphere), according to measurements of Spoor and Giles (1973).

Transfer of heat from the soil to the atmosphere at night helps protect plants such as grape vines from damage by occasional frosts (Landers and Witte, 1967; Gradwell, 1968). Conditions favouring this transfer are compact and moist soils which have high thermal conductivity and volumetric heat capacity and an absence of weeds or mulches which reduce daytime warming and night-time heat loss from the soil.

Soil temperature in areas with hot summers can become too high for the growth of roots of some plants. Irrigation applications can help keep the temperature down at critical periods by increasing evaporation from the surface. Irrigation also affects soil temperature if the temperature of the water and soil differ greatly. Management of water in relation to soil temperature is discussed by Raney and Mihara (1967).

12

Plants and soil water

The descriptions of water retention in soil, its flow through soil and the effect it can have upon physical properties such as consistence, strength of aggregates, aeration and the temperature regime are relevant to the conditions of growth of plants. The most direct effect of the moisture status of the soil upon a plant growing in it is, however, concerned with the ability of roots to absorb the amount of water that the plant needs for transpiration, and the amount of solutes that it needs for its mineral nutrition. The transpiration flux of water vapour into the atmosphere, from the plant foliage, is the final stage of a process that began as a flow of water through the soil towards the roots. The transport of water through the various stages that in total are referred to as the soil-water-plant-atmosphere continuum is the subject of this chapter. The conducting system of plants for this is described by Kramer (1983).

12.1 Absorption of water by roots

Water is absorbed by roots through or past their epidermal cells. There is some uncertainty about the actual paths that the water follows in its passage to the xylem conducting system near the axis of the root. Mees and Weatherley (1957) suggest that the main passage for water is through the cell walls and interstitial spaces, but at least some part of the pathway must go through the cells and the associated membranes, because the Casparian strip at the endodermis is considered to be impermeable to water (Woolley, 1965; Slatyer, 1967).

The root cortex has a mechanism to sieve out the solutes of the soil water not required for mineral nutrition of the plant, because the electrolyte concentration of the xylem sap is usually much less than the concentrations in the soil water whence it was derived. Since the passage of such water into the plant therefore is often against the gradient of osmotic potential (energy), its effective transport must be mediated by the consumption of metabolic energy. Water in the proto-xylem of the roots then moves axially along them to the xylem tissues of the stem through which it is conducted to the leaves. Its loss from the leaves as water vapour, which diffuses through stomatal openings, is referred to as transpiration. Some liquid water diffuses through the cuticle of leaves and stems and then evaporates, but the amount is usually a small fraction of the total, unless there are numerous damaging lesions on the leaf (Hall and Jones, 1961). Loss of water from whatever cause must make the total potential of the water in the leaves to be less than that in the soil, so that a gradient favouring upward movement of water exists through the continuous liquid system in soil, root, stem and leaf (Slatyer, 1967; Kramer, 1983).

The cohesion theory for the transmission of water through plants, so named because it requires the water column to be continuous from roots to leaves, was firmly established by the work of Dixon (1914) and many others subsequently. As discussed in Section 2.5, the tensile strength of water is adequate for this to be possible for all plants. But in trees where water has to be lifted to a considerable height, one would expect air to appear in the conducting vessels at the high suctions required in the xylem and that the continuity would be broken. There is no complete explanation of why such cavitation of the water does not happen. It certainly does in a glass capillary tube when the suction approaches 100 kPa. Adsorption of water and isolation of air bodies within by-passed cells perhaps helps to maintain continuity (Slatyer, 1967).

The total potential of the water in any part of the system can be represented, as observed in Section 2.2, by

$$\Phi = \Pi + P + Z, \qquad (12.1)$$

where Π, P and Z represent, respectively, the osmotic, pressure and gravitational potentials. These apply to water in plant tissues in the same way as in soil, so that Π is associated with the presence of solutes and P with hydrostatic pressure. When potential is expressed per unit volume of water, it has the dimensions of pressure so that P equals the hydrostatic pressure and Π is equal in magnitude but opposite in sign to osmotic pressure. Osmotic pressure (or solute suction) plays an important role, by means of the semipermeable membranes of plants, in maintaining turgor in plant tissues. Π has negative values. P is positive in turgid tissues in which the water absorbed osmotically is constrained by cell walls. It is usually negative in the conducting xylem vessels of the stem because when leaves are transpiring the water there is under tensile stress. As discussed for soils in Section 2.2, the matric potential, ψ, may be substituted for P in Equation (12.1) when dealing with water adsorbed by a solid matrix. An application will be given shortly. The place of matric potential in plant–water relations has been examined critically by Passioura (1980).

In dealing with small plants and plant tissues when gravity is unimportant, we can omit Z and the sum $\Pi + P$ is then referred to as the water potential (Slatyer, 1967). At full turgor of the plant cells, Π and P, although of different signs, have a similar magnitude so that the water potential is approximately zero. If sufficient water is lost from the cells for their pressure potential, P, to fall to zero, (i.e. to atmospheric pressure), the tissues become flaccid. The water potential is then equal to Π which will, however, have decreased somewhat because the same amount of solutes is now dissolved in less water. However the decrease in osmotic potential on dehydration may be greater than that accountable to water loss because osmotic adjustment can occur within the cell. Additional solutes are accumulated to bring this about. This adjustment helps leaves and roots to maintain turgor when the plant is under stress from water deficit or soil salinity. In this way growth can continue at a lower water potential than would otherwise be possible. Turner (1986) has discussed this fully.

The water potential of plant tissues can be measured by psychrometric methods as described in Section 3.3 with adaptations if *in situ* measurements are to be made (e.g. McBurney and Costigan, 1987). A pressure chamber method of Scholander *et al.* (1965) is also used for the study of herbaceous plants. The technique involves cutting a leaf from the plant with a sufficient length of petiole to enable the end of the petiole to protrude through the compression gasket of a pressure chamber. The leaf itself is subjected to an increasing air chamber pressure until the sap is made to flow out of the xylem elements exposed in the cut end of the petiole. The expression of the sap is usually detected accurately by use of a magnifying glass. The chamber pressure is conventionally then stated to have the magnitude of the water potential of the leaf before it was cut, on the understanding that the sap issuing at the sectioned petiole has come both from the cell protoplasm and from a restoration, by the applied pressure, of the elastic storage volume that the xylem sap occupied before the leaf was cut off. The leaf water potential obtained in this way is the sum $\Pi + \psi$ since the water is retained osmotically and by adsorption in cell walls. The osmotic potential can be measured on sap expressed from the plant tissues and placed in a psychrometer chamber, or it can be obtained by measuring its freezing point depression as discussed in Chapter 10. Full descriptions of methods for use with plants are given by Campbell (1985).

Because the passage of solutes can be controlled by the semi-permeable membranes of plant cells, Π has to be included in the total potential of soil water when considering its entry into roots, as mentioned in the introductory remarks to this chapter and further discussed in Section 10.6 on soil salinity. The decrease in total potential from soil to leaf can exceed 1 MPa in plants that are transpiring actively but it may be less than 100 kPa when they are not. Within the soil, the decrease varies with distance over which water moves to roots and inversely with water content. The decrease in the matric component of the total potential of the soil water is responsible for the flow of the soil solution towards the surface of the root. There is considerable evidence that old roots are unable to absorb water as effectively as new roots and that the region behind the root tip is particularly permeable to water. It is there too, that the roots of angiosperms are equipped with root hairs, small protuberances from the epidermal cells, that enable the fine roots to present a larger surface area for absorption of water and mineral nutrients. The locations of the strongest sinks for water absorption by the root system change during the growing season as new roots grow through the soil, old roots suberize and their attached root hairs slough off.

The flow of water to an absorbing root can be idealized in order to solve the equation for radial flow to a single root of infinite length, which can be written as

$$\frac{\partial \theta}{\partial t} = \frac{1}{r} \frac{\partial}{\partial r} \left(rD \frac{\partial \theta}{\partial r} \right), \tag{12.2}$$

where θ is the volumetric soil water content at a distance r from the root axis and D is the soil water diffusivity (see Chapter 4). Equation (12.2) is the equivalent of Equation (4.29) when the latter is written in cylindrical polar co-ordinates for two-dimensional flow that is symmetrical with respect to the cylindrical axis. Its solution is difficult because the soil water diffusivity is not an analytical function, although its dependence upon water content has been attempted so to be described by numerous authors (see Newman, 1974, for a summary).

The purpose of examining Equation (12.2) is to gain some understanding about the manner of variation of water content of the soil with radial distance from the absorbing root, and about the rate of depletion of soil water content within the zone affected by the root. If D is taken to be constant, well-known solutions obtained in the theory of conduction of heat in solids (Carslaw and Jaeger, 1959) can be put to use here. For example Gardner (1960) examined time and distance scales for absorption of water by a single, infinitely long root. Fig. 12.1 shows the influence of initial matric potential (i.e. wetness) of the soil upon the distribution of matric potential in a radial direction, for a constant rate of uptake. Gardner's work has been followed by many other attempts to give theoretical guidance for understanding the influence of root density and

312   *Plants and soil water*

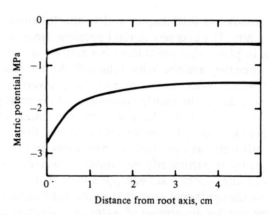

Fig. 12.1. The matric potential of the soil, derived by solution of Equation (12.2), as a function of distance from the axis of a root, after water absorption for one day at a rate of 10 mm³ mm⁻¹ day⁻¹. The effect of initial water status of the soil is shown by the larger gradient near the root in the drier soil. (After Gardner, 1960.)

soil water content upon the possible rates of extraction of water from the root zone.

Taking specific cases and assuming a constant rate of uptake of $100 \, \text{mm}^3 \, \text{mm}^{-1}$ root day^{-1} Gardner (1968) calculated that the drop of potential from soil to root was negligible when the matric potential of the water in bulk soil was near zero. The decrease was about 100 kPa when the matric potential of the bulk soil was -100 kPa and 10 MPa when the bulk soil was at -1.5 MPa. Gardner's conclusions have been criticized by Newman (1969, 1974), Lawlor (1972) and Williams (1974), on the grounds that his assumed root density was too sparse and his assumed rate of uptake too great. These authors used data obtained for plants ranging from grasses to trees to support this criticism. They lead to much lower values for the drop in matric potential from soil to root. On the basis of these parameters it appears that the drop in matric potential is unimportant where root systems are well developed until the matric potential decreases to -1.5 MPa. Most of the evidence shows resistance to flow to be mainly due to resistance in the root (Reicosky and Ritchie, 1976) or in the zone of contact between root and soil (Herkelrath, Miller and Gardner, 1977). However where root density is low and hydraulic conductivity of the unsaturated soil is low, resistance to flow within the soil may be limiting (Hulugalle and Willatt, 1983).

12.2 Availability of water

The effect of matric potential of soil water on the rate of transpiration is illustrated for a range of meteorological conditions in Fig. 12.2, from

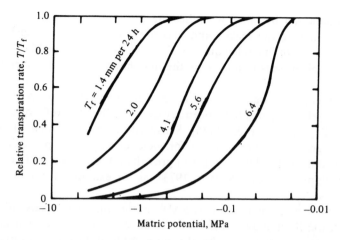

Fig. 12.2. Effect of matric potential of soil water on transpiration rate under five different evaporating conditions. Transpiration rate, T, is expressed as a fraction of the rate, T_f, obtained when water was freely available at field capacity. (After Denmead and Shaw, 1962.)

experiments on maize grown in pots in the field by Denmead and Shaw (1962). In general the availability of water for transpiration by the plant decreases as the matric potential of the soil water decreases and the trend is most marked when potential evaporation rates are highest. The rate at which water is lost from the leaves to the air can be regulated by guard cells in the stomatal openings. If these lose turgor, the stomata tend to close and so reduce the rate at which water vapour diffuses from the leaves. Turgor is lost ($P \to 0$) when the rate at which water can be supplied is insufficient to meet the evaporative loss by transpiration, and water is withdrawn from the cells. Stomata have been shown to close in various crops grown in the field at a leaf water potential ranging from about -1 MPa in beans to about -2 MPa in sorghum (Begg and Turner, 1976).

Water deficits can affect the germination of seeds and the development of plants adversely. Turgor is required for cell enlargement during growth. Also water deficits to which plants are commonly subjected reduce the rate of formation of carbohydrates in the leaves in their photosynthesis from water and carbon dioxide. Stomatal closure can contribute to this by reducing the rate of diffusion of carbon dioxide into the leaves. In this connection it may be remarked that, although loss of water can be reduced by spraying leaves with compounds that reduce transpiration, those treatments that create films or cause stomata to close also reduce photosynthesis (Begg and Turner, 1976). Broadly the effect of matric potential of the soil water on growth is similar to its effect on transpiration. The ratio T/T_f in Fig. 12.2 for the relative amount of water lost through the stomata can be taken as indicating the relative amount of carbon dioxide entering the leaves and according to Hanks and Rasmussen (1982) it is approximately equal to Y/Y_f where Y is the yield of a given crop in a given season and Y_f is the maximum yield with water freely available.

The effects that water deficits have in reducing the growth of plants and the yields of crops have been reviewed by Salter and Goode (1967) and by Begg and Turner (1976). Sensitivity to matric potential of soil water is greater in crops grown for yield of vegetative parts than for grain. It differs at different stages of growth and is greatest in tissues that are developing rapidly (Slatyer, 1973), as for example at the flowering of cereals.

While a matric potential of -1.5 MPa in the soil water is a useful guide to the lower limit of available water for most plants, it is recognized that plant water potential can be the more critical factor in availability (Lascano and van Bavel, 1984). In particular drought tolerant plants can use soil water at a matric potential of -1.5 MPa or less over long periods. They can maintain some leaf turgor ($P > 0$), although the water potential ($P + \Pi$) of the leaf is < -1.5 MPa, because the osmotic potential of their leaf sap may be as low as -20 MPa. It is much higher (about -1 to -2 MPa) in agricultural plants.

12.3 Movement of solutes to roots

Solutes reach the surface of a root by mass (or convective) flow in the stream of water that finally enters the root, and by diffusion of particular ions or molecules in the direction of decreasing concentration in the soil solution. Material dissolved in soil water can include salts that are not needed as nutrients by the plant and some selectivity can be exercised by roots with respect to these. The concentration of the liquid phase varies greatly in differing circumstances and, as discussed in Chapter 10, it can restrict water uptake by the roots when excessive salinity causes a low osmotic potential to exist in the soil water. Some aspects of solute movement that affect plant nutrition will be touched on here without, however, dealing with the chemistry of solubility and ion exchange or the physiology of ion uptake by the roots. More comprehensive accounts are given in reviews by Olsen and Kemper (1968), Barley (1970), Danielson (1972) and Nye (1979). The composition of the soil solution and the suction and displacement methods used for obtaining it are discussed by Adams (1974).

The flux density of a solute transferred by molecular or ionic diffusion in a solution is given by Fick's law:

$$q_s = -D_0(dC/dx),$$

where q_s is the quantity of the diffusing substance transferred in unit time across unit cross-sectional area normal to the x-direction, C is the concentration (quantity of the solute in unit volume of the solution) and D_0 is the diffusion coefficient with dimensions of $L^2 T^{-1}$ when the same unit of quantity (such as kg) is used for both q_s and C. D_0 has a value of 1.5×10^{-9} m^2 s^{-1} for sodium chloride at 25 °C, for example. Diffusion of solutes through a porous material such as soil is restricted to the fraction θ of the cross-sectional area that is water filled and it has to follow a tortuous path. The flux across unit cross-section of the porous material is then represented by

$$q_s = -D_p(\mathrm{d}C/\mathrm{d}x),\tag{12.3}$$

where $D_p = b\theta D_0$ for an inert solid phase and b is a tortuosity factor which probably varies with θ but is often allotted a value of about 0.6 as for gas diffusion (Section 11.3). Values of D_p range from 2×10^{-12} to 5.6×10^{-10} m^2 s^{-1} in published data for soil and soil materials assembled by Danielson (1972). In each of his materials D_p increased with θ.

The transfer of a solute carried by convective flow of the water is given by

$$q_s = q_w C,\tag{12.4}$$

where q_s is the flux density of the solute and q_w is the flux density of the water (the quantity of each crossing unit area of soil in unit time in the x-direction normal to the cross-section).

When both diffusion and convective flow contribute to the transfer of the solute, we have from Equations (12.3) and (12.4)

$$q_s = -D_p(\mathrm{d}C/\mathrm{d}x) + q_w C.\tag{12.5}$$

These are the main components in an equation given by Barber (1962) for the transfer of nutrients to the root surface in plant nutrition. The second term on the right hand side of Equation (12.5) is likely to be dominant when the plant is transpiring actively and the first when it is not. However, certain dissolved nutrients (e.g. nitrate) are probably carried principally by convection and others, adsorbed by the solid phase (e.g. phosphate), by diffusion (Olsen and Kemper, 1968). Transfer to the root system as a whole by either process is greatly influenced by root density. It is not likely to be restricted in moist surface soil under pasture or established crops with a high root density. But in subsoils and in surface soil on which seedlings are growing, low root density may impose a restriction on nutrient transfer to the plant (Barley, 1970).

While Equation (12.5) serves to illustrate the mechanics of solute movement to roots, it applies only to steady state conditions which are not usual in soils. For the nonsteady state, use is made of the continuity equation (Appendix C) which relates the rate of change in the amount of solute in a volume element of soil with the difference between the rate at which it enters and leaves the element,

$$\partial m/\partial t = -\partial q_s/\partial x,$$ (12.6)

where m is the quantity of solute per unit volume of soil and t is time. In this equation, m can be replaced by θC, where θ is the volumetric water content, and, on combining Equations (12.5) and (12.6), an equation for transient movement of solute by both diffusion and convection is obtained:

$$\frac{\partial(\theta C)}{\partial t} = \frac{\partial(D_p\partial C/\partial x)}{\partial x} - \frac{\partial(q_w C)}{\partial x}.$$ (12.7)

On putting $D_p = \theta D$ and $q_w = \theta v$ (where v is the mean interstitial velocity of water through the water-filled pore space), Equation (12.7) becomes

$$\frac{\partial C}{\partial t} = D\frac{\partial^2 C}{\partial x^2} - v\frac{\partial C}{\partial x},$$ (12.8)

if D, θ and v are assumed to be constant. The diffusion coefficient, D, can be interpreted as the coefficient appropriate to Equation (12.3) if m, the quantity of solute in unit volume of soil, is substituted for C (Nielsen *et al.*, 1972, p. 32).

Equation (12.8) has the same form as Equation (10.16) for the miscible displacement of one solution by another in a porous material, as discussed in Section 10.7. In Equation (10.16) the dispersion coefficient that takes the place of D in Equation (12.8) covers both hydrodynamic dispersion and diffusion. Hydrodynamic dispersion occurs because of differing velocities of water in different sizes of pores and in different parts of a given pore when a resident solution is being displaced by a solution of a different concentration. At high velocities, it greatly exceeds the effect of diffusion as a cause of mixing of the two solutions. But, as shown by Olsen and Kemper (1968), when the velocity is less than about 2×10^{-8} m s^{-1} there is probably little enhancement of dispersion above that due to diffusion. They consider the velocity of flow to roots to be within this range and they conclude that hydrodynamic dispersion does not greatly affect nutrient movement to roots.

12.4 Water loss from plant foliage to the atmosphere

The total potential (energy) of water substance in the soil, the plant and the atmosphere shows a systematic trend in its magnitude that has often been described in discussions of the soil-plant-atmosphere continuum (see for example Philip, 1966). The potential difference between the liquid water in the mesophyll cells which is about to evaporate into the stomatal cavity and the water vapour present in the atmospheric air close to the surface of the leaf is particularly large. That difference $(\Phi_1 - \Phi_v)$, generalized over the whole foliage canopy, can be related to the average water vapour flux away from the land surface, q_v, in a simple flux equation, thus

$$\Phi_1 - \Phi_v = rq_v, \tag{12.9}$$

where r is regarded as a resistance to the flux that includes the effect of diffusion of water vapour through the stomatal openings, diffusion of water vapour through the air layers so close to the individual leaf surfaces that within it the air flow is laminar, and the effects of turbulent transport in the lowest layers of the atmosphere up to the height where Φ_v is measured. Equation (12.9) is a formal equation, not used in calculations, but it has a conceptual utility.

The physiological processes that deliver the liquid water to the sites of evaporation (transpiration) in the plant foliage are not easy to incorporate into a theory of water loss from plants. Nor is that necessary, because the micrometeorological processes are sufficiently distinct to enable them to be considered separately. The water vapour flux into the atmosphere at its lower boundary, the plant surfaces and the soil, can be specified after the change of state has occurred.

The earliest quantitative study of evaporation appears to be due to Dalton (early in the nineteenth century), who proposed that the evaporation rate from an open water surface, such as a lake, could be described by

$$E = F(u)(e_s - e_a), \tag{12.10}$$

where e_s is the saturated water vapour pressure at the temperature of the water surface, e_a is the measured vapour pressure in the air above the lake and $F(u)$ is a transfer coefficient that serves both as a proportionality factor and as a parameter to confer dimensional homogeneity on the equation. The reciprocal of $F(u)$ is equivalent to the resistance, r, of Equation (12.9) and it may be termed a conductance. The vapour pressures, e_s and e_a, depend upon the temperature regime in which the evaporation is proceeding, which, in turn, is related to the energy budget of the environment.

The description of evaporation from the surface of the earth (land and water) in terms of the energy budget is an alternative concept to the view that evaporative flow proceeds in response to a gradient of the partial pressure of water vapour in the boundary layer. The radiant energy budget is described by

$$R = (1 - \alpha)R_s - R_1, \tag{12.11}$$

where R is the net amount of radiant energy per second, that is transformed into other forms of energy (in watts m^{-2}), R_s is the incoming short-wave radiation from the sun and the sky, R_l is the outgoing long-wave radiation emitted by the surface and α is a reflection coefficient, often termed the albedo.

The net radiation itself can be equated to all the separate forms of energy that contribute to the energy flow away from the sites of absorption of radiation; thus

$$R = H + LE + G + Q + F, \qquad (12.12)$$

where F is the photosynthetic rate and the other terms represent the fluxes of sensible heat (H) and latent heat (LE) into the atmosphere, the conduction of heat into the ground (G) and the conduction of heat into the biomass (Q). The latent heat of vaporisation of water is represented by L $(J\,kg^{-1})$ so that the evaporative flux, E, can be appropriately stated in $kg\,m^{-2}\,s^{-1}$. The magnitudes of F and Q in the energy budget overall are small (about one to two percent at most) and they can be neglected in the present context.

The radiation balance (Equation 12.11) provides a description of processes that become essentially one-dimensional when all the radiation fluxes are resolved along the perpendicular to the earth's surface. Equation (12.12) is to be interpreted in the same way, as referring to one-dimensional flows of energy.

The albedo is an important natural coefficient, the magnitude of which will vary with the height at which the radiant fluxes are measured. Viewed from a satellite, the earth's albedo varies in the range of 0.6 to 0.2, depending upon the amount of cloud-cover, proportion of land to water, ice and snow and other changing aspects of the underlying surface as seen by the sensor. Close to the earth's surface the albedo has the values summarised in Table 12.1. It can be seen, for example, that deep, clear water absorbs more short-wave radiation than any other surface and that forest absorbs more radiation than grassland and cropland.

There is a ceaseless long-wave radiation exchange between the atmosphere, the clouds and the earth's surface. The quantity designated R_l in Equation (12.11) is the net amount that radiates to outer space. It is characteristically less at night than during the day when the ground surface is heated by the sun's radiation. Fortunately it is not necessary to measure the radiation components separately, because the net radiation can be measured with a single radiometer designed to be sensitive over the full wave-length spectrum of interest (Funk, 1959).

The soil moisture status can influence, to some degree, all the quantities specified in Equations (12.11) and (12.12), with the exception of R_s. Accounts of our knowledge of micrometeorology are given in numerous

Table 12.1. *Some representative values
of albedo of the earth*

Surface	Albedo
Sand (dry)	0.30–0.40
Dark clay soil (dry)	0.14–0.20
Wet soil	0.18–0.25
Grassland	0.22–0.28
Green cereal crop	0.20–0.26
Evergreen forest	0.17–0.23
Conifer forest	0.10–0.15
Shallow water	0.08–0.15
Deep, clear water	0.04–0.07

monographs (Priestley, 1959; Thom, 1975), the scope of which can indicate to the reader that the treatment given here must be relatively brief and concise. The sensible heat and latent heat fluxes often represent together a large part of the total energy budget. They are also the two fluxes that carry energy into the lowest layers of the atmosphere and are amenable to description by the aerial gradients of temperature and water vapour concentration (specific humidity).

The flux of sensible heat can be described by the equation

$$H = -K_H \, d(\rho C_p T)/dz, \qquad (12.13)$$

where T is the temperature at height z and K_H is a transfer coefficient that depends upon the horizontal wind velocity, aerodynamic roughness of the land surface, height of measurement, and stability or instability of the atmospheric air column. The density of the air, ρ, and the heat capacity at constant pressure, C_p, (joules $kg^{-1} K^{-1}$) are substantially constant, so that Equation (12.13) becomes

$$H = -K_H \rho C_p \, dT/dz. \qquad (12.14)$$

An equation similar to Equation (12.14) can describe the flux of evaporation namely

$$E = -K_w \rho \, dq/dz, \qquad (12.15)$$

where K_w is the transfer coefficient for water vapour, analogous to K_H and influenced by the same physical conditions of the atmosphere, and q is the specific humidity ($kg \, kg^{-1}$).

The ratio of sensible heat to latent heat is the Bowen ratio, β, (Bowen, 1926) and has proved to be particularly useful in discussing the relative magnitude of the evaporative flux. It is, from Equations (12.14) and (12.15),

$$\beta = H/LE = \frac{K_H C_p \, dT/dz}{K_w L \, dq/dz}.$$ (12.16)

When the lowest layer of the atmosphere, which is our concern here, is in stable equilibrium implying that the temperature increases with height, or is in neutral equilibrium, implying that there is no tendency for air to rise or sink because buoyancy is absent, the transfer coefficients are equal to each other. Therefore

$$\beta = \frac{\gamma \, dT/dz}{dq/dz}$$

$$= \gamma \partial T/\partial q,$$ (12.17)

where γ is C_p/L.

The partial differentiation term in Equation (12.17) can be replaced by differences that are appropriate for the analysis of measurements of specific humidity and temperature, thus

$$\beta = \frac{\gamma(T_1 - T_2)}{(q_1 - q_2)},$$ (12.18)

where the subscripts mean that the temperature and the specific humidity are measured at the same two heights, z_1, and z_2. They are often about 2 m apart over short grass, the lower measurement being made at about 1 m above the grass. The combination of Equation (12.18) with Equation (12.11) suggests that the evaporation and sensible heat fluxes can be derived, if the available energy, R, can be measured.

It can be seen from examination of Equation (12.18) that β can be negative if the temperature at z_1 is less than it is at z_2 ($z_2 > z_1$). Then the sensible heat flux is directed downwards and is negative. The evaporative flux may also be negative, and if H and E are both negative, β becomes positive again. Occasions when one or both of the fluxes are negative occur when the energy budget is relatively small, during night-time when R is negative, and only rarely during daylight hours except in high latitudes.

Application of the theory just given, for short intervals of time ($\sim$ minutes) is not successful because the conduction of energy into or out of the heat storage of the ground, G, and the biomass, Q, cannot be estimated with the accuracy required. For a whole day, however, the combination of energy budget and the Bowen ratio method is often satisfactory, when mean values of $(T_1 - T_2)/(q_1 - q_2)$ for intervals of about 30 min are used to derive let us say 24 individual values of Bowen ratio. The mean Bowen ratio is then obtained by weighting according to the magnitude of the energy budget for each half-hour period.

This introductory theory must assume that the energy exchange is one dimensional, so that conditions in the horizontal are everywhere the same for a distance sufficient for the evaporating and heat-exchanging surface to be effectively of infinite extent. When energy is advected by the wind to or from the site of measurement, the mathematical description becomes more formidable and the mechanisms for transfer of energy prove difficult to quantify (Rider, Philip and Bradley, 1963). Furthermore, since the evaporative flux is largest when the evaporating surface is heated strongly by the sun's radiation, the buoyant atmosphere must be considered in detail for a complete treatment (Webb, 1965). In particular, the eddy coefficients K_H and K_W cannot always be considered to be equal, and they are certainly not equal to the eddy coefficient for transfer of momentum in the vertical, K_M, the assumed equality that has to lie at the root of similarity theory. But further consideration of the aerodynamics of the lowest layers of the atmosphere is beyond our scope.

Priestley and Taylor (1972) analysed the partitioning of the available energy in a way that suggests the following treatment. If Equation (12.16) is rewritten as

$$\frac{LE}{H} = \frac{L}{C_p}\left(\frac{\partial q}{\partial T}\right),$$
(12.19)

we can examine the significance of the term $\partial q/\partial T$. If the specific humidity is that appropriate to saturated air, q_s, at the temperature, T_0, of the evaporating surface, then $\partial q_s(T_0)/\partial T$ could be a useful partition function. A correct specification of the temperature T_0 is, however, elusive even for a short grass sward, in which shading or exposure of the leaf elements could produce only small temperature contrasts compared with more structured canopies. The specification of T_0 for an open water surface might be no more secure, because it is often considered that in calm conditions turbulent mixing of the water does not eliminate transient cool (or warm) skins at the water surface.

Suppose a mean temperature $\bar{T}$ is measured at a height a little above the surface. It could be a radiative temperature obtained by a radiation thermometer exposed to survey the evaporating surface. Priestley and Taylor suggest that the relation

$$\frac{LE}{H} = \frac{L}{C_p}\left(\frac{\partial q_s}{\partial T}\right)_{T=\bar{T}}$$
(12.20)

is approximately correct when the evaporating surface is so wet that the specific humidity of the air very close to the evaporating sites is the saturated value at the surface temperature. Then Equation (12.20) would

Plants and soil water

be appropriate to foliage recently wet by rain; probably it would be applicable to dry canopies of plant communities adequately supplied with soil moisture, and it is likely to deviate from the observed partitioning of available energy when the plant communities of a landscape are subject to drought. Equation (12.20) can be written as

$$LE/H = s/\gamma, \qquad (12.21)$$

where s is the slope of the saturated specific humidity versus temperature relationship, evaluated at $\bar{T}$. Then an exact relation can be defined as

$$LE = \alpha sR/(s+\gamma), \qquad (12.22)$$

where α is a quantity that can be examined by comparing the predictions of Equation (12.22) with observed data of latent heat flux. Priestley and Taylor remark that ... 'The quantity α may also be useful in analyzing data from unsaturated surfaces; in this case the only *a priori* expectation would be that α would have a smaller value than for saturated surfaces; indeed the ratio of the two alphas could be regarded as an index of aridity.' They gave a best estimate of 1.26 for α when the land surface was in a condition for the evaporation to be at the maximum possible rate, i.e. at potential evaporation. (The quantity α here bears no relation to the albedo α.)

Equations (12.21) and (12.22) contain the explicit prediction that the Bowen ratio, as a parameter partitioning the available energy into sensible heat and latent heat, must depend upon the characteristic temperature of the environment, through the strong dependence of s upon temperature. Fig. 12.3 shows that relationship.

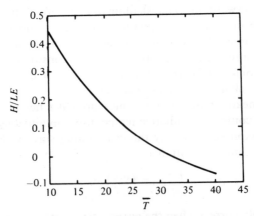

Fig. 12.3. The Bowen ratio, H/LE, plotted against the temperature $\bar{T}$ characteristic of the evaporating environment, when $\alpha = 1.26$, i.e. for a surface freely supplied with water. (After Priestley and Taylor, 1972.)

A concept similar to that expressed by Equation (12.22) was developed by Penman (1948), who sought to partition the net radiation into sensible heat and latent heat fluxes by more elaborate physical arguments than those of Priestley and Taylor. Penman's equation may be written as

$$LE = \frac{sR + \gamma F(q_s - q)}{s + \gamma},$$ (12.23)

where F is a transfer coefficient that has a role similar to $F(u)$ in Equation (12.10) and $(q_s - q)$ is the difference between the specific himidity of the air into which evaporation occurs (q) and the saturated specific humidity (q_s) of that air at the same temperature. $(q_s - q)$ therefore is the saturation deficit of the over-running air.

Penman's method of estimating evaporation, when soil moisture is non-limiting, influenced the development of ideas because of his emphasis that the net radiation as a measure of the energy available for evaporation should always occupy a prominent aspect of any physical theory (Penman, 1953). Numerous combination formulae that depend upon his ideas were offered (see Tanner, 1968 for a review) with the intent of enabling potential evaporation to be calculated using weather data that are usually observed at regular meteorological stations. There is, as yet, no proven, routine method of obtaining the actual evaporation rate if it is less than the potential rate, given by Equations (12.22) or (12.23), except by direct measurement.

Swinbank (1951) introduced the idea that the fluxes of latent heat, sensible heat and momentum could be studied by direct experiment, if the turbulent eddies could be examined closely. It is reasonable to suppose that, if the velocity of the wind is resolved into three, mutually perpendicular components, u, v, and w, the actual instantaneous velocities can be expressed as

$$u = \bar{u} + u'$$
$$v = \bar{v} + v' \quad \text{and}$$ (12.24)
$$w = w',$$

where $\bar{u}$ is the mean value of u and u' is the momentary fluctuation from the mean; similarly for v. Taking w to be the wind velocity resolved in the vertical, there can be a vertical component w' at any moment, but its long-term mean, $\bar{w}$, is zero. That is to say the durations and speeds of the air resolved vertically upwards or downwards, as the turbulent eddies of the wind pass over a fixed location, cannot produce a mean flow perpendicular to the ground.

The transport of water vapour convected by the motion of the wind can be resolved vertically too. It is momentarily $\rho w q (\text{kg m}^{-2} \text{ s}^{-1})$. Denoting the long-term mean of the water vapour flux as $\overline{\rho w q}$, Sutton (1953,

p. 63) showed that the average evaporative flux is given by

$$E = \overline{\rho w q} = \overline{(\rho w)'q'} = \rho\overline{(w'q')}, \qquad (12.25)$$

where, similarly to w', q' is the momentary fluctuation of the specific humidity away from its long-term mean, $\overline{q}$.

Equation (12.25) is the basis for design of instruments that can measure evaporation directly. The sensor for specific humidity has to have a very quick response to changes of humidity and the 'fluxatron' for water vapour eddy transport has been long delayed in its development (Dyer, Hicks and King, 1967; Raupach, 1978). The fluxatron for sensible heat, for which Equation (12.26), namely

$$H = \overline{(\rho C_p w)'T'} = \rho C_p\overline{(w'T')}, \qquad (12.26)$$

is the basis for design, has been more successful because it is easy to obtain a fine-bead thermistor that has a fast enough response. Ability to measure actual evaporation from forests and other aerodynamically rough vegetation must depend upon successful development of eddy flux apparatus, because the eddy transport coefficients are so large that the gradients of temperature or specific humidity (as in Equations (12.14) and (12.15)) are often too small to measure accurately.

12.5 The water budget of grassland

The discharge of the River Seine was stated by Perrault (1674) and Marriotte (1686) to be related to the amount of rain that fell in its drainage basin. Before the seventeenth century that fact, obvious to us now, was not understood. Perhaps the reason was that the storage of water in natural reservoirs, particularly the underground reservoir, made possible such long delays in discharge that the correlation of dry-weather flow in rivers with rainfall in preceding seasons seemed to be unlikely.

At the present day the large reservoir of water underground sometimes presents problems of measurement, particularly of change in its water storage. The direct measurement of deep percolation through the unsaturated soil zone to the water table still can hardly be done reliably in the field at large. It is customary to use lysimeters for the purpose. There have been many lysimeter installations, and of them an example that greatly influenced the popularity of the method is the lysimeter station of the Central Institute for Agricultural Research, at Wageningen in the Netherlands, commissioned in 1952 (Wind, 1958).

The Wageningen station eventually had 32 lysimeters. Each lysimeter was made of mild steel plate, in the shape of a cylindrical pot, 1 m^2 in cross-section and either 1 m or 1.5 m deep. Some were filled with soil of the field in which the lysimeters were placed. Others were filled with soils of widely varying composition to study the effects of soil type and all were planted with grass that was kept short cropped. Each pot was housed in a closely fitting well-pit so that its surface was at the same level as the field, but they could be lifted by a travelling crane for weighing.

Loss of water by drainage from a lysimeter, no matter what its design may be, can usually be arranged either by collection of the water or by measurement with a flow meter. It is the loss by evaporation that is hard to measure precisely except by weighing the container. Even then there is a limit to the size of the lysimeter imposed by the dependence of the relative sensitivity upon the weight to be determined.

Some lysimeters, designed more for water budget purposes than for observing the evaporation rate in detail, have been constructed simply by letting the container directly into the ground. The measurement of water content change in the soil above the water table is usually then done with the neutron moisture meter (Holmes and Colville, 1964), but it is not so precise as weighing. As for any installation that has to measure, as part of the experiment, evaporation from the land surface, it is important to ensure that obstructions to the wind are minimized by putting working spaces underground, or in sheds that are 100 m or more from the site.

A lysimeter should be designed to minimize the errors to which it is prone. Its surface should be as nearly as possible representative of the field in which it is placed, particularly with respect to the relative proportions of the plant species of the whole community. It should have the same heat budget as the undisturbed soil. Designers therefore try to make the gap between the weighed pot and its lined pit as small as possible. Continuous weighing mechanisms such as that developed by McIlroy (McIlroy and Angus, 1963) help to minimize traffic to the lysimeters in addition to their other obvious advantage of providing records of weight loss at small time intervals.

The water balance of an element of the soil, such as a $1 \, \text{m}^2 \times 1.5 \, \text{m}$ cylinder can be expressed as

$$P + I = E + D + A + \Delta W, \qquad (12.27)$$

where P is precipitation, I is irrigation (if any), E is evapotranspiration, D is underground drainage, A is direct or surface run-off, and ΔW is the amount of water that enters or leaves the soil water store. A lysimeter eliminates A, permits the capture of D for volume measurement and enables ΔW to be determined by weighing or by bore-hole logging with the neutron moisture meter. When P is measured by rain gauges and I by suitable experimental control the balance term is E.

Our knowledge of the yearly water budget has been gained very largely by lysimeter experiments. Fig. 12.4 shows an example of the annual evapotranspiration from grass at the location of the experimental farm

Plants and soil water

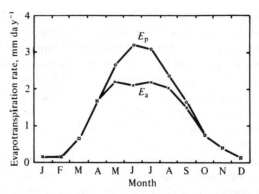

Fig. 12.4. The potential and actual evapotranspiration rate near Copenhagen, Denmark, 1955–64. E_p, potential; E_a, actual. (After Aslyng, 1965b.)

of the Royal Veterinary and Agricultural College, 20 km west of Copenhagen, Denmark (Aslyng, 1965a, b). The evaporation rates shown are characteristic of north-western Europe. The annual water budget at that location is summarized in Table 12.2, from which it can be noted that the surplus of precipitation over actual evapotranspiration (167 mm) could be quite sensitive to variations in the actual evapotranspiration.

Table 12.2. *Potential evapotranspiration, precipitation, actual evapotranspiration and cumulative water surplus (or deficit) of grassland near Copenhagen, Denmark, 1955–64 (after Aslyng, 1965b)*

Month	Potential evapotranspiration mm	Precipitation mm	Actual evapotranspiration mm	Cumulative balance mm
January	5	38	5	33
February	5	37	5	65
March	20	23	20	68
April	50	44	50	62
May	80	35	66	31
June	96	48	63	16
July	93	79	66	29
August	72	82	61	50
September	50	53	45	58
October	24	54	24	88
November	12	47	12	123
December	5	49	5	167
Total	512	589	422	167

Table 12.3. *Potential evapotranspiration, precipitation, actual evapotranspiration and cumulative water surplus (or deficit) of grassland near Mount Gambier, Australia (after Holmes and Colville, 1970b)*

Month	Potential evapotranspiration mm	Precipitation mm	Actual evapotranspiration mm	Cumulative balance mm
January	126	21	41	−20
February	87	21	38	−37
March	78	28	38	−47
April	47	52	31	−26
May	31	76	30	20
June	19	95	23	92
July	24	94	24	162
August	41	94	41	215
September	65	71	65	221
October	91	61	85	197
November	119	39	102	134
December	126	34	68	100
Total	854	686	586	100

In some contrast with the water balance near Copenhagen, (latitude 56° N), the water balance of grassland near Mount Gambier, Australia (latitude 39° S), given in Table 12.3, shows the influence of larger amounts of available energy and the lack of summer rainfall. The actual evapotranspiration during the summer is suppressed by a factor of about 0.32 in January, relative to the potential rate. At Copenhagen the same factor is about 0.66 for June. In each case the depletion of stored soil water enables the actual evapotranspiration to exceed the month's rainfall.

12.6 The water budget of forests

Penman (1963) reviewed the knowledge gained up to about 1960 concerning the influence of plant cover upon the hydrology of catchments. He suggested that water loss from forests could be somewhat larger than from grassland, but he expressed a reluctance, shared with many others, to accept a yearly evapotranspiration that could exceed the net radiation above the forest, on the assumption that it accounted for nearly all of the available energy. Indeed, there were few data to support any contrary view and none that had been obtained to that time, by acceptable micrometeorological techniques. In 1965 a seminar on forest hydrology was held in Pennsylvania. Some of the papers included in the report of

Table 12.4. *Annual precipitation, stream discharge and catchment evaporation from the Severn* (*100 per cent forested*) *and the Wye* (*grassland*) *catchments* (*after Newson, 1979*)

	Precipitation, mm yr^{-1} (P)		Stream flow, mm yr^{-1} (Q)		Evaporation, mm yr^{-1} $=(P-Q)$	
Year	Severn	Wye	Severn	Wye	Severn	Wye
1970	2485	2869	1636	2415	849	454
1971	1762	1993	797	1562	965	431
1972	2124	2131	1342	1804	782	328
1973	2380	2606	1581	2164	799	442
1974	2703	2794	1785	2320	918	474
1975	2035	2099	1213	1643	822	456
1976	1645	1736	921	1404	724	332
1977	2573	2651	1638	2236	935	325
Mean	2213	2348	1364	1944	849	405

its proceedings (Sopper and Lull, 1967) showed that hydrological processes of forests could not be described solely by the one-dimensional (vertical) fluxes of water vapour and radiation, although that had been assumed without serious error for low-growing plants and pastures. Since then a great amount of research has been done on the hydrometeorology of forests.

The recording of the water vapour flux above the canopy of a forest and the study of the associated physical processes, necessarily conducted from the platform of a tower reaching above the trees, can be done only intermittently. The synthesis of an annual water budget for a whole forest from such data requires such an amount of interpolation that other techniques are to be to be preferred for long-term studies. Experiments about hydrological processes in forests were therefore supplemented, in several countries, by fresh approaches to run-off measurements. A dense network of raingauges is required to determine the water input to the experimental catchment, whose output is measured as the discharge at a well-sited gauging station, where a single stream carries away the surplus of rainfall over evapotranspiration for the whole of the watershed.

The result of one such experiment is shown in Table 12.4, taken from a summary prepared by Newson (1979) of the comprehensive work done by the Institute of Hydrology (UK) in the Plynlimon (upland) region of Wales during the 1970s.

It can be seen, in Table 12.4, that the evaporation from the grassland catchment (Wye) was less than one half of that from the afforested

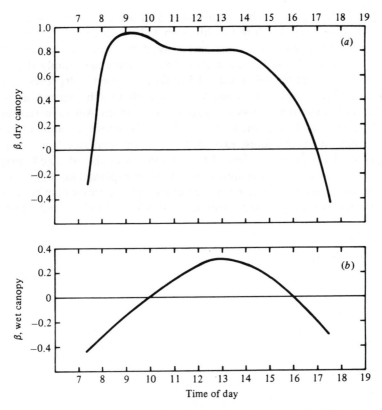

Fig. 12.5. Bowen ratios of a pine forest, when the canopy was dry (A) and when it was wet (B). Hourly values of the mean Bowen ratio from 16 days in each case, with long-fetch conditions. (After Moore, 1976.)

catchment (that part of the total Severn in evergreen spruce forest). The energy required for the evaporation of 405 mm yr^{-1} could be supplied by the net radiation absorbed by the grassland, according to ideas expressed in Equations (12.11) and (12.12). The evaporation of 849 mm yr^{-1} of water from the afforested catchment would require a large amount of supplementary energy in addition to the radiant energy. Advected sensible heat, brought to the forest by the wind, is the source of that additional energy, but it appears to be used for evaporation only when the canopy has water droplets and films upon its foliage, both during rain and after it has ceased. There is scarcely any difference in the dry-weather transpiration of grassland and the forests studied so far.

Moore (1976) employed the Bowen ratio, as a single parameter, to gain some understanding of the contrasts in evaporative regime between the wet and dry foliage conditions of a pine forest and these results are displayed in Fig. 12.5. The mean values of β are 0.12 for the wet forest and 0.74 for the dry forest, with both values calculated using as a weighting parameter the radiant energy available for conversion into the latent heat and sensible heat fluxes. These values then imply that the evaporation from the wet forest is 1.5 times that from the dry forest when the available energy is the same and is provided by the net radiation.

Moore's experiment was conducted in southern Australia (latitude 38 °S), whereas Plynlimon is at latitude 52 °N, where the annual water loss from the spruce forest exceeded the energy available from the net radiation by a factor of about 1.8. The sources of the advected energy have yet to be identified quantitatively. Further explanations and reviews of the contrasts in water use between grassland and forest may be found in papers by Calder (1982), Holmes and Wronski (1981) and Morton (1984).

Appendixes

A. SI units and some conversion factors for other units

Table A1. *Some basic and derived units of the International System of Units (SI) together with corresponding CGS units*

Physical quantity	Name of SI unit	Symbol for SI unit	Relation of CGS unit to corresponding SI unit
Basic			
Length	metre	m	$1\ cm = 10^{-2}\ m$
Mass	kilogram	kg	$1\ g = 10^{-3}\ kg$
Time	second	s	s
Temperature	kelvin	K	$x\ °C = (273.16 + x)\ K$
Electric current	ampere	A	A
Derived			
Area		m^2	$1\ cm^2 = 10^{-4}\ m^2$
Volume		m^3	$1\ cm^3 = 10^{-6}\ m^3$
Density		$kg\ m^{-3}$	$1\ g\ cm^{-3} = 10^3\ kg\ m^{-3}$
Velocity		$m\ s^{-1}$	$1\ cm\ s^{-1} = 10^{-2}\ m\ s^{-1}$
Acceleration		$m\ s^{-2}$	$1\ cm\ s^{-2} = 10^{-2}\ m\ s^{-2}$
Force	$newton = kg\ m\ s^{-2}$	N	$1\ dyne = 10^{-5}\ N$
Pressure, stress	$pascal = N\ m^{-2}$	Pa	$1\ dyne\ cm^{-2} = 10^{-1}\ Pa$
Work, energy	$joule = N\ m$	J	$1\ erg = 10^{-7}\ J$
Power	$watt = J\ s^{-1}$	W	$1\ erg\ s^{-1} = 10^{-7}\ W$
Surface tension		$N\ m^{-1}$	$1\ dyne\ cm^{-1} = 10^{-3}\ N\ m^{-1}$
Surface tension		$J\ m^{-2}$	$1\ erg\ cm^{-2} = 10^{-3}\ J\ m^{-2}$
Viscosity (dynamic)		$kg\ m^{-1}\ s^{-1}$	$1\ poise = 10^{-1}\ kg\ m^{-1}\ s^{-1}$
Electrical conductance	siemens	S	$1\ mho = 1\ S$
Electrical conductivity		$S\ m^{-1}$	$1\ mmho\ cm^{-1} = 1\ dS\ m^{-1}$

Table A2. *Prefixes and symbols that indicate multiples of basic and derived*
SI units

Fraction	Prefix	Symbol	Multiple	Prefix	Symbol
10^{-1}	deci	d	10	deka	da
10^{-2}	centi	c	10^2	hecto	h
10^{-3}	milli	m	10^3	kilo	k
10^{-6}	micro	μ	10^6	mega	M
10^{-9}	nano	n	10^9	giga	G
10^{-12}	pico	p	10^{12}	tera	T

Table A3. *Relation of other units to the basic and derived SI units*
in Table A1

Length	1 angstrom (Å)	$= 10^{-10}$ m or 100 pm
	1 micron	$= 10^{-6}$ m or 1 μm
	1 inch	$= 2.54 \times 10^{-2}$ m
Area	1 hectare	$= 10^4$ m^2
	1 acre	$= 4046$ m^2
Volume	1 litre	$= 10^{-3}$ m^3 or 1 dm^3
Mass	1 tonne	$= 10^3$ kg or 1 Mg
Pressure	1 bar	$= 100$ kPa
	1 m head of water at 20 °C	$= 9.8$ kPa
	1 standard atmosphere	$= 101.3$ kPa
	1 mm mercury at 20 °C	$= 133$ Pa

B. *Miscellaneous data including some properties of liquid water at 20 °C*

Acceleration due to gravity (approx.) $= 9.80$ m s^{-2}
Gas constant $\qquad\qquad\qquad = 8.314$ J K^{-1} mol^{-1}
Density of water $\qquad\qquad = 998.2$ kg m^{-3} (0.9982 g cm^{-3})
Surface tension of water $\quad = 72.75 \times 10^{-3}$ N m^{-1} or J m^{-2}
Viscosity of water $\qquad\quad = 1.0$ g m^{-1} s^{-1}
Heat of vaporization for water $= 2.453 \times 10^6$ J kg^{-1}
Density of mercury at 20 °C $\; = 13.546$ Mg m^{-3} or g cm^{-3}
$\ln x$ or $\log_e x$ $\qquad\qquad = 2.30258 \log_{10} x$

Additional data on water at temperatures other than 20 °C are given in Table
1.7 (density and viscosity) and Table 2.2 (surface tension). Diffusion coefficients
for oxygen and carbon dioxide in water and in air at 25 °C are given in Table 11.3.

C. *The continuity equation*

Consider the flow of a physical quantity such as an incompressible liquid through
a volume element in the form of a rectangular parallelepiped (Fig. C1) with sides
of length Δx, Δy, and Δz parallel to the axes of co-ordinates. The concentration
of the physical quantity at the point $P(x, y, z)$ at the centre of the element at

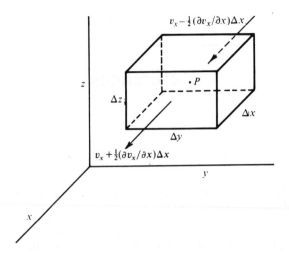

Fig. C1. The x component of velocity of flow through an element of volume.

time t is C. The velocity of flow at P is v and this has components v_x, v_y and v_z at P parallel to the x, y and z axes, respectively. The x component of velocity is approximately $v_x - \frac{1}{2}(\partial v_x/\partial x)\,\Delta x$ at the centre of one face of the element and $v_x + \frac{1}{2}(\partial v_x/\partial x)\,\Delta x$ at the other. Since $\Delta y\,\Delta z$ is the area of each of these faces, the volume of the physical quantity entering the element per unit time is

$$\left(v_x - \frac{1}{2}\frac{\partial v_x}{\partial x}\Delta x\right)\Delta y\,\Delta z,$$

and that leaving it is

$$\left(v_x + \frac{1}{2}\frac{\partial v_x}{\partial x}\Delta x\right)\Delta y\,\Delta z.$$

The rate of change in the volume contained in the element due to the x component is thus

$$-\frac{\partial v_x}{\partial x}\Delta x\,\Delta y\,\Delta z.$$

Similarly changes in the volume are caused by the y and z components of velocity and the total rate of change in volume of the physical quantity in the element is equal to

$$-\left(\frac{\partial v_x}{\partial x}+\frac{\partial v_y}{\partial y}+\frac{\partial v_z}{\partial z}\right)\Delta x\,\Delta y\,\Delta z.$$

When this expression is divided by the volume $\Delta x\,\Delta y\,\Delta z$ of the element, it represents the rate of change of concentration. Hence

$$\frac{\partial C}{\partial t}=-\left(\frac{\partial v_x}{\partial x}+\frac{\partial v_y}{\partial y}+\frac{\partial v_z}{\partial z}\right). \tag{C.1}$$

This holds exactly in the limit as Δx, Δy and Δz approach zero at point P. Equation (C.1) is written in vector notation for the divergence of v as

$$\partial C/\partial t = -\mathbf{\nabla}\cdot v. \tag{C.2}$$

References

Abramova, M. M. (1963). Movement of water vapour in soil. *Pochvovedenie* **10**, 49-63 and *Soviet Soil Sci.* 952-63.

Adams, F. (1974). Soil solution. In *The plant root and its environment*, ed. E. W. Carson, Univ. Press of Virginia, Charlottesville, pp. 441-81.

Adams, J. E., Ritchie, J. T., Burnett, E., and Fryrear, D. W. (1969). Evaporation from a simulated soil shrinkage crack, *Soil Sci. Soc. Am. Proc.* **33**, 609-13.

Ahuja, L. R., Naney, J. W., and Williams, R. D. (1985). Estimating soil water characteristics from simpler properties or limited data. *Soil Sci. Soc. Am. J.* **49**, 1100-5.

Ahuja, L. R., and Swartzendruber, D. (1972). Effect of Portland cement on soil aggregates and hydraulic properties. *Soil Sci.* **114**, 359-66.

Aitchison, G. D. (1961). Relationships of moisture stress and effective stress functions in unsaturated soils. In *Pore pressure and suction in soils*, 47-52. Butterworths, London.

Aitchison, G. D. (1973). Twenty-five years of application of soil survey principles in the practice of foundation engineering. *Geoderma* **10**, 99-112.

Aitchison, G. D., and Butler, P. F. (1951). Gypsum block moisture meters as instruments for the measurement of tension in soil water. *Aust. J. Appl. Sci.* **2**, 257-66.

Aitchison, G. D., and Donald I. B. (1956). Effective stresses in unsaturated soils. *Second Aust. N. Z. Conf. Soil Mech. Foundation Eng.* pp. 192-9.

Aitchison, G. D., and Holmes, J. W. (1953). Aspects of swelling in the soil profile. *Aust. J. Appl. Sci.* **4**, 249-59.

Alexander, L., and Skaggs, R. W. (1986). Predicting unsaturated hydraulic conductivity from the soil water characteristic. *Trans. ASAE* **29**, 176-84.

Alizai, H. U., and Hulbert, L. C. (1970). Effects of soil texture on evaporative loss and available water in semi-arid climates. *Soil Sci.* **110**, 328-32.

Allison, G. B. (1975). Estimation of the water resources of a portion of the Gambier Plain, S. A., using a new method for evaluating local recharge. *Hydrology Symp. Inst. Engnrs Aust. Armidale*, 1975, pp. 1-5.

Allison, G. B., and Barnes, C. J. (1983). Estimation of evaporation from non-vegetated surfaces using natural deuterium. *Nature* **301**, 143-5.

Allison, G. B., Barnes, C. J., and Hughes, M. W. (1983). The distribution of deuterium and ^{18}O in dry soils. 2. Experimental. *J. Hydrology* **64**, 377-97.

Allison, G. B., and Holmes, J. W. (1973). The environmental tritium concentration of underground water and its hydrological interpretation. *J. Hydrol.* **19**, 131-43.

Allison, G. B., and Hughes, M. W. (1972). Comparison of recharge to ground-water under pasture and forest using environmental tritium. *J. Hydrol.* **17**, 81-96.

Allison, G. B., and Hughes, M. W. (1974). Environmental tritium in the unsaturated zone: estimation of recharge to an unconfined aquifer. In *Isotope techniques in groundwater hydrology*, Vol. 1, IAEA Vienna, Proc. Symp. 11-15 March 1974, pp. 57-72.

Allison, G. B., and Hughes, M. W. (1977). The history of tritium fallout in southern Australia as inferred from rainfall and wine samples. *Earth & Planetary Sci. Lett.* **36**, 334-40.

Allison, L. E. (1947). Effect of microorganisms on permeability of soil under prolonged submergence. *Soil Sci.* **63**, 439-50.

Allmaras, R. R., and Dowdy, R. H. (1985). Conservation tillage systems and their adoption in the United States. *Soil and Tillage Res.* **5**, 197-222.

Allmaras, R. R., Burrows, W. C., and Larson, W. E. (1964). Early growth of corn as affected by soil temperature. *Soil Sci. Soc. Am. Proc.* **28**, 271-5.

American Society of Agricultural Engineers (1971). *Compaction of agricultural soils*, Monog. Am. Soc. Agric. Engnrs, St Joseph, Michigan.

American Society for Testing and Materials (1984). ASTM Standards, vol. 04.08, Am. Soc. Test. Mater., Philadelphia.

Anderson, D. M., Brown, R. L., and Buol, S. W. (1967). Diffusion of the dyes, eosin yellowish, bromophenol blue, and napthol green bluish, in water adsorbed by montmorillonite, *Soil Sci.* **103**, 281-7.

Anderson, V. G. (1945). Some effects of atmospheric evaporation and transpiration on the composition of natural waters in Australia. *Aust. Chem. Inst. J. and Proc.* **12**, 41-98.

Archer, J. R. (1975). Soil consistency. In *Soil physical conditions and crop production*, Min. of Agr., Fish. and Food, London, Tech. Bull. **29**, 289-97.

Arkin, G. F., and Taylor, H. M., eds. (1981). *Modifying the root environment to reduce crop stress*. Am. Soc. Agric. Engrs., St Joseph, Michigan.

Ashton, F. M. (1956). Effects of a series of cycles of alternating low and high water contents on the rate of apparent photosynthesis in sugar cane. *Pl. Physiol.* **31**, 266-74.

Aslyng, H. C. (Chairman) (1963). Soil physics terminology. *Int. Soc. Soil Sci. Bull.* **23**, 7-10.

Aslyng, H. C. (1965*a*). Weather, water balance and plant production at Copenhagen, 1955-64. *Roy. Vet. Agric. Coll. Copenhagen Yearbook* 1966, pp. 1-21.

Aslyng, H. C. (1965*b*). Evaporation, evapotranspiration and water balance investigations at Copenhagen 1955-64. *Acta Agric. Scand.* **15**, 284-300.

Atterberg, A. (1911). Die Plastizität der Tone. *Int. Mitt. Bodenk.* **1**, 10-43.

Australian Water Resources Council (1974). *Soil moisture measurement and its assessment*, Aust. Water Res. Coun. Hydrological Ser. 9, Dept. Environ. and Conserv., Canberra.

Awadhwal, N. K., and Thierstein, G. E. (1985). Soil crust and its impact on crop establishment: a review. *Soil and Tillage Res.* **5**, 289-302.

Baeumer, K., and Bakermans, W. A. P. (1973). Zero-tillage. *Adv. Agron.* **25**, 77-123.

Bagnold, R. A. (1941). *The physics of blown sand and desert dunes*, Methuen, London.

Bakhtin, P. U., Kirtbaya, Yu. K., Nikolayeva, I. N., and Volotskaya, V. I. (1969). Changes in resistance of some soils to plowing and deformations in relation to their moisture content. *Soviet Soil Sci.* 592-602. (*Pochvovedenie*, 136-46).

Bandara, B. W., and Fritton, D. D. (1986). Directional response of corn roots to physical barriers. *Plant and Soil* **96**, 359-68.

Barber, S. A. (1962). A diffusion and mass-flow concept of soil nutrient availability. *Soil Sci.* **93**, 39–49.

Barley, K. P. (1970). The configuration of the root system in relation to nutrient uptake. *Adv. Agron.* **22**, 159–201.

Barley, K. P. (1976). Mechanical resistance of the soil in relation to the growth of roots and emerging shoots. *Agrochimica* **20**, 173–82.

Bartelli, L. J. (1978). Technical classification system for soil survey information. *Adv. Agron.* **30**, 247–89.

Bateman, H. P., Naik, M. P., and Yoerger, R. R. (1965). Energy required to pulverize soil at different degrees of compaction. *J. Agric. Eng. Res.* **10**, 132–41.

Baver, L. D. (1938). Soil permeability in relation to non-capillary porosity. *Proc. Soil Sci. Soc. Am.* **3**, 52–6.

Beatty, H. J., and Loveday, J. (1974). Soluble cations and anions. In *Methods for analysis of irrigated soils.* J. Loveday ed. *Tech. Comm.* **54**, Bur. Soils, Comm. Agric. Bur., Farnham Royal, England.

Begg, J. E., and Turner, N. C. (1976). Crop water deficits. *Adv. Agron.* **28**, 161–217.

Belcher, D. J., Cuykendall, T. R., and Sack, H. S. (1950). *The measurement of soil moisture and density by neutron and gamma-ray scattering,* Tech. Dev. Rep. No. 127, Civil Aeronautics Admin., USA.

Bell, J. M., and Cameron, F. K. (1906). The flow of liquids through capillary spaces, *J. Phys. Chem.* **10**, 658–74.

Bennett, H. H. (1939). *Soil conservation,* McGraw-Hill, New York.

Bernhard, R. K., and Chasek, M. (1953). Soil density determination by means of radioactive isotopes. *Nondestructive Testing* **11**, 17–23.

Bernstein, L. (1975). Effects of salinity and sodicity on plant growth. *Ann. Rev. Phytopathol.* **13**, 295–312.

Bernstein, L., and Francois, L. E. (1973). Comparisons of drip, furrow and sprinkler irrigation. *Soil Sci.* **115**, 73–86.

Betson, R. P. (1964). What is watershed runoff? *J. Geophys. Res.* **69**, 1541–52.

Bevan, K., and Germann, P. (1982). Macropores and water flow in soils. *Water Resour. Res.* **18**, 1311–25.

Biggar, J. W., and Nielsen, D. R. (1967). Miscible displacement and leaching phenomenon. In *Irrigation of agricultural lands,* eds. R. M. Hagan, H. R. Haise, and T. W. Edminster, *Agronomy* **11**, pp. 254–74.

Biggar, J. W., and Nielsen, D. R. (1976). Spatial variability of the leaching characteristics of a field soil. *Water Resour. Res.* **12**, 78–84.

Bird, R. B., Stewart, W. E., and Lightfoot, E. N. (1960). *Transport phenomena,* J. Wiley, New York, 780 pp.

Bisal, F., and Ferguson, W. S. (1970). Effect of nonerodible aggregates and wheat stubble on initiation of soil drifting. *Can. J. Soil Sci.* **50**, 31–4.

Bisal, F., and Nielsen, K. F. (1962). Movement of soil particles in saltation. *Can. J. Soil Sci.* **42**, 81–6.

Bishop, A. W., and Blight, G. E. (1963). Some aspects of effective stress in saturated and partly saturated soils. *Geotechnique* **13**, 177–97.

Blackmore, A. V. (1973). Aggregation of clay by the products of iron: (iii) hydrolysis. *Aust. J. Soil Res.* **11**, 75–82.

Blake, G. R., and Hartge, K. H. (1986a). Bulk density. In *Methods of soil analysis,* part 1, 2nd edn, ed. A. Klute, *Agronomy* **9**, 363–75.

Blake, G. R., and Hartge, K. H. (1986b). Particle density. In *Methods of soil analysis,* part 1, 2nd edn, ed. A. Klute, *Agronomy* **9**, 377–82.

Blume, H. P. (1968). The mechanism of mottling and formation of concretions in waterlogged soils. *Z. Pflanzenernähr. Bodenk.* **119**, 124-34.

Blyth, K., and Rodda, J. C. (1973). A stream length study. *Water Resources Res.* **9**, 1454-62.

Bodman, G. B. (1949). Methods of measuring soil consistency. *Soil Sci.* **68**, 37-56.

Bodman, G. B., and Colman, E. A. (1944). Moisture and energy conditions during downward entry of water into soils. *Soil Sci. Soc. Am. Proc.* **8**, 116-22.

Bodman, G. B., and Constantin, G. K. (1965). Influence of particle size distribution in soil compaction. *Hilgardia* **36**, 567-91.

Bodman, G. B., and Rubin, J. (1948). Soil puddling. *Soil Sci. Soc. Am. Proc.* **13**, 27-36.

Bolt, G. H. (Chairman) (1976). Soil Physics terminology. *Int. Soc. Soil Sci. Bull.* **49**, 16-22.

Bolt, G. H., and Koenigs, F. F. R. (1972). Physical and chemical aspects of the stability of soil aggregates. In *Symposium on the fundamentals of soil conditioning*, ed. M. de Boodt, State Univ. Ghent, Fac. Agric. Sci. No. 37, pp. 955-73.

Boltzmann, L. (1894). On the integration of the diffusion equation with variable diffusion coefficient. *Ann. Physik. (Leipzig)* **53**, 959-64.

Bond, J. J., and Willis, W. O. (1969). Soil water evaporation: surface residue rate and placement effects. *Soil Sci. Soc. Am. Proc.* **33**, 445-8.

Bond, R. D. (1972). Germination and yield of barley when grown in a water-repellent sand. *Agron. J.* **64**, 402-3.

Bond, R. D., and Hammond, L. C. (1970). Effect of surface roughness and pore shape on water repellence of sandy soils. *Soil and Crop Sci. Soc. Florida Proc.* **30**, 308-15.

Boone, F. R., and Kuipers, H. (1970). Remarks on soil structure in relation to zero-tillage. *Neth. J. Agric. Sci.* **18**, 262-9.

Bosch, B., Guegan, B., Hubert, P., Marcé, A., Olive, P., and Siwertz, E. (1970). The cycle of tritium in the atmosphere and hydrosphere at middle latitudes since 1952. [*In French.*] *C.R. Acad. Sci. Paris* **270D**, 267-70.

Bouma, J. (1986). Using soil survey information to characterize the soil-water state. *J. Soil Sci.* **37**, 1-7.

Bouma, J., Belmans, C. F. M., and Dekker, L. W. (1982). Water infiltration and redistribution in a silt loam soil with vertical worm channels. *Soil Sci. Soc. Am. J.*, **46**, 917-21.

Bouwer, H. (1973). Renovating secondary effluent by ground water recharge with infiltration basins. In *Recycling treated municipal wastewater and sludge through forest and cropland*, eds. W. E. Sopper and L. T. Kardos, Pennsyl. State Univ. Press, pp.164-75.

Bouwer, H. (1986). Intake rate: cylinder infiltrometer. In *Methods of soil analysis*, part 1. *Physical and mineralogical methods*, 825-44. Agronomy 9, 2nd edn., ed. A. Klute.

Bouyoucos, G. J. (1927). The hydrometer as a new method for the mechanical analysis of soils. *Soil Sci.* **23**, 343-50.

Bouyoucos, G. J., and Mick, A. H. (1940). *An electrical resistance method for the continuous measurement of soil moisture under field conditions*, Mich. Agric. Exp. Sta. Tech. Bull. No. 172.

Bowen, I. S. (1926). The ratio of heat losses by conduction and by evaporation from any water surface. *Phys. Rev.* **27**, 779-87.

Bower, C. A. (1974). Salinity of drainage waters. In *Drainage for agriculture*, ed. J. van Schilfgaarde, *Agronomy* **17**, 471-87.

Bower, C. A., Ogata, G., and Tucker, J. M. (1969). Rootzone salt profiles and alfalfa growth as influenced by irrigation water salinity and leaching fraction. *Agron. J.* **61**, 783–5.

Bowman, R. S. (1984). Evaluation of some new tracers for soil water studies. *Soil Sci. Soc. Am. J.* **48**, 987–93.

Bradford, J. M. (1986). Penetrability. In *Methods of soil analysis*, part 1, 2nd edn, ed. A. Klute, *Agronomy* **9**, 463–78.

Braunack, M. V., and Dexter, A. R. (1976). Compaction of aggregate beds. In *Modification of soil structure*, eds. W. W. Emerson, R. D. Bond and A. R. Dexter, J. Wiley, Chichester, pp. 119–26.

Bresler, E. (1977). Trickle-drip irrigation: principles and application to soil-water management. *Adv. Agron.* **29**, 343–93.

Bresler, E., McNeal, B. L., and Carter, D. L. (1982). *Saline and sodic soils. Principles–dynamics–modelling.* Springer-Verlag, Berlin.

Brewer, R., (1964). *Fabric and mineral analysis of soils*, J. Wiley, New York.

Bridge, B. J., and Collis-George, N. (1973). A dual source gamma-ray traversing mechanism suitable for the non-destructive simultaneous measurement of bulk density and water content in columns of swelling soil. *Aust. J. Soil Res.* **11**, 83–92.

Briggs, L. J. (1949). A new method for measuring the limiting negative pressure of liquids. *Science,* **109**, 440.

Briggs, L. J. and Shantz, H. L. (1912). *The wilting coefficient for different plants and its indirect determination,* US Dept. Agr. Bur. Plant Ind. Bull. **230**, 1–83.

British Standards Institution (1975). *Methods of tests for soils for civil engineering purposes,* British Standard 1377, Brit. Stan. Inst., London.

Brooks, R. H., and Corey, A. T. (1966). Properties of porous media affecting fluid flow. *J. Irrigation and Drainage Div., Am. Soc. Civil Eng.* **92** (IR2), 61–88.

Bruce, R. R., and Klute, A. (1956). The measurement of soil moisture diffusivity. *Proc. Soil Sci. Soc. Am.* **20**, 458–62.

Bruce, R. R. and Klute, A. (1963). Measurement of soil moisture diffusivity from tension plate outflow data. *Proc. Soil Sci. Am.* **27**, 18–21.

Bruce-Okine, E., and Lal, R. (1975). Soil erodibility as determined by raindrop technique. *Soil Sci.* **119**, 149–57.

Brunauer, S., Emmett, P. H., and Teller, E. (1938). Adsorption of gases in multimolecular layers. *J. Am. Chem. Soc.* **60**, 309–10.

Bryan, R. B. (1968). The development, use and efficiency of indices of soil erodibility. *Geoderma* **2**, 5–26.

Buckingham, E. (1904). *Contributions to our knowledge of the aeration of soils,* US Dept. Agr. Bur. Soils Bull. No. 25.

Buckingham, E. (1907). *Studies in the movement of soil moisture,* US Dept. Agr. Bur. Soils Bull. No. 38.

Budyko, M. I. (1974). *Climate and life,* Trans. D. H. Miller, Academic Press, New York.

Buol, S. W., and Hole, F. D. (1959). Some characteristics of clay skins on peds in the B horizon of a gray-brown podzolic soil. *Soil Sci. Soc. Am. Proc.* **23**, 239–41.

Buol, S. W., Hole, F. D., and McCracken, R. J. (1973). *Soil genesis and classification,* Iowa State Univ. Press, Ames, Iowa.

Burdine, N. T. (1953). Relative permeability calculations from pore-size distribution data. *Petroleum Trans. Am. Inst. Mining Eng.* **198**, 71–8.

Burov, D. I. (1954). The effect of the content of dust-like particles in the plowed layer on the conditions of fertility of chernozem soils of the Trans-Volga. *Pochvovdenie* No. 12, 11-19.

Burrows, W. C. (1963). Characterization of soil temperature distribution from various tillage-induced microreliefs. *Soil Sci. Soc. Am. Proc.* **27**, 350-3.

Calder, I. R. (1982). Forest evaporation. Proc. Canadian Hydrology Symposium, 173-94, Nat. Res. Council, Ottawa, Ontario.

Campbell, C. A., Paul, E. A., Rennie, D. A., and McCallum, K. J. (1967). Applicability of the carbon-dating method of analysis to soil humus studies. *Soil Sci.* **104**, 217-24.

Campbell, G. S. (1985). Instruments for measuring plant water potential and its components. In *Instrumentation for environmental physiology*, B. Marshall and F. I. Woodward eds., Cambridge Univ. Press, Cambridge, England.

Campbell, G. S., and Campbell, M. D. (1982). Irrigation scheduling using moisture measurements: theory and practice. *Adv. Irrigation* **1**, 25-42.

Campbell, G. S., and Gardner, W. H. (1971). Psychrometric measurement of soil water potential: temperature and bulk density effects. *Soil Sci. Soc. Am. Proc.* **35**, 8-12.

Cannell, R. Q. (1985). Reduced tillage in north-west Europe-a review. *Soil and Tillage Res.* **5**, 129-77.

Carman, P. C. (1937). Fluid flow through granular beds. *Trans. Inst. Chem. Eng.* **15**, 150-66.

Carman, P. C. (1956). *Flow of gases through porous media*, Butterworths, London.

Carrier, G. F., Krook, M., and Pearson, C. E. (1966). *Functions of a complex variable.* McGraw-Hill, New York.

Carslaw, H. S., and Jaeger, J. C. (1959). *Conduction of heat in solids*, 2nd edn, Clarendon Press, Oxford.

Carter, D. L., Mortland, M. M., and Kemper, W. D. (1986). Specific surface. In *methods of soil analysis*, part 1, 2nd edn, ed. A. Klute, *Agronomy* **9**, 413-23.

Cary, J. W. (1965). Water flux in moist soil: thermal versus suction gradients. *Soil. Sci.* **100**, 168-175.

Cary, J. W. (1966). Soil moisture transport due to thermal gradients; practical aspects. *Proc. Soil Sci. Soc. Am.* **30**, 428-33.

Cassel, D. K. (1982). Tillage effects on soil bulk density and mechanical impedance. In *Predicting tillage effects on soil physical properties and processes*. Am. Soc. Agronomy, *Special Pub.* **44**, 45-67.

Cassel, D. K., and Klute, A. (1986). Water potential: tensiometry. In *Methods of soil analysis*, part 1, 2nd edn, ed. A. Klute, *Agronomy* **9**, 563-96.

Cassel, D. K., and Nielsen, D. R. (1986). Field capacity and available water capacity. In *Methods of soil analysis*, part 1, 2nd edn, ed. A. Klute, *Agronomy* **9**, 901-26.

Cavazza, L. (1969). Agronomic aspects of irrigation with brackish water in southern Italy. In *Value to agriculture of high-quality water from nuclear desalination*, Int. Atomic Energy Agency, Vienna, pp. 219-24.

Chahal, R. S., and Yong, R. N. (1964). Validity of the energy characteristic of soil water determined with pressure apparatus. *Nature, Lond.* **201**, 1180-1.

Chaney, K., and Swift, R. S. (1984). The influence of organic matter on aggregate stability in some British soils. *J. Soil Sci.* **35**, 223-30.

Chaney, K., and Swift, R. S. (1986). Studies on aggregate stability. 2. The effect of humic substances on the stability of re-formed soil aggregates. *J. Soil Sci.* **37**, 337-43.

Chang, Jen-hu (1958). *Ground temperature*, Vols 1 and 2, Harvard Univ., Blue Hill Met. Observatory, Milton, Mass, USA.

Chepil, W. S. (1953). Field structure of cultivated soils with special reference to erodibility by winds. *Soil Sci. Soc. Am. Proc.* **17**, 185-90.

Chepil, W. S. (1962). A compact rotary sieve and the importance of dry sieving in physical analysis. *Soil Sci. Soc. Am. Proc.* **26**, 4-6.

Chepil, W. S., and Woodruff, N. P. (1963). The physics of wind erosion and its control. *Adv. Agron.* **15**, 211-302.

Childs, E. C. (1943*a*). A note on electrical methods of determining soil moisture. *Soil Sci.* **55**, 219-23.

Childs, E. C. (1943*b*). The water table, equipotentials and streamlines in drained lands, *Soil Sci.* **56**, 317-30.

Childs, E. C. (1953). A new laboratory for the study of the flow of fluids in porous beds. *Proc. Inst. Civil Engnrs, Lond.* **2**(III), 134-41.

Childs, E. C. (1969). *An introduction to the physical basis of soil water phenomena*, J. Wiley, London.

Childs, E. C., Cole, A. H., and Edwards, D. H. (1953). The measurement of the hydraulic permeability of saturated soil *in situ*. II. *Proc. Roy. Soc. Lond.* A **216**, 72-89.

Childs, E. C., and Collis-George, N. (1950). The permeability of porous materials. *Proc. Roy. Soc. Lond.* A **201**, 392-405.

Childs, E. C., Collis-George, N., and Holmes, J. W. (1957). Permeability measurements in the field as an assessment of anisotropy and structure development. *J. Soil Sci.* **8**, 27-41.

Childs, E. C., and Tzimas, E. (1971). Darcy's law at small potential gradients. *J. Soil Sci.* **22**, 319-27.

Clarke, G. B., and Marshall, T. J. (1947). The influence of cultivation on soil structure and its assessment in soils of variable mechanical composition. *J. Coun. Sci. Ind. Res. Aust.* **20**, 162-75.

Clifford, S. M., and Hillel, D. (1986). Knudsen diffusion: the effect of small pore size and low gas pressure on gaseous transport in soil. *Soil Sci.* **141**, 289-97.

Clothier, B. E., and White, I. (1981). Measurement of sorptivity and soil water diffusivity in the field. *Soil Sci. Soc. Am. J.* **45**, 241-45.

Clothier, B. E., White, I., and Hamilton, G. J. (1981). Constant rate rainfall infiltration: field experiments. *Soil Sci. Soc. Am. J.* **45**, 245-9.

Conway, B. E. (1977). The state of water and hydrated ions at interfaces. *Adv. Colloid Interface Sci.* **8**, 91-211.

Corey, A. T., and Brooks, R. H. (1975). Drainage characteristics of soils. *Soil Sci. Soc. Am. Proc.* **39**, 251-5.

Cornish, P. M., Laryea, K. B. and Bridge, B. J. (1973). A nondestructive method of following moisture content and temperature changes in soils using thermistors. *Soil Sci.* **115**, 309-14.

Crafts, A. S., Currier, H. B., and Stocking, C. R. (1949). *Water in the physiology of the plant*, Chronica Botanica Co., Waltham, Mass.

Craig, H. (1957). Isotopic standards for carbon and oxygen and correction factors for mass-spectrographic analysis of carbon dioxide. *Geochim. e Cosmochim. Acta* **12**, 133-49.

Craig, H. (1961). Isotopic variations in meteoric waters. *Science* **133**, 1702-3.

Crank, J., and Henry, M. E. (1949*a*). Diffusion in media with variable properties: I. *Trans. Faraday Soc.* **45**, 636-42.

Crank, J., and Henry, M. E. (1949b). Diffusion in media with variable properties; II. *Trans. Faraday Soc.* **45**, 1119-28.

Criddle, W. D., and Kalisvaart, C. (1967). Subirrigation systems. In *Irrigation of agricultural lands*, eds. R. M. Hagan, H. R. Haise and T. W. Edminster, *Agronomy* **11**, 905-21.

Croney, D., and Coleman, J. D. (1953). Soil moisture suction properties and their bearing on moisture distribution in soils. *Proc. 3rd. Int. Conf. Soil Mech. Foundation Eng.* **1**, 13-18.

Croney, D., and Coleman, J. D. (1954). Soil structure in relation to soil suction (pF). *J. Soil Sci.* **5**, 75-84.

Croney, D., Coleman, J. D., and Black, W. P. M. (1958). Movement and distribution of water in soil in relation to highway design and performance. In *Water and its conduction in soils*, ed. H. F. Winterkorn, Highway Res. Board Special Rep. No. 40, Washington, D.C., pp. 226-52.

Crouch, R. J. (1976). Variation in the structural stability of soil in a tunnel eroding area. In *Modification of soil structure*, eds. W. W. Emerson, R. D. Bond and A. R. Dexter, J. Wiley, Chichester, pp. 267-74.

Curran, S. C., Angus, J., and Cockroft, A. L. (1949). Investigation of soft radiations. II. The beta spectrum of tritium. *Phil. Mag.* **40**, 53-60.

Currie, J. A. (1960). Gaseous diffusion in porous media. Part 2: Dry granular materials. *Brit. J. Appl. Phys.* **11**, 318-24.

Currie, J. A. (1961). Gaseous diffusion in porous media. Part 3: Wet granular materials. *Brit. J. Appl. Phys.* **12**, 275-81.

Currie, J. A. (1962). The importance of aeration in providing the right conditions for plant growth. *J. Sci. Food Agr.* **13**, 380-5.

Currie, J. A. (1970). Movement of gases in soil respiration. In *Sorption and transport processes in soils*, Monograph No. 37, Soc. of Chem. Ind., pp. 152-78.

Currie, J. A. (1975). Soil respiration. In *Soil physical conditions and crop production*, Min. of Ag. Fish. and Food, London, Tech. Bull. No. 29, pp. 461-68.

Currie, J. A. (1984). Gas diffusion through soil crumbs: the effects of compaction and wetting. *J. Soil Sci.* **35**, 1-10.

Czeratzki, W. (1971). The importance of ground frost in agriculture, especially in relation to soil cultivation. *Landbauforschung Völkenrode* **21**(1), 1-12.

Dallavalle, J. M. (1948). *Micromeritics: the technology of fine particles*, 2nd edn, Pitman, New York.

Danielson, R. E. (1972). Nutrient supply and uptake in relation to soil physical conditions. In *Optimizing the soil physical environment towards greater crop yield*, ed. D. Hillel, Academic Press, New York, pp. 193-221.

Danilin, A. I. (1955). The measurement of soil moisture with the help of gamma rays. [*In Russian.*] *Pochvovedenie* No. 7, 74-83. Abstract in *Soils and Ferts.* **19** (1956), 32.

Dansgaard, W. (1965). Stable isotopes in precipitation. *Tellus* **16**, 436-468.

Darcy, H. (1856), *Les fontaines publiques de la ville de Dijon*, Dalmont, Paris.

Davis, P. F., Dexter, A. R., and Tanner, D. W. (1973). Isotropic compression of hypothetical and synthetic tilths. *J. Terramechanics* **10**, 21-34.

Day, P. R. (1950). Physical basis of particle size analysis by the hydrometer method. *Soil Sci.* **70**, 363-74.

Day, P. R., and Holmgren, G. G. (1952). Microscopic changes in soil structure during compression. *Soil. Sci. Soc. Am. Proc.* **16**, 73-7.

Day, P. R., and Luthin, J. N. (1956). A numerical solution of the differential equation of flow for a vertical drainage problem. *Proc. Soil Sci. Soc. Am.* **20**, 443-7.

Debano, L. F., and Letey, J., eds (1969). Water-repellent soils. *Proc. Symp. on Water-repellent Soils*, Univ. Calif., Riverside, Calif.

De Boodt, M. (1972). Improvement of soil structure by chemical means. In *Optimising the soil physical environment towards greater crop yields*, ed. D. Hiltel, Academic Press, New York, pp. 43–55.

Delaney, A. C., and Zenchelsky, S. (1976). The organic component of wind-erosion-generated soil-derived aerosol. *Soil Sci.* 121, 146–55.

Denmead, O. T., and Shaw, R. H. (1962). Availability of soil water to plants as affected by soil moisture content and meteorological conditions. *Agron. J.* 54, 385–90.

De Roo, H. C. (1968). Tillage and root growth. In *Root growth*, ed. W. J. Whittington, Butterworths, London, pp. 339–57.

Deshpande, T. L., Greenland, D. J., and Quirk, J. P. (1968). Changes in soil properties associated with the removal of iron and aluminium oxides. *J. Soil Sci.* 19, 108–22.

de Vries, D. A. (1950). Some remarks on gaseous diffusion in soils. *Trans. 4th Int. Cong. Soil Sci.* 2, 41–3.

de Vries, D. A. (1953). Some results of field determinations of the moisture content of soil from thermal conductivity measurements. *Neth. J. Agric. Sci.* 1(2), 115–21.

de Vries, D. A. (1963). Thermal properties of soils. In *Physics of plant environment*, ed. W. R. van Wijk, North-Holland, Amsterdam, pp. 210–35.

de Wiest, R. J. M. (1966). On the storage coefficient and the equation of ground-water flow. *J. Geophys. Res.* 71, 1117–22.

Dexter, A. R. (1986a). Model experiments on the behaviour of roots at the interface between a tilled seed-bed and a compacted sub-soil. 1. Effects of seed-bed aggregate size and sub-soil strength on wheat roots. *Plant and Soil* 95, 123–33.

Dexter, A. R. (1986b). Model experiments on the behaviour of roots at the interface between a tilled seed-bed and a compacted sub-soil. 2. Entry of pea and wheat roots into sub-soil cracks. *Plant and Soil* 95, 135–47.

Dexter, A. R. (1986c). Model experiments on the behaviour of roots at the interface between a tilled seed-bed and a compacted sub-soil. 3. Entry of pea and wheat roots into cylindrical biopores. *Plant and Soil* 95, 149–61.

Dexter, A. R. and Kroesbergen, B. (1985). Methodology for determination of tensile strength of soil aggregates. *J. Agric. Engng. Res.* 31, 139–47.

Dimo, V. N., and Utkayera, V. F. (1984). Heat of wetting as one of the energy properties of soil. *Pochvovedeniye*, 1984, (2), 37–46. *Soviet Soil Sci.* 16, (1), 83–93.

Dixon, H. H. (1914). *Transpiration and the ascent of sap in plants*, Macmillan, London.

Dobrzanski, B., Witkovska, B., and Walezak, R. (1975). Soil aggregation and water-stability index. *Pol. J. Soil Sci.* 8, 3–8.

Doering, E. J. (1965). Soil-water diffusivity by the one-step method. *Soil. Sci.* 99, 322–6.

Doering, E. J., and Sandoval, F. M. (1976). Hydrology of saline seeps in the northern Great Plains. *Trans. Am. Soc. Agric. Eng.* 19, 856–61.

Downes, R. G. (1946). Tunnelling erosion in north-eastern Victoria. *J. Coun. Sci. Ind. Res. Aust.* 19, 283–92.

D'Souza, V. P. C., and Morgan, R. P. C. (1976). A laboratory study of the effect of slope steepness and curvature on soil erosion. *J. Agric. Eng. Res.* 21, 21–31.

Duley, F. L., and Kelly, L. L. (1939). Effect of soil type, slope, and surface conditions on intake of water. Univ. Nebr. Coll. Agric. *Res. Bul.* 112.

Dyal, R. S., and Hendricks, S. B. (1950). Total surface of clays in polar liquids as a characteristic index. *Soil Sci.* **69**, 421-32.

Dyer, A. J., Hicks, B. B., and King, K. M. (1967). The fluxatron—a revised approach to the measurements of eddy fluxes in the lower atmosphere. *J. Appl. Met.* **6**, 408-13.

Eagleson, P. S. (1970). *Dynamic hydrology.* McGraw-Hill, New York.

Eavis, B. W. (1972). Soil physical conditions affecting seedling root growth. 1. Mechanical impedance, aeration and moisture availability as influenced by bulk density and moisture levels in a sandy soil. *Pl. Soil* **36**, 613-22.

Edlefsen, N. E., and Anderson, A. B. C. (1943). Thermodynamics of soil moisture. *Hilgardia* **15**, 31-298.

Edwards, A. P., and Bremner, J. M. (1967). Dispersion of soil particles by sonic vibration. *J. Soil Sci.* **18**, 47-63.

Edwards, A. P., and Bremner, J. M. (1967). Microaggregates in soil. *J. Soil Sci.* **18**, 64-73.

Ehlers, W., Kopke, U., Hesse, F., and Böhm, W. (1983). Penetration resistance and root growth of oats in tilled and untilled loess soil. *Soil Tillage Res.* **3**, 261-75.

Ellison, W. D. (1947). Soil erosion studies. Pt 1. *Agric. Eng.* **28**, 145-6.

El Rayah, H. M. E., and Rowell, D. L. (1973). The influence of iron and aluminium hydroxides on the swelling of Na-montmorillonite and the permeability of a Na-soil. *J. Soil Sci.* **24**, 137-44.

Elrick, D. E., and Bowman, D. H. (1964). Note on an improved apparatus for soil moisture flow measurements. *Proc. Soil Sci. Soc. Am.* **28**, 450-2.

El-Swaify, S. A., and Emerson, W. W. (1975). Changes in the physical properties of soil clays due to precipitated aluminium and iron hydroxides: I. Swelling and aggregate stability after drying. *Soil Sci. Soc. Am. Proc.* **39**, 1056-63.

Emerson, W. W. (1954). The determination of the stability of soil crumbs. *J. Soil Sci.* **5**, 233-250.

Emerson, W. W. (1959). The structure of soil crumbs. *J. Soil Sci.* **10**, 235-44.

Emerson, W. W. (1967). A classification of soil aggregates based on their coherence in water. *Aust. J. Soil Res.* **5**, 47-57.

Emerson, W. W. (1971). Determination of the contents of clay-sized particles in soils. *J. Soil Sci.* **22**, 50-9.

Emerson, W. W. (1983). Inter-particle bonding. In *Soils: an Australian viewpoint.* Div. Soils, CSIRO, 475-98. CSIRO: Melbourne and Academic Pr.: London.

Emerson, W. W., and Bakker, A. C. (1973). The comparative effects of exchangeable calcium, magnesium and sodium on some physical properties of red-brown earth subsoils: 2. The spontaneous dispersion of aggregates in water. *Aust. J. Soil Res.* **11**, 151-7.

Engelund, F. (1951). Mathematical discussion of drainage problems. *Trans. Danish Acad. Tech. Sci.* **3**, 1-64.

Erickson, A. E. (1973). Physical changes of soils used for land application of municipal water. In *Recycling municipal sludges and effluents on land,* Proc. Joint Conf. Environment Protection Agency, US Dept. Agr. and Land Grant Colleges, pp. 39-47.

Erickson, A. E., Hansen, C. M., and Smucker, A. J. M. (1968). The influence of sub-surface asphalt barriers on the water properties and productivity of sand soils. *Trans. 9th Int. Congr. Soil Sci.* **1**, 331-7.

Erie, L. J. (1968). Management: a key to irrigation efficiency. *Proc. Am. Soc. Civ. Engnrs* **94**(IR3), 285-93.

Eriksson, E. (1965). An account of the major pulses of tritium and their effects in the atmosphere. *Tellus* 17, 118-30.

Fairbourn, M. L., and Gardner, H. R. (1975). Water-repellent soil clods and pellets as mulch. *Agron. J.* 67, 377-85.

FAO (1971). Salinity seminar, Baghdad. Report on *Methods of amelioration of saline and water-logged soils*, Food and Agricultural Organisation, Rome, Irrigation and Drainage Paper No. 7.

Farrell, D. A. and Greacen, E. L. (1966). Resistance to penetration of fine probes in compressible soils. *Aust. J. Soil Res.* 4, 1-17.

Farrell, D. A., Greacen, E. L., and Gurr, C. G. (1966). Vapour transfer in soil due to air turbulence. *Soil Sci.* 102, 305-13.

Farrell, D. A., Greacen, E. L., and Larson, W. E. (1967). The effect of water content on axial strain in a loam soil under tension and compression. *Soil Sci. Soc. Am. Proc.* 31, 445-50.

Farrer, D. M., and Coleman, J. D. (1967). The correlation of surface area with other properties of nineteen British clay soils. *J. Soil Sci.* 18, 118-24.

Fawcett, R. G. (1972). Physical aspects of swelling clay soils. In *Physical aspects of swelling clay soils*, Symp. Univ. New England, Armidale, Aust., pp. 183-5.

Feddes, R. A. (1972). Effects of water and heat on seedling emergence. *J. Hydrology* 16, 341-359.

Ferguson, H., and Gardner, W. H. (1962). Water content measurement in soil columns by gamma ray absorption. *Soil Sci. Soc. Am. Proc.* 26, 11-14.

Field, J. A., Parker, J. C., and Powell, N. L. (1984). Comparison of field- and laboratory-measured and predicted hydraulic properties of a soil with macropores. *Soil Sci.* 138, 385-96.

Fink, D. H. (1984). Paraffin-wax water-harvesting soil treatment improved with antistripping agents. *Soil Sci.* 138, 46-53.

Fisher, R. A. (1923). Some factors affecting the evaporation of water from soil. *J. Agric. Sci.* 13, 121-43.

Fisher, R. A. (1926). On the capillary forces in an ideal soil; correction of formulae given by W. B. Haines. *J. Agric. Sci.* 16, H29-505.

Foss, J. E., Wright, W. R., and Coles, R. H. (1975). Testing the accuracy of field textures. *Soil Sci. Soc. Am. Proc.* 39, 800-2.

Foster, R. C., Rovira, A. D., and Cock, T. W. (1983). *Ultrastructure of root-soil interface.* The Am. Phytopathological Soc., St. Paul, Minn.

Fountaine, E. R. (1954). Investigations into the mechanism of soil adhesion. *J. Soil Sci.* 5, 251-63.

Fountaine, E. R., and Payne, P. C. J. (1954). Causes of non-scouring in soil working implements. *Trans. 5th Int. Congr. Soil Sci.* 2, 35-45.

Fox, W. E. (1964). Cracking characteristics and field capacity in a swelling soil. *Soil Sci.* 98, 413.

Frank, H. S., and Wen, W. Y. (1957). Structural aspects of ion-solvent interaction in aqueous solutions: a suggested picture of water structure. *Discuss. Faraday Soc.* 24, 133-40.

Freebairn, D. M., and Wockner, G. H. (1986). A study of soil erosion on vertisols of the eastern Darling Downs, Queensland. 1. Effects of surface conditions on soil movement within contour bay catchments. *Aust. J. Soil Res.* 24, 135-58.

French, R. J. (1963). New facts about fallowing. *J. Agr. South Aust.* 67, 42-8.

Fritton, D. D., Kirkham, D., and Shaw, H. R. (1967). Soil water and chloride redistribution under various evaporation potentials. *Soil Sci. Soc. Am. Proc.* 31, 599-603.

Frydman, S. (1964). The applicability of the Brazilian (indirect tension) test to soils. *Aust. J. Appl. Sci.* **15**, 335-43.

Funk, J. P. (1959). Improved polythene-shielded net radiometer. *J. Scient. Instr.* **36**, 267-70.

Gardner, H. R., and Danielson, R. E. (1964). Penetration of wax layers by cotton roots as affected by some physical conditions. *Soil Sci. Soc. Am. Proc.* **28**, 457-60.

Gardner, H. R., and Hanks, R. J. (1966). Evaluation of the evaporation zone in soil by measurement of heat flux. *Soil Sci. Soc. Am. Proc.* **30**, 425-8.

Gardner, W., and Kirkham, D. (1952). Determination of soil moisture by neutron scattering. *Soil Sci.* **73**, 391-401.

Gardner, W. H. (1986). Water content. In *Methods of soil analysis*, part 1, 2nd edn, ed. A. Klute, *Agronomy* **9**, 493-544.

Gardner, W. H., and Calissendorff, C. (1967). Gamma ray and neutron attenuation in measurement of soil bulk density and water content. In *Isotope radiation techniques in soil physics and irrigation studies*, IAEA, Vienna, pp. 101-13.

Gardner, W. R. (1956). Calculation of capillary conductivity from pressure plate outflow data. *Soil Sci. Soc. Am. Proc.* **20**, 317-20.

Gardner, W. R. (1959). Solutions of the flow equation for the drying of soils and other porous media. *Soil Sci. Soc. Am. Proc.* **23**, 183-7.

Gardner, W. R. (1960). Dynamic aspects of water availability to plants. *Soil Sci.* **89**, 63-73.

Gardner, W. R. (1968). Availability and measurement of soil water. In *Water deficits and plant growth*, Vol. 1, ed. T. T. Kozlowski, Academic Press, New York, pp. 107-35.

Gardner, W. R., and Fireman, M. (1958). Laboratory studies of evaporation from soil columns in the presence of a water table. *Soil Sci.* **85**, 244-9.

Gardner, W. R., Hillel, D., and Benyamini, Y. (1970). Post-irrigation movement of soil water: 2. Simultaneous redistribution and evaporation. *Water Resources Res.* **6**, 1148-53.

Gat, J. R. (1970). Environmental isotope balance in Lake Tiberias. In *Isotope hydrology 1970*, Proc. Symp. Vienna, 9-13 March 1970, IAEA, Vienna, pp. 109-27.

Gavrilov, N. F. (1961). Effect of cracks on soil-moisture content. *Pochvovedenie* No. 7, 118-21.

Gee, G. W., and Bauder, J. W. (1986). Particle-size analysis. In *Methods of soil analysis*, part 1, 2nd edn, ed. A. Klute, *Agronomy* **9**, 383-41.

George, K. P. (1970). Theory of brittle fracture applied to soil cement. *J. Soil Mech. Foundation Div., Proc. Am. Soc. Civ. Eng.* **96**, 991-1010.

Gibbs, R. J., Matthews, M. D., and Link, D. A. (1971). The relationship between sphere size and settling velocity. *J. Sed. Petrol.* **41**, 7-18.

Gill, W. R., and McCreery, W. F. (1960). Relation of size of cut to tillage tool efficiency. *Agric. Eng.* **41**, 372-4.

Gill, W. R., and Vanden Berg, G. E. (1967). *Soil dynamics in tillage and traction*, Agric. Handb. No. 316, US Dept. Agr.

Gingrich, J. R., and Russell, M. B. (1957). A comparison of the effects of soil moisture tension and osmotic stress on root growth. *Soil Sci.* **84**, 185-94.

Gliński, J., and Stepniewski, W. (1985). *Soil aeration and its role for plants*. CRC Press, Boca Raton, Florida.

Grable, A. R. (1966). Soil aeration and plant growth. *Adv. Agron.* **18**, 57-106.

Gradwell, M. W. (1966). Soil moisture deficiencies in puddled pastures. *N.Z.J. Agric. Res.* **9**, 127-36.

Gradwell, M. W. (1968). The effect of grass cover on overnight heat losses from soil and some factors influencing the thermal conductivities of soils. *N.Z.J. Sci.* 11, 284–310.

Grant, C. D., Kay, B. D., Groenvelt, P. H., Kidd, G. E., and Thurtell, G. W. (1985). Spectral analysis of micropenetrometer data to characterize soil structure. *Can. J. Soil Sci.* 65, 789–804.

Greacen, E. L. (1960a). Water content and soil strength. *J. Soil Sci.* 11, 313–33.

Greacen, E. L. (1960b). Aggregate strength and soil consistence. *Trans. 7th Int. Congr. Soil Sci.* 1, 256–64.

Greacen, E. L., ed. (1981). Soil water assessment by the neutron method. C.S.I.R.O., Australia.

Greacen, E. L. (1986). Root reponse to soil mechanical properties. *Trans.* 13th Congress, Int. Soc. Soil Sci., Hamburg, 5, 20–47.

Greacen, E. L., Barley, K. P., and Farrell, D. A. (1968). The mechanics of root growth in soils, with particular reference to the implications for root distribution. In *Root growth*, ed. W. J. Whittington, Butterworths, London, pp. 256–68.

Greacen, E. L., Correll, R. L., Cunningham, R. B., Johns, G. G., and Nicolls, K. D. (1981). Calibration. In *Soil water assessment by the neturon method*, ed. E. L. Greacen, 50–81. C.S.I.R.O., Australia.

Greacen, E. L., and Oh, J. S. (1972). Physics of root growth. *Nature New Biol.* 235, 24–5.

Greacen, E. L., and Schrale, G. (1976). The effect of bulk density on neutron meter calibration. *Aust. J. Soil Res.* 14, 159–169.

Greb, B. W., Smika, D. E., and Black, A. L. (1970). Water conservation with stubble mulch fallow. *J. Soil Water Cons.* 25, 58–62.

Green, R. E., Ahuja, L. R., and Chong, S. K. (1986). Hydraulic conductivity, diffusivity and sorptivity of unsaturated soils: field methods. In *methods of soil analysis*, part 1. *Physical and mineralogical methods*, 771–98. Agronomy 9, 2nd edn., ed. A. Klute.

Green, W. H., and Ampt, G. A. (1911). Studies on soil physics: I. The flow of air and water through soils. *J. Agric. Sci.* 4, 1–24.

Greene-Kelly, R. (1962). Charge densities and heats of immersion of soil clay minerals. *Clay Minerals Bull.* 5, 1–8.

Greenland, D. J. (1965). Interaction between clays and organic compounds in soils: Part 2, adsorption of soil organic compounds and its effect on soil properties. *Soils and Fert.* 28, 521–32.

Greenland, D. J., and Quirk, J. P. (1962). Surface areas of soil colloids. *Trans. Int. Soc. Soil Sci. Comm.* IV and V, NZ, pp. 79–87.

Greenland, D. J., Rimmer, D., and Payne, D. (1975). Determination of the structural stability class of English and Welsh soils, using a water coherence test. *J. Soil Sci.* 26, 294–303.

Greenwood, D. J. (1975). Measurement of soil aeration. In *Soil physical conditions and crop production*, Min. of Agr. Fish. and Food, London, Tech. Bull. No. 29, 261–72.

Greminger, P. J., Sud, Y. K., and Nielsen, D. R. (1985). Spatial variability of field-measured, soil-water characteristics. *Soil. Sci. Soc. Am. J.* 49, 1075–82.

Grim, R. E. (1968). *Clay mineralogy*, 2nd edn, McGraw, New York.

Groenevelt, P. H., de Swart, J. G., and Cesler, J. (1969). Water content measurement with 60 keV gamma ray attenuation. *Bull. Int. Ass. Sci. Hydrol.* 14(2), 67–78.

Gupta, I. C., and Abichandani, C. T. (1968). Salt composition of some saline water irrigated soils of Western Rajasthan. *J. Indian Soc. Soil Sci.* **16**, 305–13.

Gupta, S. C., Farrell, D. A., and Larson, W. E. (1974). Determining effective soil water diffusivities from one-step outflow experiments. *Proc. Soil Sci. Soc. Am.* **38**, 710–16.

Gupta, S. C., and Larson, W. E. (1979). Estimating soil water retention characteristics from particle size distribution, organic matter percent, and bulk density. *Water Resour. Res.* **15**, 1633–5.

Gurr, C. G. (1962). Use of gamma rays in measuring water content and permeability in unsaturated columns of soil. *Soil Sci.* **94**, 224–9.

Gurr, C. G., and Jakobsen, B. (1978). Gamma probe for measurement of field bulk density and water content. In *Modification of soil structure*, eds. W. W. Emerson, R. D. Bond, and A. R. Dexter, J. Wiley, Chichester, pp. 127–33.

Gurr, C. G., Marshall, T. J., and Hutton, J. T. (1952). Movement of water in soil due to a temperature gradient. *Soil Sci.* **74**, 335–45.

Guy, H. P. (1969). Laboratory theory and methods for sediment analysis. In *Techniques of water-resources investigations of the US Geological Survey*, Book 5, Chap. 1., US Govt Print. Off., Washington, D.C.

Hadas, A. (1975). Drying of layered soil columns under nonisothermal conditions. *Soil Sci.* **119**, 143–8.

Hadas, A., and Wolf, D. (1983). Energy efficiency in tilling dry, clod forming soil. *Soil and Tillage Res.* **3**, 47–59.

Hadas, A., and Wolf, D. (1984a). Refinement and reevaluation of the drop-shatter soil fragmentation. *Soil and Tillage Res.* **4**, 237–49.

Hadas, A., and Wolf, D. (1984b). Soil aggregates and clod strength dependence on size, cultivation, and stress load rates. *Soil Sci. Soc. Am. J.* **48**, 1157–64.

Hagan, R. M., Haise, H. R., and Edminster, T. W., eds. (1967). *Irrigation of agricultural lands, Agronomy* **11**.

Haines, W. B. (1923). The volume changes associated with variation of water content in soil. *J. Agric. Sci.* **13**, 296–310.

Haines, W. B. (1925). Studies in the physical properties of soils. (2) A note on the cohesion developed by capillary forces in an ideal soil. *J. Agric. Sci.* **15**, 529–35.

Hall, D. M., and Jones, R. L. (1961). Physiological significance of surface wax on leaves. *Nature, Lond.* **191**, 95–6.

Halvorson, A. D., and Rhoades, J. D. (1976). Field mapping soil conductivity to delineate dryland saline seeps with four electrode technique. *Soil Sci. Soc. Am. J.* **40**, 571–5.

Hanks, R. J., and Rasmussen, V. P. (1982). Predicting crop production as related to plant water stress. *Adv. Agron.* **35**, 193–215.

Haridason, M., and Jensen, R. D. (1972). Effect of temperature on pressure head-water content relationship and conductivity of two soils. *Soil Sci. Soc. Am. Proc.* **36**, 703–8.

Harris, R. F., Chesters, G., and Allen, O. N. (1966), Dynamics of soil aggregation. *Adv. Agron.* **18**, 107–69.

Harris, W. L. (1971). The soil compaction process. In *compaction of agricultural soils*, Am. Soc. Agric. Engnrs, St Joseph, Michigan, pp. 9–44.

Harrold, L. L., Peters, D. B., Dreibelbis, F. R., and McGuiness, J. L. (1959). Transpiration evaluation of corn grown on a plastic covered lysimeter. *Soil Sci. Soc. Am. Proc.* **23**, 174–8.

Hartge, K. H., and Bohne, H. (1984). Mechanism of stabilization of clay aggregates by quicklime. *Pochvovdeniye* 1984 (6), 47-53. *Soviet Soil Sci.* **16** (3), 94-101.

Haverkamp, R., and Parlange, J-.Y. (1986). Predicting the water-retention curve from particle-size distribution: 1. Sandy soils without organic matter. *Soil Sci.* **142**, 325-39.

Hénin, S., Féodoroff, A., Gras, R., and Monnier, G. (1960). *Le profil cultural. Principes de physique du sol.* Soc. d'Editions des Ingenieurs Agricoles, Paris.

Hénin, S., and Santamaria, R. (1975). Observation comparée sur le comportment des fragments de terre soumis à l'action de l'eau de différentes solutions et de l'alcool. *Science du Sol* pp. 171-82.

Herdan, G. (1960). *Small particle statistics*, Butterworth Sci. Pub., London.

Herkelrath, W. N., Miller, E. E., and Gardner, W. R. (1977). Water uptake by plants. 2. The root contact model. *J. Soil Sci. Soc. Am.* **41**, 1039-43.

Hillel, D. (1971). *Soil and water*, Academic Press, New York.

Hillel, D., and Berliner, P. (1974). Waterproofing surface-zone soil aggregates for water conservation. *Soil Sci.* **118**, 131-5.

Hodgson, J. M., ed. (1974). *Soil survey field handbook*, Tech. Monog. No. 5., Soil Survey, Rothamsted Exp. Sta., Harpenden, England.

Hofmann, V., Endell, K., and Wilm, D. (1933). Kristallstruktur und Quellung von Montmorillonit. *Zeit. Krist.* **86**, 340-8.

Holmes, J. W. (1955). Water sorption and swelling of clay blocks. *J. Soil Sci.* **6**, 200-8.

Holmes, J. W. (1960). Water balance and the water-table in deep sandy soils of the Upper South-East, South Australia. *Aust. J. Agric. Res.* **11**, 970-88.

Holmes, J. W., and Colville, J. S. (1964). The use of the neutron moisture meter and lysimeters for water balance studies. *Trans. 8th Int. Congr. Soil Sci.* **2**, 445-54.

Holmes, J. W., and Colville, J. S. (1970*a*). Forest hydrology in a karstic region of southern Australia. *J. Hydrol.* **10**, 59-74.

Holmes, J. W., and Colville, J. S. (1970*b*). Grassland hydrology in a karstic region of southern Australia. *J. Hydrol.* **10**, 38-58.

Holmes, J. W., Greacen, E. L., and Gurr, C. G. (1960). The evaporation of water from bare soils with different tilths. *Trans. 7th Int. Congr. Soil Sci.* **1**, 188-94.

Holmes, J. W., Taylor, S. A., and Richards, S. J. (1967). Measurement of soil water. In *Irrigation of agricultural lands*, eds. R. M. Hagan, H. R. Haise and T. W. Edminster, *Agronomy* **11**, pp. 275-303.

Holmes, J. W. and Wronski, E. B. (1981). The influence of plant communities upon the hydrology of catchments. *Agric. Water Management* **4**, 19-34.

Hooghoudt, S. B. (1940). Contributions to the knowledge of some physical characteristics of soil. 7. General considerations of the problem of drainage and infiltration by means of parallel drains, furrows and canals. [*In Dutch.*] *Versl. Landb. Ond.* **46**, 515-707.

Horne, R. A. (1968). The structure of water and aqueous solutions. *Surv. Prog. Chem.* **4**, 1-43.

Horton, R. E. (1940). An approach towards a physical interpretation of infiltration capacity. *Soil Sci. Soc. Am. Proc.* **5**, 399-417.

Hubert, P., Marcé, A., Olive, P., and Siwertz, E. (1970). Tritium study of the dynamics of underground water. *C.R. Acad. Sci. Paris* D **270**, 908-11.

Hudson, N. (1981). *Soil conservation*, Cornell Univ. Pr., Ithaca, N.Y., 2nd edn.

Hulugalle, N. R., and Willatt, S. T. (1983). The role of soil resistance in determining water uptake by plant root systems. *Aust. J. Soil Res.* **21**, 571-4.

Hutton, J. T. (1955). A *method of particle size analysis of soil*, Div. Rep. No. 11/55, Div. Soils, Comm. Sci. Ind. Res. Org. Australia.

IAEA (1968). *Guidebook on nuclear techniques in hydrology*, IAEA Tech. Rep. No. **91**, IAEA, Vienna.

IAEA (1981). *Stable isotope hydrology*. IAEA, Vienna.

IAHS (1974). Mathematical models in hydrology. *Symposium of Warsaw, July 1971* IAHS Publ. No. 100, IAHS-UNESCO-WMO.

Idso, S. B., Reginato, R. J., Jackson, R. D., Kimball, B. A., and Nakayama, F. S. (1974). The three stages of drying of a field soil. *Soil Sci. Soc. Am. Proc.* **38**, 831-7.

Ingles, O. G., and Metcalf, J. B. (1972). *Soil stabilization: principles and practice*, Butterworths, Sydney.

Innes, W. B. (1957). Use of a parallel plate model in calculation of pore size distribution. *Analyt. Chem.* **29**, 1069-73.

Jackson, R. D. (1973). Diurnal changes in soil water content during drying. In *Field soil water regime*, Spec. Publ. No. 5, Soil Sci. Soc. Am., Madison, Wisconsin, pp. 37-55.

Jaeger, J. C. (1964). *Elasticity, fracture and flow*, Methuen, London.

Jahnke, E., and Emde, F. (1945). *Tables of functions with formulae and curves*, Dover Publications, New York.

Janert, H. (1934). The application of heat of wetting measurements to soil research problems. *J. Agric. Sci.* **24**, 136-50.

Jenny, H., and Grossenbacher, K. (1963). Root-soil boundary zones as seen in the electron microscope. *Soil Sci. Soc. Am. Proc.* **27**, 275-7.

Johnston, J. R., and Hill, H. O. (1945). A study of the shrinkage and swelling properties of rendzina soils. *Soil Sci. Soc. Am. Proc.* **9**, 24-9.

Kalinin, Ya D. (1969). Experiment in the leaching of saline land in southern Kazakhstan. *Soviet Hydrology: Selected Papers*, **2**, 201-9.

Kanwar, J. S. (1976). Soil and water management—the key to production in rainfall agriculture of semi-arid tropics. *J. Indian Soc. Soil Sci.* **24**, 230-9.

Kaufman, S., and Libby, W. F. (1954). The natural distribution of tritium. *Phys. Rev.* **93**, 1337-44.

Kay, B. D., Grant, C. D., and Groenvelt, P. H. (1985). Significance of ground freezing on soil bulk density under zero tilling. *Soil Sci. Soc. Am. J.* **49**, 973-8.

Kaye, G. W. C., and Laby, T. H. (1973). *Tables of physical and chemical constants*, 14th edn, Longman, London.

Keen, B. A. (1931). *The physical properties of the soil*, Longmans, Green and Co., London.

Keisling, T. C., Davidson, J. M., Weeks, D. L., and Morrison, R. D. (1977). Precision with which selected soil physical parameters can be estimated. *Soil Sci.* **124**, 241-8.

Kemper, W. D., and Koch, E. J. (1966). *Aggregate stability of soils from western United States and Canada*, US Dept. Agr. Tech. Bull. No. 1355.

Kemper, W. D., Olsen, J. S., and Hodgdon, A. (1975). Irrigation method as determinant of large pore persistence and crust strength of cultivated soils. *Soil Sci. Soc. Am. Proc.* **39**, 519-23.

Kemper, W. D., and Rosenau, R. C. (1986). Aggregate stability and size distribution. In *Methods of soil analysis*, part 1, 2nd edn, ed. A. Klute, *Agronomy* **9**, 425-42.

Kersten, S. (1949). *Thermal properties of soils.* Bull. Univ. Minnesota, Inst. Tech. Eng. Exp. Sta. No. **28**.

Kimball, B. A., Jackson, R. D., Nakayama, F. S., Idso, S. B. and Reginato, R. J. (1976a). Soil-heat flux determination: temperature-gradient method with computed thermal conductivities. *Soil Sci. Soc. Am. J.* **40**, 25-8.

Kimball, B. A., Jackson, R. D., Reginato, R. J., Nakayama, F. S., and Idso, S. B. (1976b). Comparison of field-measured and calculated soil-heat fluxes. *Soil Sci. Soc. Am. J.* **40**, 18-25.

Kirby, C. F. (1968). *Irrigation with waste water at Board of Works Farm, Werribee,* Water Res. Found. Aust. Rep. No. 25, Kingsford, N.S.W., Australia.

Kirkham, D. (1945). Proposed method for field measurement of permeability of soil below the water table. *Soil Sci. Soc. Am. Proc.* **10**, 58-68.

Kirkham, D. (1958). Seepage of steady rainfall through soil into drains. *Trans. Am. Geophys. Union* **39**, 892-908.

Kirkham, D., and Feng, C. L. (1949). Some tests of the diffusion theory, and laws of capillary flow in soils. *Soil Sci.* **67**, 29-40.

Kirkham, D., and Powers, W. L. (1972). *Advanced soil physics,* Wiley-Interscience, New York.

Klute, A. (1952). A numerical method for solving the flow equation for water in unsaturated materials. *Soil Sci.* **73**, 105-16.

Klute, A. (1986). Water retention: laboratory methods. In *Methods of soil analysis,* part 1, 2nd edn, ed. A. Klute, *Agronomy* **9**, 635-62.

Klute, A., ed. (1986). *Methods of soil analysis,* part 1, 2nd edn., *Agronomy* **9**.

Klute, A., and Dirksen, C. (1986). Hydraulic conductivity and diffusivity: laboratory methods. In *Methods of soil analysis,* part 1, 2nd edn, ed. A. Klute, *Agronomy* **9** 687-734.

Kolyasev, F. E., and Gupalo, A. I. (1958). *On the correlation of heat and moisture properties of soils,* Publication No. 629 (Special Report No. 40) Highway Res. Bd, Nat. Acad. Sci., Washington, pp. 106-12.

Kononova, M. M. (1966). *Soil organic matter,* 2nd English edn, Pergamon, Oxford.

Kopp, E. (1966). Das Grundprinzip der Polygonbildung im Mechanismus der Gefügeentstehung. *Z. Pflanzenernähr. Bodenk.* **112**, 201-12.

Kostiakov, A. N. (1932). On the dynamics of the coefficient of water-percolation in soils and on the necessity of studying it from a dynamic point of view for purposes of amelioration. *Trans. 6th Comm. Int. Soc. Soil Sci. Russian* Pt A, pp. 15-21. Also *Pochvovedenie* No. 3, 293-7 [*in Russian*].

Kovda, V. A., van den Berg, C. and Hagan, R. M. (eds) (1973). *Irrigation, drainage and salinity, an international source book,* UNESCO/FAO, Hutchinson, London.

Kozeny, J. (1927). Über kapillare Leitung des Wassers im Boden. *Sb. Akad. Wiss. Wien, Math-naturw. Kl.* Abt. IIa, **136**, 271-306.

Kramer, P. J. (1983). *Water relations of plants.* Academic Press, New York.

Krumbein, W. C., and Pettijohn, F. J. (1938). *Manual of sedimentary petrography,* D. Appleton-Century Co., New York.

Lafeber, D. (1965). The graphical representation of planar pore patterns in soils. *Aust. J. Soil Res.* **3**, 143-64.

Lal, R. (1974). Soil temperature, soil moisture and maize yield from mulched and unmulched tropical soils. *Plant and Soil* **40**, 129-43.

Lambe, T. W., and Whitman, R. V. (1969). *Soil Mechanics,* J. Wiley, New York.

Landers, J. N., and Witte, K. (1967). Irrigation for frost protection. In *Irrigation of agricultural lands,* eds. R. M. Hagan, H. R. Haise and T. W. Edminster, *Agronomy* **11**, pp. 1037-57.

Lang, A. R. G. (1967). Osmotic coefficients and water potentials of sodium chloride solutions from 0 to 40°C. *Aust. J. Chem.* **20**, 2017-23.

Larson, W. E., and Allmaras, R. R. (1971). Management factors and natural forces as related to compaction. In *Compaction of agricultural soils*, Am. Soc. Agric. Engnrs, St Joseph, Michigan, pp. 367-427.

Lascano, R. J., and van Bavel, C. H. M. (1984). Root water uptake and soil water distribution: test of an availability concept. *Soil Sci. Soc. Am. J.* **48**, 233-7.

Lauritzen, C. W. (1948). Apparent specific volume and shrinkage characteristics of soil materials. *Soil Sci.* **65**, 155-79.

Lauritzen, C. W., and Stewart, A. J. (1941). Soil-volume changes and accompanying moisture and pore-space relationships. *Soil Sci. Soc. Am. Proc.* **6**, 113-16.

Lawlor, D. W. (1972). Growth and water use of *Lolium perenne*. 1. Water transport. II. Plant growth. *J. Appl. Ecol.* **9**, 79-105.

Lawrence, G. P. (1977). Measurement of pore sizes in fine-textured soils: a review of existing techniques. *J. Soil Sci.* **28**, 527-40.

Laws, J. O. (1941). Measurement of fall-velocity of water-drops and raindrops. *Trans. Am. Geophys. Union* **22**, 709-21.

Lemon, E. R., and Erickson, A. E. (1952). The measurement of oxygen diffusion in the soil with a platinum microelectrode. *Soil Sci. Soc. Am. Proc.* **16**, 160-3.

Letey, J. (1985). Relationship between soil physical factors and crop production. *Adv. Soil Sci.* **1**, 277-94.

Lewis, M. R. (1937). The rate of infiltration of water in irrigation practice. *Trans. Am. Geophys. Union (18th Meeting)*, Pt 2 Sect. Hydrol., pp. 361-8.

Libby, W. F. (1963). Moratorium tritium geophysics. *J. Geophys. Res.* **68**, 4484-94.

Loehr, R. C. (1974). *Agricultural waste management*, Academic Press, New York.

Loveday, J. (1976). Relative significance of electrolyte and cation exchange effects when gypsum is applied to a sodic clay soil. *Aust. J. Soil Res.* **14**, 361-71.

Low, A. J. (1972). The effect of cultivation on the structure and other physical characteristics of grassland and arable soils. *J. Soil Sci.* **23**, 363-80.

Low, A. J. (1973). Soil structure and crop yield. *J. Soil Sci.* **24**, 249-59.

Low, P. F. (1961). Physical chemistry of clay-water interactions *Adv. Agron.* **13**, 269-327.

Low, P. F., and White, J. L. (1970). Hydrogen bonding and polywater in clay-water systems. *Clays and Clay Minerals* **18**, 63-6.

Lusczynski, N. J. (1961). Head and flow of ground-water of variable density. *J. Geophys. Res.* **66**, 4247-56.

Luthin, J. N., ed. (1957). *Drainage of agricultural lands. Agronomy*, 7.

Luthin, J. N., and Kirkham, D. (1949). A piezometer method for measuring permeability of soil *in situ* below a water table. *Soil Sci.* **68**, 34-58.

Lynch, J. M., and Bragg, E. (1985). Microorganisms and soil aggregate stability. *Adv. Soil Sci.* **2**, 133-71.

Maasland, M. (1959). Water table fluctuations induced by intermittent recharge. *J. Geophys. Res.* **64**, 549-59.

McBurney, T., and Costigan, P. A. (1987). Plant water potential measured continuously in the field. *Plant and Soil* **97**, 145-9.

McCalla, T. M. (1944). Water-drop method of determining stability of soil structure. *Soil Sci.* **58**, 117-21.

McCalla, T. M., and Army, T. J. (1961). Stubble mulch farming. *Adv. Agron.* **13**, 125-96.

McCown, R. L. Jones, R. K., and Peake, D. C. I. (1985). Evaluation of a no till, tropical legume ley farming strategy. In *Agro-research for the semi-arid tropics*:

352   *References*

North-west Australia, ed. R. C. Muchow. Univ. Queensland Press, Brisbane, 450–69.

McGuiness, J. L., Dreibelbis, F. R., and Harrold, L. L. (1961). Soil moisture measurements with the neutron method supplement weighing lysimeter. *Soil Sci. Soc. Am. Proc.* **25**, 339–42.

McIlroy, I. C., and Angus, D. E. (1963). The Aspendale multiple weighed lysimeter installation. *Comm. Sci. Ind. Res. Org. Aust. Div. Met. Physics Tech. Paper* **14**.

McIntyre, D. S. (1958a). Permeability measurements of soil crusts formed by raindrop impact. *Soil Sci.* **85**, 185–9.

McIntyre, D. S. (1958b). Soil splash and formation of surface crusts by raindrop impact. *Soil Sci.* **85**, 261–6.

McIntyre, D. S. (1970). The platinum microelectrode method for soil aeration measurement. *Adv. Agron.* **22**, 235–83.

McIntyre, D. S., and Loveday, J. (1974a). Bulk density. In *Methods for analysis of irrigated soils*, ed. J. Loveday, Tech. Comm. No. 54, Comm. Bur. Soils, Comm. Agric. Bur., Farnham Royal, England, pp. 38–42.

McIntyre, D. S., and Loveday, J. (1974b). Particle size analysis. In *Methods for analysis of irrigated soils*, ed. J. Loveday, Tech. Comm. No. 54, Comm. Bur. Soils, Comm. Agric. Bur., Farnham Royal, England, pp. 88–99.

McIntyre, D. S. and Sleeman, J. R. (1982). Macropores and hydraulic conductivity in a swelling soil. *Aust. J. Soil Res.* **20**, 251–4.

MacRae, R. J., and Mehuys, G. R. (1985). The effect of green manuring on the physical properties of temperate-area soils. *Adv. Soil Sci.* **3**, 71–94.

McQuade, C. V., Hartley, R. E. R., and Young, G. J. (1981). Land use and erosion in the catchment of the Pekina reservoir, South Australia. *Soil and Water Management Rep.* **4/81**, Dept. Agric., South Australia.

Macmillan, W. D. (1958). *The theory of the potential.* Dover Publications, New York, reprint of first publication.

Marriotte, E. (1686). *A treatise on the motion of water and other fluids*, English trans. J. T. Desaguliers, 1718, London.

Marshall, C. E. (1964). *The physical chemistry and mineralogy of soils*, vol. 1: *Soil materials*, J. Wiley, New York.

Marshall, T. J. (1947). *Mechanical composition of soil in relation to field descriptions of texture*, Bull. No. 224, Coun. Sci. Ind. Res. Aust.

Marshall, T. J. (1956). A plummet balance for measuring the size distribution of soil particles. *Aust. J. Appl. Sci.* **7**, 142–7.

Marshall, T. J. (1957). Permeability and the size distribution of pores. *Nature, Lond.* **180**, 664–665.

Marshall, T. J. (1958). A relation between permeability and size distribution of pores. *J. Soil Sci.* **9**, 1–8.

Marshall, T. J. (1959a). *Relations between water and soil*, Tech. Comm. No. 50 Bur. Soils, Comm. Agric. Bur., Farnham Royal, England.

Marshall, T. J. (1959b). Diffusion of gases through porous media. *J. Soil Sci.* **10**, 79–82.

Marshall, T. J. (1962a). The nature, development and significance of soil structure. *Trans. Int. Soc. Soil Sci. Comm. IV and V*, NZ, pp. 243–57.

Marshall, T. J. (1962b). Permeability equations and their models. In *Interaction between fluids and particles*, 3rd Congr. Eur. Fed. Chem. Eng., pp. 299–303. (Inst. Chem. Engnrs, London).

Marshall, T. J., and Gurr, C. G. (1954). Movement of water and chlorides in relatively dry soil. *Soil Sci.* **77**, 147–52.

Marshall, T. J., and Quirk, J. P. (1950). Stability of structural aggregates of dry soil. *Aust. J. Agric. Res.* 1, 266-75.

Marshall, T. J., and Stirk, G. B. (1949). Pressure potential of water moving downward into soil. *Soil Sci.* 68, 359-70.

Mazurak, A. P., and Ramig, R. E. (1962). Aggregation and air-water permeability in a chernozem soil cropped to perennial grasses and fallow-grain. *Soil Sci.* 94, 151-7.

Mees, G. C., and Weatherley, P. E. (1957). The mechanisms of water absorption by roots. 2. The role of hydrostatic pressure gradients across the cortex. *Proc. Roy. Soc. Lond.* B 147, 381-91.

Megaw, H. D. (1973). *Crystal structures: a working approach*, W. B. Saunders Co., Philadelphia.

Meiri, A. (1984). Plant response to salinity. Experimental methodology and application to the field. In *Soil salinity under irrigation. Processes and management*. I. Shainberg and J. Shalhevet eds. Springer-Verlag, Berlin.

Mering, J. (1946). On the hydration of montmorillonite. *Trans. Faraday Soc.* B 42, 205-19.

Merrill, S. D., and Rawlins, S. L. (1972). Field measurement of soil water potential with thermocouple psychrometers. *Soil Sci.* 113, 102-9.

Middleton, H. E. (1930). *Properties of soils which influence soil erosion*. US Dept. Agr. Tech. Bull. No. 178.

Mielke, L. N., Doran, J. W., and Richards, K. A. (1986). Physical environment near the surface of plowed and no-tilled soil. *Soil and Tillage Res.* 7, 355-66.

Mihara, Y. (1951). *Raindrops and soil erosion*. Bull. Nat. Inst. Agric. Sci. Japan, Series A, No. 1.

Miller, E. E., and Elrick, D. E. (1958). Dynamic determination of capillary conductivity extended for non-negligible membrane impedance. *Proc. Soil Sci. Soc. Am.* 22, 483-6.

Miller, R. D. (1980). Freezing phenomena in soils. In *Applications of Soil Physics*. D. Hillel, Acad. Press, New York, 254-99.

Millington, R. J. (1959). Gas diffusion in porous media. *Science* 130, 100-2.

Millington, R. J., and Quirk, J. P. (1959). Permeability of porous media. *Nature, Lond.* 183, 387-8.

Millington, R. J., and Quirk, J. P. (1961). Permeability of porous solids. *Trans. Faraday Soc.* 57, 1200-6.

Milthorpe, F. L. and Moorby, J. (1974 and 1979). *An introduction to crop physiology*, Cambridge Univ. Press. London.

Moench, A. F., and Evans, D. D. (1970). Thermal conductivity and diffusivity of soil using a cylindrical heat source. *Soil Sci. Soc. Am. Proc.* 34, 377-81.

Moore, C. J. (1976). Eddy flux measurements above a pine forest. *Quart. J. R. Met. Soc.* 102, 913-918.

Moore, R. E. (1939). Water conduction from shallow water tables. *Hilgardia* 12, 383-426.

Morton, F. I. (1984). What are the limits on forest evaporation? *J. Hydrology* 74, 373-98.

Moss, A. J. and Green, P. (1983). Movement of solids in air and water by raindrop impact. Effects of drop-size and water-depth variations. *Aust. J. Soil Res.* 21, 257-69.

Mualem, Y. (1976). A new model for predicting the hydraulic conductivity of unsaturated porous media. *Water Resources Res.* 12, 513-22.

Mualem, Y., (1984). A modified dependent-domain theory of hysteresis. *Soil Sci.* 137, 283-91.

354 References

Mualem, Y. (1986). Hydraulic conductivity of unsaturated soils: prediction and formulas. In *Methods of soil analysis*, part 1, 2nd edn, ed. A. Klute, *Agronomy* **9**, 799–823.

Mullins, C. E., and Panayiotopoulos, K. P. (1984). The strength of unsaturated mixtures of sand and kaolin and the concept of effective stress. *J. Soil Sci.* **35**, 459–68.

Murray, R. S., and Quirk, J. P. (1980). Clay-water interactions and the mechanism of soil swelling. *Colloids and Surfaces* **1**, 17–32.

Myers, L. E. (1967). Recent advances in water harvesting. *J. Soil Water Cons.* **22**, 95–7.

Nagpal, N. K., Boersma, L., and De Backer, L. W. (1972). Pore size distribution of soils from mercury intrusion porosimeter data. *Soil Sci. Soc. Am. Proc.* **36**, 264–7.

National Academy of Sciences (1974). *More water for arid lands. Promising technologies and research opportunities*, National Research Council, Washington, D.C.

Nerpin, S. V., and Chudnovskii, A. V. (1967). *Physics of the soil*, trans. Israel Program for Scientific Translations, Jerusalem.

Neumann, H. H., and Thurtell, G. W. (1972). A Peltier cooled thermocouple dewpoint hygrometer for *in situ* measurement of water potentials. In *Psychrometry in water relations research*, eds R. W. Brown and B. P. van Haveren, Utah Agric. Exp. Sta., Utah State Univ., Logan, pp. 103–12.

Newman, E. I. (1969). Resistance to water flow in soil and plant. II. A review of experimental evidence on the rhizosphere resistance. *J. Appl. Ecol.* **6**, 261–272.

Newman, E. I. (1974). Root and soil water relations. In *The plant root and its environment*, ed. E. W. Carson, Univ. Press of Virginia, Charlottesville, pp. 363–440.

Newson, M. D. (1979). The results of ten years' experimental study on Plynlimon, mid-Wales, and their importance for the water industry. *J. Inst. Water Engrs. and Scientists*, **33**, 321–33.

Nichols, M. L., Reed, I. F., and Reaves, C. A. (1958). Soil reaction to plowshare design. *Agric. Eng.* **39**, 336–9.

Nielsen, D. R., Jackson, R. D., Cary, J. W., and Evans, D. D., eds (1972). *Soil water*. Am. Soc. Agron., Madison, Wisconsin.

Nir, A. (1964). On the interpretation of tritium age measurements of ground-water. *J. Geophys. Res.* **69**, 2589–95.

Nir, A., Kruger, S. T., Lingenfelter, R. E., and Flamm, E. J. (1966). Natural tritium. *Rev. Geophys.* **4**, 441–56.

Norrish, K. (1954). The swelling of montmorillonite. *Discuss. Faraday Soc.* **18**, 120–34.

Norrish, K. (1972). Forces between clay particles. Proc. 4th Int. Clay Conf., Madrid, pp. 375–83.

Norrish, K., and Pickering, J. G. (1983). Clay minerals. In *Soils: an Australian viewpoint*. Div. Soils, CSIRO, 281–308. CSIRO, Melbourne; Academic Press, London.

Norrish, K., and Tiller, K. G. (1976). Subplasticity in Australian soils: 5. Factors involved and techniques. *Aust. J. Soil Res.* **14**, 273–89.

Northcote, K. H. (1979). *A factual key for the recognition of Australian soils*, 4th edn, Comm. Sci. Ind. Res. Org. and Rellim Tech. Pub., Glenside, South Australia.

Nutter, W. L. (1973). The role of soil water in the hydrologic behavior of upland basins. *Soil Sci. Soc. Am. Special Publ.* **5**, 181–93.

Nye, P. H. (1979). Diffusion of ions and uncharged solutes in soils and soil clays. *Adv. Agron.* **31**, 225–72.

Nyhan, J. W. (1976). Influence of soil temperature and water tension on the decomposition rate of carbon-14 labeled herbage. *Soil Sci.* **121**, 288–93.

O'Brien, F. E. M. (1948). The control of humidity by saturated salt solutions. *J. Sci. Instrum. Phys. Ind.* **25**, 73–6.

Olsen, S. R., and Kemper, W. D. (1968). Movement of nutrients to plant roots. *Adv. Agron.* **20**, 90–151.

Orchiston, H. D. (1954). Adsorption of water vapour. 2. Clays at 25°C. *Soil Sci.* **78**, 463–80.

Pak, K. P., and Tsyurupa, I. G. (1976). Achievements of Soviet soil science in the reclamation of solonetzes. *Geoderma* **15**, 119–30.

Panabokke, C. R., and Quirk, J. P. (1956). Effect of initial water content on stability of soil aggregates in water. *Soil Sci.* **83**, 185–95.

Parra, M. A., and Romero, G. C. (1980). On the dependence of salt tolerance of beans (*Phaseolus vulgaris* L.) on soil water matric potentials. *Plant and Soil* **56**, 3–16.

Passioura, J. B. (1977). Determining soil water diffusivities from one-step outflow experiments. *Aust. J. Soil Res.* **15**, 1–8.

Passioura, J. B. (1980). The meaning of matric potential. *J. Exp. Bot.* **31**, 1161–9.

Payne, P. C. J., and Fountaine, E. R. (1952). A field method of measuring the shear strength of soils. *J. Soil Sci.* **3**, 136–44.

Pearson, F. J. (1965). Use of $^{13}C/^{12}C$ ratios to correct radio carbon ages of materials initially diluted by limestone. *6th Int. Conf. Radiocarbon and Tritium Dating*, Seattle, 7–11 June 1965, pp. 357–66.

Pearson, F. J., and Hanshaw, B. B. (1970). Sources of dissolved carbonate species in groundwater and their effects on carbon-14 dating. In *Isotope hydrology 1970*, Proc. Symp. Vienna, 9–13 March 1970, IAEA, Vienna, pp. 271–86.

Pearson, F. J., and Swarzenki, W. V. (1974). ^{14}C evidence for the origin of arid region groundwater, northeast province, Kenya. In *Isotope techniques in groundwater hydrology*, Vol. 2, IAEA, Vienna, Proc. Symp. 11–15 March 1974, pp. 95–109.

Peck, A. J. (1960). Change of moisture tension with temperature and air pressure: theoretical. *Soil Sci.* **89**, 303–10.

Peck, A. J., and Hurle, D. H. (1973). Chloride balance of some farmed and forested catchments in southwestern Australia. *Water Resources Res.* **9**, 648–57.

Peck, A. J., Johnston, C. D., and Williamson, D. R. (1981). Analyses of solute distributions in deeply weathered soils. *Agric. Water Manage.* **4**, 83–102.

Penman, H. L. (1940). Gas and vapour movements in the soil. 1. The diffusion of vapours through porous solids. *J. Agric. Sci.* **30**, 437–62.

Penman, H. L. (1941). Laboratory experiments on evaporation from fallow soil. *J. Agric. Sci.* **31**, 454–65.

Penman, H. L. (1948). Natural evaporation from open water, bare soil and grass, *Proc. Roy. Soc. Lond.* A **193**, 120–46.

Penman, H. L. (1953). The physical basis of irrigation control. *Rep. 13th Int. Horticult. Congr.* pp. 913–24.

Penman, H. L. (1963). *Vegetation and hydrology*, Tech. Comm. No. 53, Comm. Bur. Soils, Comm. Agric. Bur., Farnham Royal, England.

Penner, E. (1970). Thermal conductivity of frozen soils. *Can. J. Earth Sci.* **7**, 982–7.

Perrault, P. (1674). *The origin of springs*, English trans. A. La Rocque, 1967, Hafner Publishing Co., New York.

Perroux, K. M., Smiles, D. E., and White, I. (1981). Water movement in uniform soils during constant-flux infiltration. *Soil Sci. Soc. Am. J.* **45**, 237–40.

Phene, C. J., Hoffman, G. J., and Rawlins, S. L. (1971). Measuring soil matric potential *in situ* by sensing heat dissipation within a porous body. 1. Theory and sensor construction. *Soil Sci. Soc. Am. Proc.* **35**, 27–33.

Phene, C. J., Hoffman, G. J., and Austin, R. S. (1973). Controlling automated irrigation with soil matric potential sensor. *Trans. Am. Soc. Agric. Eng.* **16**, 773–6.

Philip, J. R. (1954). An infiltration equation with physical significance. *Soil Sci.* **77**, 153–7.

Philip, J. R. (1957a). The theory of infiltration. 1. The infiltration equation and its solution. *Soil Sci.* **83**, 345–57.

Philip, J. R. (1957b). Numerical solution of equations of the diffusion type with diffusivity concentration-dependent II. *Aust. J. Phys.* **10**, 29–42.

Philip, J. R. (1957c). The theory of infiltration: 4. Sorptivity and algebraic infiltration equations. *Soil Sci.* **84**, 257–64.

Philip, J. R. (1957d). Evaporation, and moisture and heat fields in the soil. *J. Meteor.* **14**, 354–66.

Philip, J. R. (1966). Plant water relations: some physical aspects. *Ann. Rev. Pl. Physiol.* **17**, 245–68.

Philip, J. R. (1969). Moisture equilibrium in the vertical in swelling soils. 1. Basic theory. *Aust. J. Soil Res.* **7**, 99–120.

Philip, J. R. (1971a). Hydrology of swelling soils. In *Salinity and water use*, eds. T. Talsma and J. R. Philip, Macmillan Press, London, pp. 95–107.

Philip, J. R. (1971b). Limitations on scaling by contact angle. *Soil Sci. Soc. Am. Proc.* **35**, 507–9.

Philip, J. R. (1973). On solving the unsaturated flow equation: 1. The Flux-concentration relation. *Soil Sci.* **116**, 328–35.

Philip, J. R. (1977). Unitary approach to capillary condensation and adsorption. *J. Chem. Phys.* **66**, 5069–75.

Philip, J. R., and de Vries, D. A. (1957). Moisture movement in porous materials under temperature gradients. *Trans. Am. Geophys. Union* **38**, 222–32.

Poulovassilis, A. (1962). Hysteresis of pore water, an application of the concept of independent domains. *Soil Sci.* **93**, 405–12.

Pound, C. E., and Crites, R. W. (1973). Characteristics of municipal effluents. In *Recycling municipal sludges and effluents on land*, Proc. Joint Conf. Environment Protection Agency, US Dept. Agr. and Land Grant Colleges, pp. 49–59.

Priestley, C. H. B. (1959). *Turbulent transfer in the lower atmosphere*, Univ. Chicago Press, Chicago.

Priestley, C. H. B., and Taylor, R. J. (1972). On the assessment of surface heat flux and evaporation using large-scale parameters. *Mon. Weather Rev.* **100**, 81–92.

Proctor, R. R. (1933). Fundamental principles of soil compaction. *Eng. News Record* **3**, 286–9.

Purcell, W. R. (1949). Capillary pressures—their measurement using mercury and the calculation of permeability therefrom. *Trans. Am. Soc. Min. Engnrs* **186**, 39–46.

Quirk, J. P., and Schofield, R. K. (1955). The effect of electrolyte concentration on soil permeability. *J. Soil Sci.* **6**, 163–78.

Raats, P. A. C. (1975). Distribution of salts in the root zone. *J. Hydrol.* **27**, 237–48.

Raats, P. A. C., and Gardner, W. R. (1974). Movement of water in the unsaturated zone near a water table. In *Drainage for agriculture*, ed. J. van Schilfgaarde, *Agronomy*, **17**, pp. 311–57.

Rab, A. R., Willatt, S. T., and Olsson, K. A. (1987). Hydraulic properties of a duplex soil determined from in situ measurments. *Aust. J. Soil Res.* **25**, 1–7.

Raney, F. C., and Mihara, Y. (1967). Water and soil temperature. In *Irrigation of agricultural lands*, eds. R. M. Hagan, H. R. Haise and T. N. Edminster, *Agronomy* **11**, pp. 1024–36.

Ratliff, L. F., Ritchie, J. T., and Cassel, D. K. (1983). Field-measured limits of soil water availability as related to laboratory-measured properties. *Soil Sci. Soc. Am. J.* **47**, 770–5.

Raupach, M. R. (1978). Infrared fluctuation hygrometry in the atmospheric boundary layer. *Quart. J. Roy. Met. Soc.* **104**, 309–22.

Rawlins, S. L., and Campbell, G. S. (1986). Water potential: thermocouple psychrometry. In *Methods of soil analysis*, part 1, 2nd edn, ed. A. Klute, *Agronomy* **9**, 597–618.

Reeve, R. C., and Bower, C. A. (1960). Use of high-salt waters as a flocculant and source of divalent cations for reclaiming sodic soils. *Soil Sci.* **90**, 139–44.

Reicosky, D. C. and Ritchie, J. T. (1976). Relative importance of soil resistance and plant resistance in root water absorption. *Soil Sci. Soc. Am. J.* **40**, 293–97.

Revut, V. I. (1973). Possibility of controlling the thermal regime of soils by mulching. *Soviet Soil Sci.* 117–22. (*Pochvovedenie* 141–6.)

Reynolds, W. D., Elrick, D. E., and Topp, G. C. (1983). A re-examination of the constant-head well permeameter method for measuring saturated hydraulic conductivity above the water table. *Soil Sci.* **136**, 250–68.

Rhoades, J. D. (1984). Principles and methods of monitoring soil salinity. In *Soil salinity under irrigation-processes and management*. I. Shainberg and J. Shalhevet eds. Springer-Verlag, Berlin.

Rhoades, J. D., Oster, J. D., Ingvalson, R. D., Tucker, J. M., and Clark, M. (1974). Minimizing the salt burdens of irrigation drainage waters. *J. Environ. Quality*, **3**, 311–16.

Richards, L. A. (1928). The usefulness of capillary potential to soil moisture and plant investigators. *J. Agric. Res.* **37**, 719–42.

Richards, L. A. (1931). Capillary conduction of liquids through porous mediums. *Physics* **1**, 318–33.

Richards, L. A. (1941). A pressure membrane extraction apparatus for soil solution. *Soil Sci.* **51**, 377–86.

Richards, L. A. (1949). Methods of measuring soil moisture tension. *Soil Sci.* **68**, 95–112.

Richards, L. A. (1953). Modulus of rupture as an index of crusting of soil. *Soil Sci. Soc. Am. Proc.* **17**, 321–3.

Richards, L. A. (1960). Advances in soil physics. *Trans. 7th Int. Congr. Soil Sci.* **1**, 67–79.

Richards, L. A., and Fireman, M. (1943). Pressure-plate apparatus for measuring sorption and transmission by soils. *Soil Sci.* **56**, 395–404.

Richards, L. A., and Ogata, G. (1958). Thermocouple for vapour pressure measurements in biological and soil systems at high humidity. *Science* **128**, 1089–90.

Richards, L. A., and Weaver, L. R. (1943). Fifteen atmosphere percentage as related to the permanent wilting percentage. *Soil Sci.* **56**, 331–9.

Richards, S. J. (1938). Soil moisture calculations from capillary tension records. *Soil Sci. Soc. Am. Proc.* **3**, 37–64.

Richardson, S. J. (1976). Effect of artificial weathering on the structural stability of a dispersed silt soil. *J. Soil Sci.* **27**, 287–94.

Rider, N. E., Philip, J. R., and Bradley, E. F. (1963). The horizontal transport of heat and moisture: a micrometeorological study. *Quart. J. Roy. Met. Soc.* **89**, 507–31.

Rightmire, C. T., and Hanshaw, B. B. (1973). Relationship between the carbon isotope composition of soil CO_2 and dissolved carbonate species in groundwater. *Water Resources Res.* **9**, 958–67.

Rijtema, P. E. (1959). Calculation of capillary conductivity from pressure plate outflow data with non-negligible membrane impedence. *Neth. J. Agric. Sci.* **7**, 209–15.

River Murray Commission (1970). *Murray Valley salinity investigations,* Canberra, Australia.

Robinson, R. A., and Stokes, R. H. (1959). *Electrolyte solutions,* 2nd edn, Butterworths, London.

Rodda, J. C. (1976). *Facets of hydrology.* J. Wiley, London.

Rogowski, A. S. (1964). Strength of soil aggregates, Ph.D. thesis, Iowa State Univ., Ames, Iowa.

Rogowski, A. S., Moldenhauer, W. C., and Kirkham, D. (1968). Rupture parameters of soil aggregates. *Soil Sci. Soc. Am. Proc.* **32**, 720–4.

Rollins, R. L., Spangler, M. G. and Kirkham, D. (1954). Movement of soil moisture under a thermal gradient. *Highway Res. Bd Proc., Washington* **33**, 492–508.

Rolston, D. E. (1986). Gas diffusivity. In *Methods of soil analysis,* part 1, 2nd edn, ed. A. Klute, *Agronomy,* **9**, 1089–102.

Roos, S. A. A., Awadalla, E. A., and Khalaf, M. A. (1976). Use of gypsum and water with high salt content in reclamation of sodic soil. *Z. Pflanzenernähr Bodenk.* **6**, 725–30.

Rose, D. A. (1977). Hydrodynamic dispersion in porous materials. *Soil Sci.* **123**, 277–83.

Rose, C. W. (1966). *Agricultural physics,* Pergamon Press, Oxford.

Rose, C. W. (1968). Water transport in soil with a daily temperature wave. 1. Theory and experiment. *Aust. J. Soil Res.* **6**, 31–44.

Ross, P. J., Williams, J., and McCown, R. L. (1985). Soil temperature and the energy balance of vegetative mulch in the semi-arid tropics. 1. Static analysis of the radiation balance. *Aust. J. Soil Res.,* **23**, 493–514.

Rovira, A. D., and Greacen, E. L. (1957). The effect of aggregate disruption on the activity of microorganisms in the soil. *Aust. J. Agric. Res.* **8**, 659–73.

Rowan, J. N. (1971). *Salting on dryland farms in north-western Victoria,* Soil Conservation Authority (Victoria), Melbourne.

Royer, J. M., and Vachaud, G. (1975). Field determination of hysteresis in soil-water characteristics. *Soil Sci. Soc. Am. Proc.* **39**, 221–3.

Rozanov, B. G. (1984). Survey of major soil classifications in some foreign countries. *Pochvovedeniye,* 1984, (1), 5–16. *Soviet Soil Sci.* **16**, (1), 60–71.

Rubin, J., and Steinhardt, R. (1964). Soil water relations during rain infiltration. 3. Water uptake at incipient ponding. *Soil Sci. Soc. Am. Proc.* **28**, 614–20.

Russell, R. S. (1977). *Plant root systems,* McGraw-Hill, Maidenhead, England.

Russell, R. S., and Goss, M. J. (1974). Physical aspects of soil fertility—the response of roots to mechanical impedance. *Neth. J. Agric. Sci.* **22**, 305–18.

Sallberg, J. R. (1965). Shear strength. In *Methods of soil analysis*, eds. C. A. Black, D. D. Evans, J. L. White, L. E. Ensminger and F. E. Clark, *Agronomy* 9, 431-47.

Salter, P. J., and Goode, J. E. (1967). *Crop response to water at different stages of growth*, Res. Rev. No. 2, Comm. Bur. Hort. and Plantation Crops, Comm. Agric. Bur., England.

Sauerbeck, D. R., and Johnen, B. G. (1976). Root formation and decomposition during plant growth. In *Int. Symp. Soil Organic Matter Studies*, Brunswick, Fed. Rep. Germany, IAEA SM-211/16, IAEA.

Scharpenseel, H. W. (1973). Messung der natürlichen C-14 Konzentration in der organischen Substanz von recenten Böden. Eine Zwischenbilanz. *Z. Pflanzenernähr. Bodenk.* 133, 241-63.

Scheidegger, A. E. (1974). *The physics of flow through porous media.* Univ. Toronto Press, 3rd edn.

Schindler, U., Bohne, K., and Sauerbrey, R. (1985). Comparison of different measuring and calculating methods to quantify the hydraulic conductivity of unsaturated soil. *Z. Pflanzenernaehr. Bodenk.* 148, 607-17.

Schneider, E. C., and Gupta, S. C. (1985). Corn emergence as influenced by soil temperature, matric potential, and aggregate size distribution. *Soil Sci. Soc. Am. J.* 49, 415-22.

Scholander, P. F., Hammel, H. T., Bradstreet, E. D., and Hemmingsen, E. A. (1965). Sap pressure in vascular plants. *Science* 148, 339-46.

Scholefield, D., and Hall, D. M. (1985). Constricted growth of grass roots through rigid pores. *Plant and Soil* 85, 153-62.

Scott, G. J. T., Webster, R., and Nortcliff, S. (1986). An analysis of crack pattern in clay soil: its density and orientation. *J. Soil Sci.* 37, 653-68.

Scotter, D. R. (1976). Liquid and vapour transfer in soil. *Aust. J. Soil Res.* 14, 33-41.

Scotter, D. R., and Horne, D. J. (1985). The effect of mole drainage on soil temperatures under pasture. *J. Soil Sci.* 36, 319-27.

Shainberg, I., and Kemper, W. D. (1966). Hydration status of adsorbed cations. *Soil Sci. Soc. Am. Proc.* 30, 707-13.

Shalhavet, J. (1973). Irrigation with saline water. In *Arid zone irrigation*, eds. B. Yaron, E. Danfors and Y. Vaadia, Chapman and Hall, London, pp. 263-76.

Shalhevet, J. (1984). Management of irrigation with brackish water. In *Soil salinity under irrigation. Processes and management.* I. Shainberg and J. Shalhevet eds. Springer-Verlag, Berlin.

Shalhevet, J. and Hsiao, Th. C. (1986). Salinity and drought. A comparison of their effects on osmotic adjustment, assimilation, transpiration and growth. *Irrig. Sci.* 7, 249-64.

Shanmuganathan, R. T., and Oades, J. M. (1983). Modification of soil physical properties by addition of calcium compounds. *Aust. J. Soil Res.* 21, 285-300.

Sharma, M. L., Gander, G. A., and Hunt, C. G. (1980). Spatial variability of infiltration in a water-shed. *J. Hydrology* 45, 101-22.

Shcherbakov, R. A. (1985). Model of the hysteresis of water retention by soils. *Soviet Soil Sci.* 17, (4), 103-11. *Pochvovedeniye*, 1985, (8), 54-60.

Shirazi, M. A., and Boersma, L. (1984). A unifying quantitative analysis of soil texture. *Soil Sci. Soc. Am. J.* 48, 142-7.

Sills, I. D., Aylmore, L. A. G., and Quirk, J. P. (1974). Relationship between pore size distribution and physical properties of clay soils. *Aust. J. Soil Res.* 12, 107-17.

360   *References*

Singer, C. J., Holmyard, E. J., Hall, A. R., and Williams, T. I., eds (1957). *A history of technology*, Vol. 3, Chapter 12: Land drainage and reclamation, Clarendon Press, Oxford.

Skidmore, E. L., Fisher, P. S., and Woodruff, N. P. (1970). Wind erosion equation: computer solution and application. *Soil Sci. Soc. Am. Proc.* **34**, 931-5.

Skidmore, E. L., and Powers, D. H. (1982). Dry soil-aggregate stability: energy-based index. *Soil Sci. Soc. Am. J.* **46**, 1274-9.

Slager, S., and Koenigs, F. F. R. (1964). Particle size analysis of soils by means of a torsion balance and a plummet. *Sedimentology* **3**, 240-52.

Slatyer, R. O. (1967). *Plant-water relationships*, Academic Press, London.

Slatyer, R. O. (1973). The effect of internal water status on plant growth, development and yield. *Ecology and Conservation*, **5**, UNESCO, Paris, pp. 177-91.

Slichter, C. S. (1899). Theoretical investigations of the motion of ground water. *US Geol. Surv. 19th Ann. Rep. 1897-98*, pp. 295-384.

Smalley, I. J. (1970). Cohesion of soil particles and the intrinsic resistance of simple soil systems to wind erosion. *J. Soil Sci.* **21**, 154-61.

Smart, P. (1975). Soil microstructure. *Soil Sci.* **119**, 385-93.

Smiles, D. E., and Bond, W. J. (1982). An approach to solute movement during unsteady, unsaturated water flow in soils. In *Prediction in water quality*, E. M. O'Loughlin and P. Cullen eds., Aust. Academy of Sci., Canberra, 265-87.

Smith, D. B., Wearn, P. L., Richards, H. J., and Rowe, P. C. (1970). Water movement in the unsaturated zone of high and low permeability strata by measuring natural tritium. In *Isotope hydrology 1970*, Proc. Symp. Vienna, 9-13 March 1970, IAEA Vienna, pp. 73-87.

Smith, D. D., and Wischmeier, W. H. (1962). Rainfall erosion. *Adv. Agron.* **14**, 109-48.

Smith, K. A. (1980). A model of anaerobic zones in aggregated soils, and its potential application to estimates of denitrification. *J. Soil Sci.* **31**, 263-77.

Snyder, V. A., and Miller, R. D. (1985*a*). Tensile strength of unsaturated soils. *Soil Sci. Soc. Am. J.* **49**, 58-65.

Snyder, V. A., and Miller, R. D. (1985*b*). A pneumatic fracture method for measuring the tensile strength of unsaturated soils. *Soil Sci. Soc. Am. J.* **49**, 1369-74.

Soane B. D. (1967). Dual energy gamma-ray transmission for coincident measurement of water content and dry bulk density of soil. *Nature, Lond.* **214**, 1273-4.

Soane, B. D. (1970). The effects of traffic and implements on soil compaction. *J. and Proc. Inst. Agric. Engnrs* **25**, 115-26.

Soane, B. D. Blackwell, P. S., Dickson, J. W., and Painter, D. J. (1981). Compaction by agricultural vehicles: a review. 1. Soil and wheel characteristics. *Soil and Tillage Res.* **1**, 207-37.

Soane, B. D., and Pidgeon, J. D. (1975). Tillage requirement in relation to soil physical properties. *Soil Sci.* **119**, 376-84.

Soil Survey Staff (1975), *Soil taxonomy*, Agric. Handb. No. 436, Soil Cons. Ser., US Dept. Agr. Washington, D.C.

Sopper, W. E., and Lull, H. W. (1967). *Forest hydrology*, Proc. Seminar Pennsylvania State Univ., Sept. 1965, Pergamon, London.

Sowers, G. F. (1965). Consistency. In *Methods of soil analysis*, eds. C. A. Black, D. D. Evans, J. L. White, L. E. Ensminger and F. E. Clark, *Agronomy* **9**, pp. 391-9.

Spanner, D. C. (1951). The Peltier effect and its use in the measurement of suction pressure, *J. Exp. Bot.* **2**, 145-68.

References 361

Spinks, J. W. T., Lane, D. A., and Torchinsky, B. B. (1951). A new method for moisture determination in soil. *Can. J. Tech.* **29**, 371-4.

Spoor, G., and Giles, D. F. H. (1973). Effect of cultivations on raising spring soil temperatures for germination with particular reference to maize. *J. Soil Sci.* **24**, 392-8.

Sposito, G., and Mattigod, S. V. (1977). On the chemical foundation of the sodium adsorption ratio. *Soil Sci. Soc. Am. J.* **41**, 323-9.

Sprent, P. (1969). *Models in regression and related topics*, Methuen, London.

SRWSC (1969). *Land drainage, reclamation and groundwater utilization*, Proc. Conf. March 1969, State Rivers and Water Supply Commission, Melbourne, Victoria, Australia.

Stamp, L. D., ed. (1961). *A history of land use in arid regions*, UNESCO, Paris.

Staple, W. J. (1960). Significance of fallow as a management technique in continental- and winter-rainfall climates. In *Plant-water relationships in arid and semi-arid conditions*, UNESCO, Paris, *Arid Zone Research* **15**, 205-14.

Sternberg, L., and De Niro, M. J. (1983). Isotopic composition of cellulose from C_3, C_4 and CAM plants growing near one another. *Science*, **220**, 947-9.

Stirk, G. B. (1954). Some aspects of soil shrinkage and the effect of cracking upon water entry into the soil. *Aust. J. Agric. Res.* **5**, 279-90.

Stocker, P. T. (1975). Diffusion and diffuse cementation in lime and cement stabilised clayey soils—chemical aspects. *Aust. Road Res.* **5**(9), 6-47.

Stoops, G. (1973). Optical and electron microscopy. A comparison of their principles and their use in micropedology. In *Soil microscopy*, ed. G. K. Rutherford, Proc. 4th Working Meeting on Soil Micromorphology, Kingston, Canada, pp. 101-18.

Sutton, O. G. (1953). *Micrometeorology*, McGraw-Hill, New York.

Swanson, C. L. W. and Peterson, J. B. (1942). The use of the micrometric and other methods for the evaluation of soil structure. *Soil Sci.* **53**, 173-85.

Swartzendruber, D. (1969). The flow of water in unsaturated soils. In *Flow through porous media*, ed. R. J. M. de Wiest, Academic Press, New York, pp. 215-92.

Swinbank, W. C. (1951). A measurement of vertical transfer of heat and water vapour and momentum in the lower atmosphere with some results. *J. Meteorol.* **8**, 135-45.

Szabolcs, I., ed. (1971). *European solonetz soils and their reclamation*, Akademiai Kiado, Budapest.

Szabolcs, I., Darab, K., and Varallyay, G. (1972). Prediction and prevention of secondary salinization of irrigated soils on the Hungarian lowland. *Pochvovedenie* **1**, 115-124, or *Soviet Soil Sci.* 113-21.

Takagi, S. (1960). Analysis of the vertical, downward flow of water through a two-layered soil. *Soil Sci.* **90**, 98-103.

Talsma, T. (1963). The control of saline groundwater. Thesis for the degree of Doctor in Land Technology, University of Wageningen. *Reprint of Bulletin of Univ. of Wageningen* **63**(10), 1-68.

Talsma, T. (1969). *In situ* measurement of sorptivity. *Aust. J. Soil Res.* **17**, 269-76.

Talsma, T. (1977). Measurement of the overburden component of total potential in swelling field soils. *Aust. J. Soil Res.* **15**, 95-102.

Talsma, T., and Hallam, P. M. (1980). Hydraulic conductivity measurement in forest catchments. *Aust. J. Soil Res.* **18**, 139-48.

Tanaka, D. L. (1985). Chemical and stubble-mulch fallow influences on seasonal soil water contents. *Soil Sci. Soc. Am. J.* **49**, 728-33.

362   References

Tanner, C. B. (1967). Measurement of evaporation. In *Irrigation of agricultural lands*, eds. R. M. Hagan, H. R. Haise, and T. W. Edminster, *Agronomy* 11, 534–74.

Tanner, C. B. (1968). Evaporation of water from plants and soil, in *Water deficits and plant growth*, ed. T. T. Kozlowski, Academic Press, New York, pp. 73–106.

Tanner, C. B., and Hanks, R. J. (1952). Moisture hysteresis in gypsum moisture blocks. *Soil Sci. Soc. Am. Proc.* 16, 48–51.

Taylor, C. B. (1978). Interlaboratory comparison of low-level tritium measurements in water. *Int. J. Appl. Radiation and Isotopes* 29, 39–48.

Taylor, D. W. (1948). *Fundamentals of soil mechanics*. J. Wiley, London.

Taylor, G. (1953). Dispersion of soluble matter in solvent flowing slowly through a tube. *Proc. Roy. Soc. Lond.* A219, 186–203.

Taylor, H. M. (1974). Root behavior as affected by soil structure and strength. In *The plant root and its environment*, ed. E. W. Carson, Univ. Press of Virginia, Charlottesville, pp. 271–91.

Taylor, H. M., and Ratliff, L. F. (1969). Root elongation rates of cotton and peanuts as a function of soil strength and soil water content. *Soil Sci.* 108, 113–19.

Temperley, H. N. V. (1947). Behaviour of water under hydrostatic tension: Pt 3. *Phys. Soc. Lond. Proc.* 59, 199–208.

Terzaghi, K. von (1931). The influence of elasticity and permeability on the swelling of two-phase systems. In *Colloid chemistry*, Vol. 3 *Technological applications*, ed. J. Alexander, The Chemical Catalog Co., New York, pp. 65–88.

Terzaghi, K. (1943). *Theoretical soil mechanics*, J. Wiley, New York.

Theng, B. K. G. (1982). Clay-polymer interactions: summary and perspectives. *Clays Clay Min.* 30, 1–10.

Thiem, G. (1907). Water-table levels and free flow through gravelly deposits. *J. Gasbeleucht. Wasserversorg.* 50, 377–82.

Thom, A. S. (1975). Momentum, mass and heat exchange of plant communities. In *Vegetation and the atmosphere*, Vol. 1, ed. J. L. Monteith, Academic Press, London, pp. 57–109.

Thomas, M. D. (1928). Aqueous vapour pressure of soils: 4. influence of replaceable bases. *Soil Sci.* 25, 485–93.

Tisdall, J. M., and Oades, J. M. (1982). Organic matter and water stable aggregates in soils. *J. Soil Sci.* 33, 141–64.

Tisdall, J. M., Olsen, K. A., and Willoughby, P. (1984). Soil structural management and production in a non-cultivated peach orchard. *Soil Tillage Res.* 4, 165–74.

Toksöz, S., and Kirkham, D. (1961). Graphical solution and interpretation of a new drain spacing formula. *J. Geophys. Res.* 66, 509–16.

Topp, G. C. and Davis, J. L. (1985). Measurement of soil water content using time-domain-reflectometry (T.D.R.): a field evaluation. *Soil Sci. Soc. Am. J.* 49, 19–24.

Topp, G. C., and Miller, E. E. (1966). Hysteretic moisture characteristics and hydraulic conductivities for glass-bead media. *Soil Sci. Soc. Am. Proc.* 30, 156–62.

Topp, G. C., Davis, J. L. and Annan, A. P. (1982). Electromagnetic determination of soil water content using TDR: II Evaluation of installation and configuration of parallel transmission-lines. *Soil Sci. Soc. Am. J.* 46, 678–84.

Towner, G. D. and Childs, E. C. (1972). The mechanical strength of unsaturated porous material. *J. Soil Sci.* 23, 481–98.

Troeh, F. R., Jaˑro, J. D., and Kirkham, D. (1982). Gaseous diffusion equations for porous materials. *Geoderma* **27**, 239-53.

Tschapek, M., Scoppa, C. O., and Wasowski, C. (1978). The surface tension of soil water. *J. Soil Sci.* **29**, 17-21.

Turner, N. C. (1986). Crop water deficits: a decade of progress. *Adv. Agron.* **39**, 1-51.

Unger, P. W., and Parker, J. J. (1976). Evaporation reduction from soil with wheat, sorghum and cotton residues. *Soil Sci. Soc. Am. J.* **40**, 938-42.

U.S. Salinity Laboratory Staff (1954). Diagnosis and improvement of saline and alkali soils. US Dept. Agric. Handb. No. 60.

Utomo, W. H., and Dexter, A. R. (1981*a*). Tilth mellowing. *J. Soil Sci.* **32**, 187-201.

Utomo, W. H., and Dexter, A. R. (1981*b*). Soil friability. *J. Soil Sci.* **32**, 203-13.

van Bavel, C. H. M. (1950). Mean weight-diameter of soil aggregates as a statistical index of aggregation. *Soil Sci. Soc. Am. Proc.* **14**, 20-3.

van Bavel, C. H. M. (1951). A soil aeration theory based on diffusion. *Soil Sci.* **72**, 33-46.

van Beers, W. F. J. (1963). *The auger hole method. A field measurement of the hydraulic conductivity of soil below the water table*, Int. Inst. Land Reclamation and Improvement, Bull. No. 1 (1958), Wageningen, The Netherlands.

van Deempter, J. J. (1949). Results of mathematical approach to some flow problems connected with drainage and irrigation. *Appl. Sci. Res.*, Holland, A **2**, 33-54.

van Genuchten, M.Th. (1980). A closed-form equation for predicting the hydraulic conductivity of unsaturated soils. *Soil Sci. Soc. Am. J.* **44**, 892-8.

van Olphen, H. (1975). Water in soils. In *Soil components*, ed. J. E. Gieseking, Vol. 2, Chapter 15, Springer-Verlag, Berlin.

van Olphen, H. (1977). *An introduction to clay colloid chemistry*, 2nd edn, J. Wiley, Interscience, New York.

van Schilfgaarde, J., ed. (1974*a*). *Drainage for agriculture, Agronomy,* **17**.

van Schilfgaarde, J., (1974*b*). Saturated flow theory and its application. Non-steady flow to drains. In *Drainage for agriculture*, ed. J. van Schilfgaarde, *Agronomy* **17**, 245-70.

van Schilfgaarde, J., Frevert, R. K., and Kirkham, D. (1954). A tile drainage field laboratory. *Agric. Eng.* **35**, 474-8.

Veihmeyer, F. J., and Hendrickson, A. H. (1927). Soil moisture conditions in relation to plant growth. *Pl. Physiol.* **2**, 71-82.

Veihmeyer, F. J., and Hendrickson, A. H. (1949). Methods of measuring field capacity and permanent wilting percentage of soils. *Soil Sci.* **68**, 75-94.

Veihmeyer, F. J., and Hendrickson, A. H. (1955). Does transpiration decrease as soil moisture decreases? *Trans. Am. Geophys. Union* **36**, 425-8.

Vershinin, P. V., Mel'nikova, M. K. Michurin, B. N., Moshkov, B. S., Poyasov, N. P., and Chudnovskii, A. F. (1959). *Fundamentals of agrophysics*, Trans. Israel Program for Scientific Translations, Jerusalem, 1966.

Viets, F. G. (1972). Effective drought control for successful dryland agriculture. In *Drought injury and resistance in crops*, Crop Sci. Soc. Publ. No. 2, Madison, Wisconsin, pp. 57-76.

Vinten, A. J. A., and Nye, P. H., (1985). Transport and deposition of dilute colloid suspensions in soils. *J. Soil Sci.* **36**, 531-41.

Visser, W. C. (1969). An empirical expression for the desorption curve. In *Water in the unsaturated zone*, eds P. E. Rijtema and H. Wassink, UNESCO, Paris/IASH, Gentbrugge, pp. 329-35.

Vogel, J. C. (1967). Investigation of groundwater flow with radiocarbon. In *Isotopes in hydrology*, Proc. Symp. Vienna, 14–18 November 1966, IAEA and IUGG, IAEA, Vienna, pp. 355–69.

Vomocil, J. A. (1954). *In situ* measurements of soil bulk density. *Agric. Eng.* **35**, 651–4.

Voorhees, W. B., Allmaras, R. R. and Johnson, C. E. (1981). Alleviating temperature stress. In *Modifying the root environment to reduce crop stress.* G. F. Arkin and H. M. Taylor eds., Am. Soc. Agric. Engr., St Joseph, Missouri, 217–66.

Wadleigh, C. H., and Ayres, A. D. (1945). Growth and biochemical composition of bean plants as conditioned by soil moisture tension and salt concentration. *Pl. Physiol.* **20**, 106–32.

Walker, P. H., and Costin, A. B. (1971). Atmospheric dust accession in south-eastern Australia. *Aust. J. Soil. Res.* **9**, 1–5.

Walker, P. H., Woodyer, K. D., and Hutka, J. (1974). Particle size measurement by Coulter Counter of very small deposits and low suspended sediment concentrations of streams. *J. Sed. Petrology* **44**, 673–9.

Wallace, A., and Wallace, G. A. (1986). Effects of very low rates of synthetic soil conditioners on soils. *Soil Sci.* **141**, 324–7.

Wallace, A., Wallace, G. A., Abouzamzam, A. M., and Cha, J. W. (1986). Soil tests to determine application rates for polymeric soil conditioners. *Soil Sci.* **141**, 390–4.

Wang, J. S. Y., and Narasimhan (1985). Hydrologic mechanisms governing fluid flow in a partially saturated, fractured, porous medium. *Water Resources Res.* **21**, 1861–74.

Ward, R. C. (1975). *Principles of hydrology.* 2nd edn, McGraw-Hill, London.

Waring, S. A., Fox, W. E., and Teakle, L. J. H. (1958). Fertility investigations on the black earth wheatlands of the Darling Downs, Queensland. 1. Moisture accumulation under short fallow. *Aust. J. Agric. Res.* **9**, 205–16.

Water Quality Criteria Committee (1972). *Water quality criteria 1972*, Nat. Acad. Sci., Washington, D.C.

Watson, K. K. (1959). The application of drainage theory to the radius and spacing of drain channels. *J. Agric. Eng. Res.* **4**, 222–8.

Watson, K. K., Reginato, R. J., and Jackson, R. D. (1975). Soil water hysteresis in a field soil. *Soil Sci. Soc. Am. Proc.* **39**, 242–6.

Weaver, H. A., and Jamison, V. C. (1951). Effects of moisture on tractor tire compaction of soil. *Soil Sci.* **71**, 15–23.

Webb, E. K. (1965). Aerial microclimate. *Met. Monog., Am. Meteorol. Soc.* **6**, 27–58.

Webster, R. (1985). Quantitative spatial analysis of soil in the field. In *Adv. in Soil Sci.* **3**, 1–70.

Werner, P. W. (1957). Some problems in non-artesian groundwater flow. *Trans. Am. Geophys. Union* **38**, 511–18.

Wesseling, J. (1973). Sub-surface flow into drains. In *Theories of field drainage and watershed runoff*, Int. Inst. Land Reclamation and Improvement, Publ. No. 16, vol. 2, Wageningen, The Netherlands.

Wesseling, J. (1974). Crop growth and wet soils. In *Drainage for agriculture*, ed. J. van Schilfgaarde, *Agronomy*, **17**, 7–37.

West, E. S. (1952). A study of the annual soil temperature wave. *Aust. J. Sci. Res.* A**5**, 303–14.

Weymouth, J. H., and Williamson, W. O. (1953). The effects of extrusion and some other processes on the microstructure of clay. *Am. J. Sci.* **251**, 89–107.

White, I., Smiles, D. E., and Perroux, K. M. (1979). Absorption of water by soil: The constant flux boundary condition. *Soil. Sci. Soc. Am. J.* **43**, 659–64.

White, R. E. (1985). The influence of macropores on the transport of dissolved and suspended matter through soil. *Adv. Soil Sci.* **3**, 95–120.

Wiersum, L. K. (1957). The relationship of the size and structural rigidity of pores to their penetration by roots. *Pl. Soil* **9**, 75–85.

Wilcox, L. V., and Durum, W. H. (1967). Quality of irrigation water. In *Irrigation of agricultural lands*, eds. R. M. Hagan, H. R. Haise, and T. W. Edminster, *Agronomy* **11**, 104–22.

Willatt, S. T. and Pullar, D. M. (1984). Changes in soil physical properties under grazed pastures. *Aust. J. Soil Res.* **22**, 343–8.

Williams, J. (1974). Root density and water potential gradients near the plant root. *J. Exp. Bot.* **25**, 669–674.

Williams, J. and Sinclair, D. F. (1981). Accuracy, bias and precision. In: *Soil water assessment by the neutron method*, ed. E. L. Greacen, pp. 35–49. C.S.I.R.O., Australia.

Williams, J., Prebble, R. E., Williams, W. T., and Hignett, C. T. (1983). The influence of texture, structure and clay mineralogy on the soil moisture characteristic. *Aust. J. Soil Res.* **21**, 15–32.

Willis, W. O., Haas, H. J., and Robins, J. S. (1963). Moisture conservation by surface and subsurface barriers and soil configuration under semi-arid conditions. *Soil Sci. Soc. Am. Proc.* **27**, 577–80.

Willoughby, P., and Willatt, S. T. (1981). Penetration of pea radicles into saturated soil cores related to organic matter amendments. *Aust. J. Soil Res.* **19**, 343–53.

Wilson, D. J., and Ritchie, A. I. M. (1986). Neutron moisture meters: the dependence of their response on soil parameters. *Aust. J. Soil Res.* **24**, 11–23.

Wind, G. P. (1955). A field experiment concerning capillary rise of moisture in a heavy clay soil. *Neth. J. Agric. Sci.* **3**, 60–9.

Wind, G. P. (1958). The lysimeters in the Netherlands. *Comm. Hydrolog. Res. T.N.O. Proc. Inf.* No. 3, 164–228.

Wires, K. C. (1984). The Casagrande method versus the drop-cone penetrometer method for the determination of liquid limit. *Can. J. Soil Sci.* **64**, 297–300.

Wischmeier, W. H. (1970). Relation of soil erosion to crop and soil management. *Proc. Int. Water Erosion Symp., Int. Comm. Irrig. Drainage, Prague* **2**, 201–20.

Wischmeier, W. H. (1976). Use and misuse of the universal soil loss equation. *J. Soil Water Cons.* **31**, 5–9.

Wischmeier, W. H., and Mannering, J. V. (1969). Relation of soil properties to its erodibility. *Soil Sci. Soc. Am. Proc.* **33**, 131–7.

Wischmeier, W. H., and Smith, D. D. (1978). *Predicting rainfall-erosion losses-a guide to conservation planning.* Agric. Handb. No. 537, US Dept. Agr.

Wood, W. E. (1924). Increase in salt in soil and streams following the destruction of native vegetation. *J. Roy. Soc. W. Aust.* **10**, 35–47.

Woodruff, N. P., and Siddoway, F. H. (1965). A wind erosion equation. *Soil Sci. Soc. Am. Proc.* **29**, 602–8.

Woolley, J. T. (1965). Radial exchange of labelled water in intact maize roots. *Pl. Physiol.* **40**, 711–17.

Wyllie, M. R. J., and Gregory, A. R. (1953). Formation factors of unconsolidated porous media. Influence of particle size and effect of cementation. *Petroleum Trans., Am. Inst. Min. Engnrs* **198**, 103–9.

Wyllie, M. R. J., and Gregory, A. R. (1955). Fluid flow through unconsolidated porous aggregates. Effect of porosity and particle size on Kozeny-Carman constants. *Ind. Eng. Chem.* **47**, 1379–88.

Wyllie, M. R. J., and Spangler, M. B. (1952). Application of electrical resistivity measurements to problem of fluid flow in porous media. *Bull. Am. Ass. Petroleum Geologists,* **36**, 359–403.

Yaron, D., Bielorai, H., Shalhevet, J., and Gavish, Y. (1972). Estimation procedures for response functions of crops to soil, water content and salinity. *Water Resources Res.* **8**, 291–300.

Yoder, R. E. (1936). A direct method of aggregate analysis of soils and a study of the physical nature of erosion losses. *J. Am. Soc. Agron.* **28**, 337–51.

Yong, R. N., and Warkentin, B. P. (1975). *Soil properties and behaviour,* Elsevier, Amsterdam.

Young, R. A., and Wiersma, J. L. (1973). The role of rainfall impact in soil detachment and transport. *Water Resources Res.* **9**, 1629–36.

Youngs, E. G. (1968). Shape factors for Kirkham's piezometer method for determining the hydraulic conductivity of soil *in situ,* for soils overlying an impermeable floor or infinitely permeable stratum. *Soil Sci.* **106**, 235–7.

Youngs, E. G., and Poulovassilis, A. (1976). The different forms of moisture profile development during redistribution of soil water after infiltration. *Water Resources Res.* **12**, 1007–12.

Zegelin, S. J., and White, I. (1982). Design for a field sprinkler infiltrometer. *Soil Sci. Soc. Am. J.* **46**, 1129–33.

Zimmermann, U., Ehhalt, D, and Munnich, K. O. (1967). Soil-water movement and evapotranspiration: changes in the isotopic composition of the water. In *Isotopes in hydrology,* Proc. Symposium held in Vienna, 14–18 Nov. 1966, (IAEA 1967), 567–585.

Zimmermann, U., Munnich, K. O., and Roether, W. (1967). Downward movement of soil moisture traced by means of hydrogen isotopes. *Geophys. Monog.* No. 11, ed. G. E. Stout, *Am. Geophys. Union* **28**.

Zingg, A. W., and Hauser, V. L. (1959). Terrace benching to save potential run-off for semi-arid land. *Agron. J.* **51**, 289–92.

Index